Muddy Ground

JOHN WILLIAM NELSON

Muddy Ground

Native Peoples, Chicago's Portage,
and the Transformation of a Continent

The University of North Carolina Press *Chapel Hill*

Set in Arno Pro by Westchester Publishing Services
Manufactured in the United States of America

Library of Congress Cataloging-in-Publication Data
Names: Nelson, John William, author.
Title: Muddy ground : Native peoples, Chicago's portage, and the transformation
    of a continent / John William Nelson.
Other titles: David J. Weber series in the new borderlands history.
Description: Chapel Hill : The University of North Carolina Press, [2023] |
    Series: The David J. Weber series in the new borderlands history | Includes
    bibliographical references and index.
Identifiers: LCCN 2023008565 | ISBN 9781469675190 (cloth ; alk. paper) |
    ISBN 9781469675206 (pbk ; alk. paper) | ISBN 9781469675213 (ebook)
Subjects: LCSH: Portages—Illinois—Chicago—History. | Portages—Great Lakes
    Region (North America)—History. | Indian trails—Illinois—History. | Indian
    trails—Great Lakes Region (North America)—History. | Indians of North
    America—Illinois—History. | Indians of North America—Great Lakes Region
    (North America)—History. | United States—Race relations—History.
Classification: LCC F548.4 .N45 2023 | DDC 977.3/11—dc23/eng/20230301
LC record available at https://lccn.loc.gov/2023008565

Cover illustration: *Chicago in 1820* (Chicago: Chicago Lithographing Co., 1867).
Courtesy of the Library of Congress Prints and Photographs Division.

For Shruti

Contents

Illustrations

Muddy Ground

Murky Waters and Muddy Ground
Continental Conquests and the View from Chicago

portage 3 a: *the carrying of boats or goods overland from one body of
water to another or around an obstacle (such as a rapids)*
b: *the route followed in making such a transfer*[1]

The canoes coasted along the sand dunes of southern Lake Michigan as the
voyageurs sang a French boat song. All in the party wore their finest clothes
and the boats flew flags in anticipation of their destination. The small flotilla
of bateaux and canoes, manned by an assortment of fur traders and their Indigenous wives and children, had traveled twenty days along the lakeshore. In
the coming months, members of this "Illinois Brigade" would disperse across
the river valleys and wetlands south of the Great Lakes to exchange their merchandise for animal pelts. They acquired this valuable commodity on behalf
of the American Fur Company. By spring, they would reassemble and return
to Michilimackinac, at the confluence of Lake Michigan and Lake Huron, to
report their successes and collect their wages. For now, they were preparing
to disembark near the ancient gateway between the Illinois Country, to the
south of them, and the Great Lakes waterways upon which they had spent
the past three weeks. Many in the party had made this trip countless times
before, spanning two of the great watersheds of the continent across a few
miles of muddy terrain. But the sixteen-year-old apprentice, Gurdon Hubbard, would record the experience with fresh eyes as he passed over Chicago's
portage for the first time.[2]

The party of canoemen, traders, and family members made landfall along
the Chicago River on October 1, 1818. Hubbard had grown up in Montreal
and spent the past several months traveling the water routes of Canada and
the northern Great Lakes, learning his chosen profession as a fur trader. But
landing at Chicago, he and his fellow adventurers entered a new and wholly
unfamiliar environment. He "gazed in admiration on the first prairie" of his
life, with its "waving grass," wildflowers, open "oak woods" and abundant
wildlife. The sheer distance he could see in this open landscape south of Lake

Michigan left him "spell-bound."[3] Chicago stood as a portal between the biome of the forest-lined Great Lakes watershed and the tallgrass prairies of the Illinois Country and Mississippi drainage. Chicago's portage consisted of a short but muddy pathway connecting the Chicago and Des Plaines Rivers across the narrowest continental divide separating the Gulf of St. Lawrence and the Gulf of Mexico. Waters from the Chicago River flowed sluggishly into Lake Michigan, while the Des Plaines rolled on to the Illinois River and, eventually, the Mississippi. For centuries, Indigenous travelers had used the overland "carrying place" through Chicago to pass their boats from one watershed to the other, and in the 150 years or so before Hubbard's time, Europeans had been learning to do the same from Native guides, allies, and kin.

But Hubbard's admiration at Chicago's unique geography would turn to grumbling after a few days spent along the shoreline repairing boats. Hubbard and his compatriots had to face the daunting prospect of hauling their boats and supplies across the portage pathway. When they made their portage, there was "sufficient water to float an empty boat" through the waterlogged terrain between the two rivers. The divide between the watersheds was only ever a few feet above water level and often flooded. To tackle this amphibious space, Hubbard's brigade divided the labor of the portage between three groups. Some of the party hauled goods and supplies to the western terminus of the overland pathway in packs weighing ninety pounds each. Others stayed in the boats, using poles to ease the vessels through the mud and vegetation. As those aboard pushed with their poles, "others waded in the mud alongside," frequently sinking to their waists. While they toiled, swarms of mosquitoes beleaguered them. Those working through the murky water faced further attacks from leeches, or "blood suckers," which they had to remove from their bodies each night in camp. Though they labored from "early dawn to dark," it still took Hubbard's brigade of experienced traders "three consecutive days of such toil" to get all their boats and supplies across the divide.[4] But once they made it across, the unbroken waters of the Des Plaines, Illinois, and Mississippi Rivers lay open to them, as did thousands of wetlands, feeder streams, tributaries, and other vital riverways from the Ohio to the Missouri.

Despite the demanding work described by Hubbard, the overland portage through Chicago remained the most effective means of traversing the continent between the Mississippi and Great Lakes waterways through the early nineteenth century. By 1818, Chicago operated as a small cross-cultural hub with transcontinental implications. Hubbard came to Chicago in the company of French Canadian voyageurs, their Anishinaabe wives, and their mixed-heritage, or *métis*, children. At Chicago, he interacted with American officers and enlisted men at the recently rebuilt Fort Dearborn. He dined at the table of John Kinzie, a local Anglophone trader with ties to British Canada, under the roof of the oldest trading house at Chicago, built during the eighteenth century by Jean Baptiste Point de Sable, a French-speaking free Black man. Though regarded as a far-flung frontier outpost by most Americans during the period, Chicago was already an entangled crossroads with a deep and multicultural history.[5]

To operate in this cosmopolitan space, Hubbard had to master both French and Indigenous languages, as well as customs. During his first year as a fur trader in the Illinois Country, he worked to learn Neshnabémwen, the language spoken by many of his Potawatomi companions and clients. He relied on the region's Indigenous women to supply his crews with maize and other farm produce, while he traded gunpowder and tobacco with Native men for meals of wild game and fish. He briefly took an Indigenous wife, Watseka, "in the fashion of the country," hoping to ingratiate himself into the familial networks of the local Anishinaabeg.[6] Like generations of French and British adventurers before him, he relied on Native expertise when it came to negotiating with Indigenous leaders, collecting pelts from Native suppliers, and judging the dangers of various water routes. The realm Hubbard entered as an apprentice of the fur trade was a world in motion. And that world of movement followed a geographic logic that Native peoples had perfected in the waterborne interior of North America for centuries.

But the Chicago that Hubbard experienced in 1818 and recollected years later would transform in the coming decades. The geographic rationale that had made the Chicago portage a space of interaction and mobility since before European contact faced a new challenge from the United States, and a particular American notion of how a settled landscape should function. Indigenous nations confronted pressures from not only white settlers and land speculators but especially the U.S. government. The American state sought to overhaul the geography and environments of places like Chicago that for centuries had operated as sites of Indigenous power and colonial consternation. When U.S. officials and incoming settlers finally did bring about a conquest of Chicago in the 1830s, they did so only by transforming the geography and ecology of the local area through a series of infrastructure projects—canal diggings, grid surveys, swamp drainages, and harbor improvements that undercut the geographically based power of Chicago's Anishinaabeg. This new conquest of Chicago's amphibious space aimed to eradicate the Indigenous geographic logic of the portage. Only after extensive state investment and intensive alterations to the landscape did Americans succeed in creating a usable site of empire at Chicago. The overhaul of Chicago's portage geography rendered the space unrecognizable from Hubbard's descriptions in a matter of decades.[7]

The U.S. government and its officials already had ambitions to transform Chicago by the time Hubbard arrived. The year before, a young congressman, John C. Calhoun, had proposed that Americans' last great adversary was the North American continent itself, and he declared that the United States "must conquer space." Calhoun had risen as part of a new cohort of American nationalist politicians in the aftermath of the War of 1812 who argued that the federal government needed to take a more active role in building roads, canals, and other transportation networks across the nation. This generation of leaders argued for such projects to bolster American economics, settlement, and national security. Calhoun deployed the language of war because he fully believed that a "perfect system" of state-backed infrastructure had the potential to conquer the North American continent more effectively than any military campaign ever had. Government investment in "internal improvements," as these infrastructure projects were called at the time, had the power to reshape the geography of the interior and bring spaces that had previously eluded U.S. control under the dominion of the American republic once and for all.[8]

John C. Calhoun's conquest of "space" went hand in hand with concurrent U.S. attempts to subjugate Native peoples across the continent. U.S. officials, eastern business interests, and incoming settlers worked to reshape the Chicago area into a more legible geography from which they could profit.[9] In this way, the story of Chicago brings together two important but often separate threads of American history, demonstrating how internal improvements during the era of the early republic fit within the wider process of Indigenous dispossession, environmental degradation, and the U.S. conquest of the American interior.[10] The overhauling of Chicago's portage geography into a manageable landscape for U.S. control and settler development explain the shift from a site of Indigenous hegemony to a hub of American commerce, settlement, and state power by the mid-nineteenth century. As U.S. officials reconceptualized their vision for Chicago's local space and leveraged state capacity to conform Chicago to such imaginings, they directly undermined the power of Native peoples at the portage.[11] In this way, the American state demonstrated its ability to deploy physical force on two distinct but reinforcing fronts—first, as a violent overwriting of Chicago's geography through infrastructure projects, and second, as a state-backed effort to eradicate the region of its Indigenous population.[12] The transformation of Chicago from an Indigenous canoe portage into a bustling settler city by the later nineteenth century represented both a subjugation of the local environment and the dispossession of Chicago's Native peoples. A two-pronged conquest shifted the balance of power away from Chicago's Anishinaabeg and undergirded U.S. success at the local, regional, and even continental levels.

This study traces that trajectory and transformation at Chicago, unearthing a new explanation for how Indigenous peoples lost power in the face of an expanding United States. Where other scholars have pointed to military victories, diplomatic shifts, racial violence, economic growth, burgeoning populations, and settler colonialism itself as the cause for this change, I argue that the transformation of the physical geography at places like Chicago bolstered U.S. success in colonizing the peoples of the American interior.[13] Far from a mere military takeover, this conquest proved more insidious, and exponentially more effective, at empowering the United States while undermining Indigenous authority in the borderlands of American expansion. By the 1830s, the United States had gained the upper hand across the Great Lakes and other regions between the Appalachian Mountains and the Mississippi River, forcibly expelling many Native peoples across the Mississippi and leaving others to parse out ways to survive in a totally overhauled landscape of U.S. governance.[14] By undermining the local environments of Indigenous power, the United States changed not simply the rules of the game but the board on which the game was played. Whereas Native peoples had leveraged their home geographies for centuries to stave off colonization and dictate the terms of engagement at places like Chicago, by the 1830s, the United States had effectively conquered both spaces and peoples across vast swaths of North America. This two-pronged conquest by the United States during the era of the early republic becomes more evident when we draw our focus onto particular sites of contact and colonization, such as Chicago's portage. Concentrating at the local level—from the mud up, as it were—we can trace the vital role that geographic control played in mediating European-Indigenous relations for several centuries and mark the catastrophic fallout for Native peoples when the United States eventually transformed local environments and waterways to bring order to its frontier regions.

This book's contention is that geographic realities mattered in the histories of contact and encounter between North America's Indigenous peoples and incoming Europeans. The spaces where Native peoples and Europeans met, traded, talked, fought, loved, and lived together had a direct impact on the ways these disparate peoples interacted over the crucial early centuries of American colonization. The ground on which cross-cultural encounter occurred was more than just a metaphor—the physical spaces of interaction shaped the terms of these relations. To demonstrate this, I focus on a particularly small patch of earth that connected the waterborne regions of the Mississippi drainage and the Great Lakes watershed—the Chicago portage. This narrow strip of land between watersheds featured an overland pathway, or portage, where travelers could carry canoes and supplies between the Chicago and Des Plaines Rivers. Depending on the time of year and the water levels, the portage could stretch anywhere from a couple hundred yards to several miles, yet it remained a point of connection throughout the seventeenth, eighteenth, and early nineteenth centuries. Studying the portage through this long stretch of time, I trace the ways Native peoples and Europeans understood and harnessed the local landscape during their dealings with one another. Chicago's geographic position as a borderland between regions, watersheds, ecological zones, and European and Indigenous realms of power made it a strategic site where multiple Indigenous and European regimes attempted to assert their dominance. The local borderlands geography shaped the ways Native peoples and Europeans consolidated power at Chicago and across the continent.[15]

Yet this is not a story of geographic determinism. Quite the opposite, in fact. While on maps, Chicago's geography seemed promising to European adventurers, Catholic missionaries, and colonial officials from early French expansion through the U.S. republic, the environmental reality proved frustratingly difficult for those looking to harness Chicago's geographic potential on the ground. The stormy waters of Lake Michigan and its winter ice floes proved a challenge to navigation northward, while varying water levels in the Des Plaines and Illinois Rivers complicated travel downriver. Meanwhile, Chicago's local wetland landscape bogged down movement over the muddy portage, and spring flooding further discouraged permanent settlement in the vicinity.

Portages like Chicago were spaces of frequent unease. These passageways between dry land and open water were disorienting for the uninitiated. Given Chicago's varying water levels, the portage never looked the same twice; the takeout point could be at changing locations along the watercourse, and the pathway itself varied in length. People lost themselves in the reeds, mud, tall grass, and twisting sloughs. The physical act of carrying a canoe (often over one's head) left portaging crews unsure of their next step, let alone their surroundings. Gear was heavy. The potential for injuries, high. Hernias, twisted ankles, pulled muscles, and broken bones could spell death for the unwary traveler. Portage paths were spaces of vulnerability as likely sites of ambush, and Chicago's muddy path offered few routes of escape. At no other point in otherwise waterborne voyages through the Great Lakes region were travelers more at risk, more on edge, and more bewildered than when they passed through these amphibious places of transit.

From the seventeenth century forward, then, the bulk of this story traces the disjoint between visions for Chicago based on its strategic geography and the local ecological and political landscape that left such aspirations unrealized. Chicago's terrain challenged, rather than promised, imperial ambitions for territorial empire in the west, despite the beneficial mobility provided by the water routes of the region. Though various groups saw potential for geographic power at Chicago—the Illinois, the French, the Weas, the Meskwakis, the British, the Anishinaabeg, and the Americans—only the Anishinaabeg succeeded in harnessing the landscape and bringing Chicago under localized control for any extended period. Chicago's utility as a site of Euro-American settlement and state control only came about later in the nineteenth century, after U.S. officials and citizens proved willing and able to renovate the area's environmental realities by draining marshes, plowing the wet prairies, constructing an artificial harbor, and digging a canal to regulate waterflow between the Great Lakes and the Mississippi drainage. When U.S. citizens enjoyed the urban boom of Chicago as a city, it was largely in spite of, and not because of, its geographic reality.[16] Only by significantly overhauling an Indigenous landscape did Euro-Americans gain the upper hand along Chicago's waterways and portages.

This book began as a work on portages, not a history of Chicago. Or rather, it began as an exploration of how people understood, used, and sought to control a geography of incredible mobility in the preindustrial era. At its most abstract, it is a story of how space and movement shape power dynamics. I hope it is still that, though along the way it has illuminated understudied aspects of Chicago's Indigenous past and offered a new interpretation for thinking about how empires failed, and the settler republic of the United States succeeded, at conquest in the American interior. As significant as these broader stakes are, I did not choose Chicago for Chicago's sake. I reasoned that to make sense of a system of such dizzying motion—maritime canoe routes, overland carrying places, material exchanges, and vying Native and European interests—it would be more manageable to stake out one small vantage point within the wider network and watch the succession of movement as it sped by.

Throughout the story, I weave in some of the more iconic and illustrative travel narratives through Chicago's portage route. I have tried to incorporate them into the analysis to show the methods for travel via canoe and portage and to highlight the experiences of those who passed along Chicago's route. Readers should note the striking continuity of the system of travel from our earliest written descriptions of the portages through the 1800s. For almost two centuries of written history, the process for connecting the waters of the Great Lakes with the Mississippi watershed followed an ancient Indigenous precedent.

Portages were ubiquitous throughout early America as nodes that linked waterways. As devices of Native American ingenuity, these geographic features combined with the portability of lightweight Indigenous-built canoes allowed for waterborne networks of mobility that stretched from the sub-Arctic to the Gulf Coast and could even transcend continental divides, like the one at Chicago. As crossover points, sites of contact, and bottlenecks of otherwise waterborne movement, portages, or "carrying places," featured prominently in the calculations of Native peoples as well as incoming Europeans.[17]

In that spirit, readers should not view the history of Chicago's portage as exceptional. The contours of the contest over Chicago could as easily apply to other intersections of movement and cross-cultural encounter throughout the Great Lakes and beyond. Indeed, from the Creek Country of the American Southeast to Iroquoia in upstate New York, similar processes played out as the United States leveraged state power to overhaul landscapes and usurp Indigenous networks of mobility—portages, river fords, mountain passes, trails, and more. Across the continent, the process of rewriting Native environments became a defining feature of the early republic that served to bolster its power and open new pathways for American encroachments into Indigenous homelands.[18]

Chicago simply provides one of the best and most illustrative examples of such a process, while also offering two other important takeaways. A focus on Chicago allows us to recover, in the clarity of a fixed space, an Indigenous amphibious geography of the interior that operated across waterways and relied on crafted geographic devices—portages—as a means to knit the continent together. It also highlights the history of a site that has often been ignored in older histories of the Great Lakes and *Pays d'en Haut*. Standard histories of the region during the colonial era spend little time on Chicago, but this is not because Chicago was an insignificant place. Rather, because it persisted as a space of overlapping Indigenous authority for so long, it has remained outside the purview of historical studies that fixate on European outposts which generated their own archival footprints. Retracing Chicago's pre-nineteenth-century history requires the creative piecing together of far-flung materials—travelogues, archaeological finds, oral traditions, trade ledgers, and imperial correspondence, to name just a few—to make sense of a site where, for the most part, Europeans failed in their schemes of empire. Understanding these failures, the Indigenous power that persisted, and the eventual ways the U.S. government undermined Native authority at the portage, allows us a new way of understanding how localized geography shaped the story of contact and conquest on the North American continent.[19]

This study begins in the seventeenth century, just before the era of contact between Indigenous peoples and Europeans in the Great Lakes. Native peoples had harnessed the fluid geography of the North American interior for centuries to enhance their mobility throughout the midcontinent. Portages, or carrying places, served as the geographic linchpins to this fluid system of interconnected lakes and rivers. Along these routes, Native peoples traveled, traded, and warred in lightweight birchbark canoes specifically built for portability. Navigating overland between water routes, Native Americans established a complex system of travel throughout the region, which enabled the seasonal mobility required for their mixed economies of trade, hunting, fishing, and agriculture. These routes also facilitated the movement of people, goods, ideas, and diplomatic influence across the region during the long era before European contact.

The distinct geography of Chicago made it a crucial link in this network of waterborne mobility. The glaciers that formed Lake Michigan had originally created the low spot between the Chicago River and the Mississippi drainage approximately 14,500 years before, when excessive runoff from the melting ice drained from the southern end of the lake toward the Illinois River valley.[20] Even in the later historical period, those who crossed the low-lying continental divide at Chicago derisively called the area "Mud Lake." When Gurdon Hubbard traversed the portage in 1818, he lamented that the pathway was often more "lagoon" than either land or water. As late as the nineteenth century, the spring melts of ice and snow flooded the marshy ground separating the Des Plaines and Chicago Rivers and connected the bodies of water in a confusing slough of rushing water, sometimes flowing south into the Illinois River and sometimes north into Lake Michigan. The watery variance of this space, while difficult to map for later European and U.S. officials, meant that if Native peoples timed their travels right, they could avoid the arduous labor of carrying their canoes and supplies between the rivers, and simply slip through the muddy waters of the spring floods without leaving their canoes in a kind of makeshift waterborne passage.[21]

Chicago functioned as a thoroughfare of Indigenous movement for a wide array of Native peoples from the seventeenth century forward. Before contact, the archaeological record shows that people of the Oneota culture inhabited the river valleys and prairies around Chicago. Archaeologists believe these Oneota inhabitants were related to the Ho-Chunk, or Winnebago people of Wisconsin today. By the time French observers arrived at Chicago's portages, a host of Native travelers including Miamis, Potawatomis, Odawas, Illinois, Shawnees, Mascoutens, Kickapoos, and Meskwakis used the carrying place. Waves of settlement followed, first with the Miami-speaking Weas and then various Anishinaabe migrants—mostly Potawatomis—who resettled the area in the mid-eighteenth century and successfully commanded the portage into the 1820s. In the following pages, I sometimes take James Merrell's suggestion in referring to Native peoples around Chicago as *settlers* and their towns as *settlements*.[22] This should not confuse the reader. More than some rhetorical table turning, *settlement* is the nearest descriptor for the dynamic and enduring ways Native peoples occupied the Chicago region and harnessed its strategic space throughout the period. The successive waves of assertive Indigenous peoples that used Chicago's space of movement offer compelling counterpoints to more well-known European colonization efforts in the Great Lakes. All these periods of meaningful occupation by Native peoples demonstrate the vibrant history of the Indigenous peoples of the lower Great Lakes and their active roles in controlling, reshaping, and cultivating the landscapes and waterways that undergirded their power in the region.[23]

While Native peoples were active settlers of Chicago, I do not mean to imply that they were interlopers or latecomers. Chicago's region was and is claimed as the homeland of many Indigenous nations. Such overlapping claims are corroborated by the historical record as well as in oral traditions of the region's Native nations. Chicago's portage as an Indigenous route of travel certainly dates back to the precontact era, and even our earliest written descriptions of the trade routes through the wetlands demonstrate that the area was utilized by Anishinaabeg and Ho-Chunk peoples from the north, Kickapoo, Mascouten, and Meskwaki peoples from the west, Illinois peoples to the south, and various Miami-speaking groups from several directions. All these peoples, from their earliest history, had some connection to the important crossroads of Chicago. Even the groups who never lived locally held a claim to its importance as a space of movement.[24] To make sense of such Indigenous diversity through Chicago's space and time, I have included "A Guide to the Peoples of Chicago's Portages," as a sort of glossary. Here, readers can find brief descriptions of various Native peoples that shaped Chicago's history and read my clarifications of the terms I have used for respective groups throughout the book. My hope is that this guide will help readers make sense of the incredible vibrancy of Great Lakes Indigenous peoples and their dynamic histories at Chicago.

Tracing Chicago's story of cross-cultural encounter over such an extended time frame allows us to move beyond the dichotomies of Indigenous persistence versus the totality of settler conquest. In this way, the following pages do not seek to argue whether Indigenous power persisted at Chicago, or Native survivance succeeded into the present. Of course it did—in formidable ways through the 1830s, and in dynamic, subversive ways ever since. There are thousands of Native people living in Chicago today and their experiences have been well documented as examples of how U.S. conquest never completely eliminated an Indigenous presence in the Midwest or elsewhere.[25] The experience of Natives, Europeans, and Americans at Chicago's portage demonstrates the ebbs and flows of Indigenous power at the local, regional, and continental levels but also illuminates the ultimate grounds on which Chicago's Anishinaabeg became divested of their power, and lands, along Lake Michigan's shoreline.

By focusing on the waterborne mobility that Chicago's portage facilitated across the continental divide, the project reestablishes the centrality of waterways to the narrative of Indigenous power and U.S. conquest in the continental interior, demonstrating how local geography and environmental factors—specifically the wetlands, riverways, and lakeshore of Chicago's waterscapes—remain essential to understanding Indigenous-European relations. This project reveals that even deep within the interior of the continent, waterborne travel facilitated colonial incursions and fostered Indigenous resistance.[26] Two interconnected spheres of waterborne movement—one constructed by Indigenous people in the continental interior, the other fostered by an expanding imperial, maritime world in the Atlantic basin—came together along the muddy ground of Chicago's portage. By examining cross-cultural interactions and movement through Chicago's strategic water route, this project bridges those disparate spheres of historical analysis, studying both the Indigenous power of the continental interior in conjunction with the waterborne connections of a maritime Atlantic World.[27]

When Europeans first arrived in the Great Lakes region, they marveled at the labyrinth of interconnected waterways that made up the heart of the continent. The French, who came to the Great Lakes' western shores during the mid-seventeenth century, became the first Europeans to explore these inland, freshwater expanses—a geographic feature wholly distinct from prior European experience.[28] Early adventurers like Louis Jolliet and René-Robert Cavelier, Sieur de La Salle conceptualized an interior maritime network connecting French interests from the St. Lawrence River to the Gulf of Mexico, linked by overland portages at several key locations—notably Niagara and Chicago. As these early French imperialists observed the region's Indigenous peoples plying the waterways of the interior in their carefully constructed canoes, they came to envision how their own maritime designs for Atlantic empire might map onto the watery geography of the Great Lakes and Mississippi River watersheds.

While the French projected their understandings of a maritime empire onto the region, those geopolitical aspirations failed to materialize at places like Chicago. Despite the promising geography, the actual terrain of the portages proved difficult to manage at the local level. The French hoped to bring Chicago into their imperial, commercial, and evangelical sway. But their attempts foundered as early efforts to build a Jesuit mission and trading hub collapsed in the face of Haudenosaunee campaigns, and, later, Meskwaki resistance around the portage. The ecological unsuitability of the site as a wetland undermined European efforts to garrison the space too. Successive waves of French and British officials struggled to bring the portages under control, and their efforts continued to unravel over the course of the eighteenth century. The local reality of the area frustrated their geographic visions. Warfare, the wetland landscape, and the constant movement of people and goods through the portage made Chicago a paradoxically busy thoroughfare while belonging to no particular group for many decades. Thus, the Chicago portage had the contradictory role of providing a pragmatic route of travel used by Native peoples, fur traders, and imperial officials alike, while remaining a frustratingly fluid space for those seeking to control its mobility. Partly a no-man's-land beyond the reach of imperial control, and partly a neutral passage where many different people could traverse the continental interior, Chicago stymied both contemporary empire builders and subsequent historians who have tried to understand its role in the Great Lakes borderland.[29]

Beginning in the 1740s, groups of Anishinaabeg expanding southwestward from their homelands in the St. Joseph and Grand River valleys, as well as Michilimackinac, came to Chicago as part of a wider growth and expansion of Anishinaabewaki—the homeland of the Anishinaabeg. These Native peoples brought a geographic appreciation for Chicago's portages. They also came with experience, having successfully navigated and controlled other sites of waterborne connection throughout the Great Lakes for centuries prior. During the latter part of the eighteenth century, while much of the region became engulfed in warfare and the political fallout of a collapsing French regime, a rising British imperial presence, and a subsequent American colonial revolt, the Anishinaabeg around Chicago's portage leveraged their home geography to expand their influence in the borderlands between Lake Michigan and the Illinois Country.

As the United States extended claims over Chicago's territory in the 1790s, the portages became a renewed site of contest. American officials approached the Great Lakes hoping to control sites like Chicago as clear-cut terrestrial space. Like the French and British before them, early American officials' hopes withered on the actual ground of Chicago. As American authorities grappled to control the region, eastern notions of organized settlement across open lands failed to account for the porous and highly mobile nature of the portage. Chicago's portage, neither wholly terrestrial nor wholly marine, muddied the waters for American officials who held a neat model for how empire should work on an expanding, land-based frontier. Despite the erection of Fort Dearborn in 1803, a wide panoply of people continued to orient their trade, travel, and livelihoods along the rivers and portage route that provided natural access to British Canada to the north and Spanish territories to the south and served as a conduit for Indigenous seasonal migrations. Illicit trade across national borders and the cyclical transience of local Anishinaabeg became a constant complaint of American agents and officers at Chicago. On maps, American officials had hoped a fort would curb movement between the Great Lakes and the midcontinent, clamping down on the unmonitored waterborne mobility of Chicago's route. In reality, however, Chicago locals—both Europeans and Native peoples—challenged U.S. hegemony, harnessing the portage and its waterborne mobility to navigate around the American presence at the mouth of the Chicago River. The tension between local interests and official U.S. control came to a head in 1812, at the outset of war with Britain, when the area's Anishinaabeg attacked the retreating American garrison and subsequently burned Fort Dearborn to the ground.

When U.S. officials managed to regain their control over Chicago after the War of 1812, they continued to face a number of imperial and local challenges stemming from divergent interpretations of Chicago's geography. In the following decades, U.S. officials set about rendering the waters and wetlands of the portage into a legible terrain that organized Chicago's space into a site which state officials could effectively govern.[30] A revamped, state-sponsored geography, more than any military defeat or settler land rush, made the U.S.

conquest of Chicago possible. While a handful of Anglo-American individuals—Hubbard among them—made the transition and helped Chicago's space remain relevant in both geographic regimes, many thousands of Native people, Francophone traders, and mixed-race individuals suffered. The environment, too, faced the degradation that went along with government-funded progress and localized efforts at overhauling the waterways and landscapes of Chicago's space.

In the end, this study also demonstrates the methodological value of localized history. The development of Chicago as a crossroads and meeting point between various Indigenous peoples, European empires, and U.S. officials played out on varying regional, national, continental, and even global scales. Yet to truly comprehend both cross-cultural interaction and competition at Chicago's portage, the following pages often focus on history at the community or even interpersonal level. Calhoun could wax vaguely about the conquest of "space" in the halls of Congress; many historians since have followed this broad, sweeping trajectory to explain U.S. expansion across the continent.[31] But to understand the fundamental nature of such a process, this study recenters our examination of that conquest on a particular patch of ground where Native peoples encountered, collaborated with, and contested European and American colonization efforts.[32] Early North America was indeed vast, but the crucial interactions between people groups and their environments that shaped the continent's history took place at intimate localities like Chicago's muddy portage.[33] Only by concentrating on shifting dynamics at such a localized level can historians ever hope to make sense of the wider transformations that shaped North America's human and nonhuman history from early contact into our modern era.

A Guide to the Peoples of Chicago's Portages

Anishinaabeg: The Anishinaabe peoples (plural Anishinaabeg) are a culturally and linguistically similar assemblage of Native nations originating in the northern and central Great Lakes. Sometimes referred to as the "Three-Fires" Confederacy within older histories, they include the *Bodéwadmi (Potawatomi)*, *Odawa (Ottawa)*, and more northerly *Ojibwe (Chippewa)*. They are all Algonquian-speaking groups, though the language of the Potawatomis of southern Lake Michigan, *Neshnabémwen*, varies from the *Anishinaabemowin* of Odawas and Ojibwes. The Anishinaabeg played a significant role in Chicago's history from earliest written records onward. In the seventeenth century, Anishinaabe traders operated as brokers between Great Lakes networks of exchange and the Illinois peoples of the Mississippi watershed, using the portages at Chicago as a major route in their travels. In the eighteenth century, Anishinaabe migrants began making Chicago a more permanent home, with Anishinaabe-dominated towns springing up near the portages by the 1740s. While members of all three major groups of the Anishinaabeg called Chicago home, Potawatomi leadership came to define the political identity of these towns by the end of the eighteenth century. The Chicago-area Anishinaabeg would continue to be called "the united tribes of Ottawas, Chipawas, and Potawotomees" in treaties with the United States throughout the era of the early republic.[1] Despite land cession treaties like those at Chicago in 1821 and 1833 that aimed to dispossess Native peoples of their homelands in the region, some Anishinaabeg continued to live in the Chicago region even after the expulsion efforts of the U.S. government in the 1830s. Thousands of Anishinaabe people continue to reside across the Great Lakes region today.[2]

Dakota (Oceti Sakowin/Sioux): Though never residents of the area surrounding Chicago's portages, Siouan-speaking Dakota people played a leading role in shaping the Indigenous political geography of the Illinois Country and upper Mississippi River valley from the era of precontact through the 1830s. They resided to the west of the Great Lakes, but they loomed as potential trade partners and allies to the French. They also allied with Indigenous groups in the Chicago area, including the Meskwakis at various times during the Fox Wars. At other times, they posed a serious threat as aggressors in the

Mississippi and Illinois River valleys, warring against Illinois and Miami peoples. The Wea, for instance, abandoned townsites around Chicago early in the eighteenth century because of, among other things, their fears of Dakota strength.[3]

Haudenosaunee (Iroquois): The Haudenosaunee, familiar as the Iroquois in many histories of early America, formed a powerful league of five nations including the **Seneca**, **Cayuga**, **Onondaga**, **Oneida**, and **Mohawk** before European contact in the lands south of Lake Ontario. A sixth nation from the southeast, the **Tuscarora**, joined the confederacy in the early eighteenth century. Haudenosaunees shaped events at Chicago even though their homeland remained far removed from the western Great Lakes. In the seventeenth century, armed with muskets from Dutch, and later, English traders at Albany, the Haudenosaunee launched a series of military campaigns against other Iroquoian and Algonquian peoples across the region in what historians have come to call the Mourning Wars. Haudenosaunee warriors raided to take captives, earn prestige for themselves and their communities, and possibly at times, to seize furs and other trade goods from Native peoples allied to the French in the Great Lakes. The Haudenosaunee campaigns eventually expanded into the Lake Michigan basin and Chicago's portages offered these expeditions an easy route for raiding against the Illinois Country downstream.[4] Haudenosaunees campaigning through Chicago threatened both Native peoples of the area, like the Mascoutens, Weas, Meskwakis, and Kickapoos, as well as French adventurers passing through the portages.

Ho-Chunk (Winnebago): A Siouan-speaking people who resided in the southern portions of Wisconsin during the postcontact era. French sources typically referred to them as the *Puans* or *Puants* because of their association with Green Bay (*The Stinking Bay*). According to very early European sources, the Ho-Chunk suffered a series of brutal wars against other Indigenous groups of the Great Lakes and Illinois Country, including a costly conflict against the Illinois sometime in the seventeenth century. There are documented Ho-Chunk journeys through Chicago's portages well into the nineteenth century, though they lived, for the most part, well north of the area. The Ho-Chunk are also relevant to the localized history of Chicago because they are the probable descendants of **Oneota** people whose townsites have been recovered and studied by archaeologists across northern Illinois and southern Wisconsin. The Oneota sites of Chicago are most often associated with what archaeologists designate as Fisher and Huber phases of artifacts. From recent excavations, it appears that Oneota people of the precontract era settled in similar strategic sites near possible portages and river confluences as later successors did during the seventeenth and eighteenth centuries.[5]

Illinois (Inʋca/Inohka/Illiniwek): This collection of Illinois/Miami-speaking groups moved into the Illinois Country sometime in the early seventeenth century, likely originating closer to Lake Erie or the upper Ohio River valley. By the time of French contact, there were five main identifiable subgroups, but there may have been as many as twelve to thirteen smaller divisions of Illinois people living along the river valleys of the region at the time. While historians have often referred to the collective groups as the "Illinois Confederacy," this large body of allied and interrelated peoples did not exhibit the same overarching governmental organization as formal confederacies such as the Iroquois. Their prowess as diplomats, warriors, merchants, and slave traders nevertheless made them the dominant political entity in the Illinois River drainage from the seventeenth into the early eighteenth century. Their relevance to French regional ambitions and alliances was so vital that many early maps refer to Lake Michigan as the Lake of the Illinois. Their large population centers made them particularly notable. The **Kaskaskia,** most famously, settled the Grand Village that attracted as many as 6,000 residents at its height in the late seventeenth century. Kaskaskian merchants traveled through the Chicago portages as early as 1674 on trade ventures back and forth from the Great Lakes, and in 1677 a delegation of eighty Kaskaskias encamped along the Des Plaines River to greet Father Claude Jean Allouez upon his arrival. Clearly during this period, Chicago's portages fell within the Grand Village's sphere of influence, though there are no historical accounts of local Kaskaskian settlements at Chicago. The other group of Illinois with notable relevance to the Chicago area are the **Peoria,** who lived periodically in the central Illinois River valley around Lake Peoria (at the time, called Lake Pimiteoui). The Peoria maintained villages as far north as Starved Rock and claimed the upper Illinois River valley as part of their hunting district as late as the 1760s, though by that time it was contested by incoming Anishinaabeg as well as Mascoutens, Sauks, Meskwakis, and Kickapoos. Interestingly, another group of Illinois, the **Michigamea,** had a prominent leader named Chikagou during the early eighteenth century. While this evinces a common origin word in the Illinois/Miami language, there is no explicit connection between Chikagou the Indigenous leader and Chicago the place.[6]

Kickapoo: The Kickapoo are an Algonquian-speaking group with connections to several geographic areas of the Great Lakes, including parts of southeastern Michigan, southern Wisconsin, the Wabash River valley, and northern Illinois. At the time of French contact, Kickapoos were living in composite towns north and west of Chicago and closely associated with Mascoutens, Sauks, Meskwakis, and occasionally Miami-speaking groups. Though the Kickapoo never held exclusive local control over Chicago, they lived in the immediate vicinity in adjacent river valleys. Unlike Anishinaabeg and more northerly groups of Iroquoian and Algonquian peoples, they remained less tied to canoe travel, but they likely were still aware of Chicago's portage routes and used them at various times. The Kickapoo faced threats of attack from both the Haudenosaunee and Illinois during the late seventeenth and early eighteenth centuries and initially allied themselves with the Meskwakis during the Fox Wars. As pressure against the Meskwakis mounted in the 1720s, they sought a separate peace with the French and their Indigenous allies. The Kickapoo cooperated with the Meskwakis, Sauks, and Anishinaabeg moving into the region during the second half of the eighteenth century, but by 1800, most Kickapoos dwelled along the Wabash River valley of Indiana. Eventually, the Kickapoo would adopt groups of Mascoutens into their society. The Kickapoo people faced ongoing pressure from the United States and its citizens during the prolonged Removal era, and their subsequent odysseys scattered groups of Kickapoo to Kansas, Oklahoma, Texas, and Mexico.[7]

Mascouten (Mathkoutench): A central Algonquian group variably recorded as the "Fire Nation" or the "Prairie People" in early sources. Like their linguistic and cultural associates, the Kickapoo, the Mascoutens maintained connections with geographic locations on both sides of Lake Michigan, as well as parts of what would become Illinois and Indiana. The Mascoutens faced threats of attack from both the Haudenosaunee and Illinois during the late seventeenth and early eighteenth centuries, and initially allied with the Kickapoo and Meskwaki during the Fox Wars. By the 1720s, they coordinated with the Kickapoo to negotiate a separate peace with the French and their Indigenous allies. The Mascoutens cooperated with the Meskwakis, Kickapoos, Sauks, and Anishinaabeg moving into the Chicago area during the second half of the eighteenth century. The Mascoutens lived in towns adjacent to the Chicago route of travel, and many lived in the composite Indigenous town known as *La Fourche* at the forks of the Kankakee and Des Plaines Rivers as late as the 1770s. They also had a town near the confluence of the Iroquois and Kankakee Rivers. By the turn of the nineteenth century, most Mascoutens had combined with settlements of Kickapoos and Miami-speaking peoples along the Wabash River valley, eventually joining the Prairie Band of the Kickapoo.[8]

Menominee (Mamaceqtaw/Folles Avoines): During the colonial period, the Menominees lived in central Wisconsin, sometimes in composite townsites alongside Ojibwes or Ho-Chunks. While they were called the "*Folles Avoines*" nation by early French observers for their use of wild rice (*Zizania aquatica*), they were not the only group to utilize this resource in the western Great Lakes. Few Menominees had a direct connection to the Chicago area, though they did rely on the more northerly portage route through the Fox and Wisconsin Rivers to span the Great Lakes and Mississippi watersheds. Mamaceqtaw is the autonym used by some Menominee people today.

Meskwaki (Fox/Renards/Outagamis): The Red Earth People, as they call themselves in their own Algonquian language, resided in portions of what would become Wisconsin, Illinois, and Iowa during most of the colonial era. Their own oral history connects them to more eastern Great Lakes geography, and for a brief period in the early eighteenth century they joined in the resettlement projects of Antoine de la Mothe Cadillac around Detroit. The Meskwakis also suffered from captive raids perpetrated by the Illinois and Haudenosaunees during the seventeenth century. Their experience throughout the French period of colonization was marred by a series of conflicts with the French but especially other Algonquian groups that sought to exclude the Meskwakis from the French alliance system. The Meskwakis countered French and Indigenous threats to their existence by blockading portage routes like those through Chicago and the Wisconsin and Fox Rivers. By the 1720s, these hostilities had escalated to the point where French colonial officials were actively engaged in campaigns of extermination against the Meskwakis. The Meskwakis, however, relied on alliances and kin ties with other groups in the Chicago borderlands—first the Mascouten and Kickapoo, and later the Sauks and even Anishinaabeg—to secure protection and stave off annihilation. During the 1740s, Meskwaki, Sauk, and Anishinaabe refugees settled into composite towns around Chicago. The Meskwakis would eventually associate closely with the Sauks, and both groups maintained towns throughout northwestern Illinois, southwestern Wisconsin, and west of the Mississippi River in Iowa. After the Black Hawk War of 1832, in which the "British Band" of Sauks, Meskwakis, and Kickapoos were targeted by U.S. forces, the Meskwakis were forcibly removed from east of the Mississippi River.[9]

Métis: A term used throughout the seventeenth, eighteenth, and nineteenth centuries for people of mixed Indigenous and European ancestry in the Great Lakes region and elsewhere. Métis people played crucial roles at Chicago and throughout the region as cultural mediators between Native communities and European empires and were critical to the flourishing fur trade. In many cases, Great Lakes métis families remained tied to the fur trade, spoke French (even after French imperial withdrawal from the region), and practiced varying Catholic traditions, sometimes alongside Indigenous beliefs. The multilingual and bicultural reality of métis individuals has led some scholars to also refer to them as "creoles," though this risks confusion with the term *creole* in other contexts. Métis families made up a large portion of the population at trade centers like Chicago, Green Bay, Mackinac, Peoria, and Prairie du Chien during the early decades of the nineteenth century. In the 1830s, amid the expulsion of Native peoples, many of Chicago's métis population faced the difficult choice between removing west of the Mississippi along with their Indigenous kin or remaining in Illinois in attempts to assimilate into the burgeoning white American society. While *Métis* was and is used as an overarching term for those in the Great Lakes region who trace mixed ancestry, it should not be confused with the Métis Nation of Manitoba, Saskatchewan, and Alberta, a formally recognized Indigenous people of Canada who likewise trace a dual heritage of Indigenous and European descent. I use *métis* as a descriptor, written in lowercase, throughout the book to distinguish these Great Lakes métis populations from the recognized First People group of Canada.[10]

Miami (*Twightwee/Myaamia/Myaamiaki/Mihtohseeniaki*): Miami peoples are closely related culturally and linguistically to the Illinois people but remained distinct, politically and diplomatically, for much of the colonial American period. Miami-speakers, of various subgroups, inhabited towns stretching from southern Wisconsin across northern Illinois and Indiana at the time of French contact in the late seventeenth century. Early Miami-speakers who may have used Chicago's portages or even lived in the area before clear archival records included the **Pepikokia** and **Piankeshaw**, though French reports about the early locations of these groups remain vague. The **Wea** settled around Chicago in large towns in the last decade of the seventeenth century and maintained villages near the portages, along the Des Plaines and Calumet Rivers through the early years of the eighteenth century. The Wea abandoned Chicago in the early decades of the eighteenth century as the portage route through the area made their towns vulnerable to amphibious attacks from other Indigenous groups. They also may have moved east at this time, along with other Miamis, to gain better access to English traders in the Ohio valley. Miamis usually appear in British colonial documents as some variation of "Twightwee," a garbled reference to the sandhill crane, an important bird within Miami cosmology.[11] Despite having withdrawn from the local area, Miamis still claimed Chicago as part of their historic homeland as late as 1795 at the Treaty of Greenville. A contingent of American-allied Miami scouts participated in the Battle of Fort Dearborn in 1812, accompanying the retreating U.S. garrison. Mihtohseeniaki, Myaamia, and Myaamiaki are all variations of the autonym that Miami peoples use for themselves.[12]

Osage (Ni-u-kon-ska): The People of the Middle Waters lived well south and west of Chicago by the colonial period, having migrated from the Ohio valley at some point before French contact. They are a Siouan-speaking people, and historically, they were distinguished by Europeans as two separate groups, the "**Little Osages**" and the "**Great Osages**." Much of their power base relied on their geographic position near the confluence region of the central Mississippi River valley. They only factor into Chicago history in a limited way, however. They were often the target of Chicago-based Anishinaabe raids down the Illinois River from the mid-eighteenth century into the early years of the nineteenth century. In earlier eras, some Osages traveled as far north as the Illinois settlements at Starved Rock in trade and diplomatic ventures.[13]

Sauk (Oθaakiiwaki/Sac): The Sauk, or Yellow Earth people, are an Algonquian group closely related to the Meskwakis through language, culture, and kinship ties. During the colonial period, they lived in portions of southern and central Wisconsin, though their precontact homelands may have been as far east as Lake Huron's Saginaw Bay. At various times during the colonial era, they lived in proximity to groups of Meskwakis, Menominees, Kickapoos, Mascoutens, and Anishinaabeg and were formally allied with the Meskwakis in various wars. Alongside Meskwaki refugees and Anishinaabe kin, some Sauks settled locally in Chicago during the 1740s, and these towns remained composite inter-Indigenous centers for several decades at least before developing under Potawatomi leadership. After the Black Hawk War of 1832, in which Black Hawk, a Sauk leader, and his "British Band" of followers was targeted by U.S. forces, the Sauk were forcibly removed from east of the Mississippi River.[14]

Shawnee (Ša·wano·ki/Shawanoe): The Shawnees have been described as "the greatest Travellers in America" by both colonial commentators and modern historians.[15] In the years after European contact, these people migrated across much of the American Southeast as well as the Ohio Country, Great Lakes region, and mid-Atlantic Seaboard in efforts to mitigate the threats posed by European colonization, endemic warfare, and epidemic diseases. This diaspora led some Shawnees into the orbit of the Grand Village of the Illinois and the colonial enterprises of the French adventurer René-Robert Cavelier, Sieur de La Salle. As such, though no Shawnees likely lived at Chicago, they shaped the Indigenous and colonial history of Illinois as a force to be reckoned with during the seventeenth century. During the eighteenth century, it was rumors of a Shawnee relocation to Chicago that first alerted Anthony Wayne to the dangers of the portage as a site of Indigenous resistance. Shawnee actions continued to influence events at Chicago into the early nineteenth century, when the Shawnee brothers, Tecumseh and Tenskwatawa, spread their message of Nativism from their bases at Greenville, Ohio, and then Prophetstown, Indiana. Many of Chicago's Anishinaabeg found this message of Indigenous revitalization appealing in the years leading up to the War of 1812. These included notable Indigenous and métis residents of the Chicago area, such as Shabbona, Main Poc, and Billy Caldwell, who all allied with the British and fought alongside Tecumseh's forces in Michigan and Canada. While many of the Anishinaabeg who fought at the Battle of Fort Dearborn were no doubt inspired by the Shawnee brothers' message, the attack against the U.S. garrison was not launched in direct coordination with Tecumseh's orders, as some earlier historians speculated.[16]

Urban Indians: A term used to designate Indigenous people—Native Americans and First Peoples—who live in urban areas like modern-day Chicago. While many urban Indians continue to have direct ties to tribal governments and specific Indigenous nations, others do not, having moved to urban centers in the wake of land dispossession or disenrollment, or in search of economic opportunities. Community organizations within urban spaces serve as cultural focal points for many urban Indigenous populations, and the National Urban Indian Family Coalition addresses the needs of urbanized Indigenous peoples nationwide. Today, Chicago is home to thousands of Native peoples. Chicagoland includes an Indigenous population of over 65,000 people, representing 175 different tribes, bands, and nations. This makes Chicago the third-largest urban Indigenous metropolis in the United States. While some of these people hail from nations that historically claim Chicago as homeland, others have migrated from various parts of North America and the world to call Chicago home. The flourishing of local Native communities in the twenty-first century is a reminder that Chicago remains a thoroughly Indigenous space into the present.[17]

Wendat (Huron/Wyandot): The Wendat are a collection of Iroquoian-speaking peoples originating from the Georgian Bay region of eastern Lake Huron. Wendat people faced threats from French missionary activity, the spread of European diseases, and Haudenosaunee attacks beginning in the mid-seventeenth century. These various pressures forced many Wendat to disperse farther west into the Great Lakes, where they sought refuge among Algonquian-speaking peoples and secured a lasting alliance with the French. Wyandots are a subset of Wendat who migrated into lands that would become the states of Michigan and Ohio during the eighteenth century. A group of Wendat warriors arrived at Chicago in the winter of 1730–31 as part of a campaign of the Fox Wars. Along with allied Haudenosaunees and Anishinaabeg, these Wendats constructed a fort near the portages as a refuge for the winter and a convalescence center for their sick and injured combatants.[18] It is also likely that other incidents of Wendat presence and travel through the portages went undocumented in European sources during the seventeenth and eighteenth centuries.[19]

Additional Resources

Much information for the above descriptions is drawn from primary sources and targeted secondary readings on Chicago and the Great Lakes region. Some useful and current reference works that provide information for Native people groups across North America include Frederick Hoxie's *Encyclopedia of North American Indians* (Boston: Houghton Mifflin, 1996) and the incomplete *Handbook of North American Indians*, 17 vols. (Washington, D.C.: Smithsonian Institution, 1978), edited by William C. Sturtevant. An instructive assessment of the northeastern volume of this series is William Starna's "After the Handbook: A Perspective on 40 Years of Scholarship since the Publication of the Handbook of North American Indians, Volume 15, Northeast," *New York History* 98, no. 1 (2017): 112–146. For Indigenous Chicago, see the ongoing work and new initiatives at the Newberry Library's D'Arcy McNickle Center for American Indian and Indigenous Studies. A dated but still helpful ethnohistorical overview of many of the regional Native peoples is W. Vernon Kinietz's *The Indians of the Western Great Lakes, 1615–1760* (Ann Arbor: University of Michigan Press, 1965); it is especially valuable as a road map to primary sources. The best work on the human geography of the

Indigenous Great Lakes during the colonial period is Helen Hornbeck Tanner and Adele Hast, *Atlas of Great Lakes Indian History* (Norman: University of Oklahoma Press, 1986). Many tribal governments now also provide updated, collaborative information on their history, language resources, and current affairs on official websites. For some starting points when it comes to methodological approaches to Indigenous history and studies, see Nancy Shoemaker, *Clearing a Path: Theorizing the Past in Native American Studies* (New York: Routledge, 2002); Patricia Kay Galloway, *Practicing Ethnohistory: Mining Archives, Hearing Testimony, Constructing Narrative* (Lincoln: University of Nebraska Press, 2006); Albert Hurtado et al., *Major Problems in American Indian History,* 3rd ed. (Stamford, CT: Cengage Learning, 2014); Chris Andersen and Jean M. O'Brien, eds., *Sources and Methods in Indigenous Studies* (London: Routledge, 2016).

CHAPTER ONE

Openings

From Ecotone to Borderland

With mud and timber, a colony of beavers divided the waters of the continent. It was sometime in the early seventeenth century, in the Chicago wetlands near the southwest shoreline of Lake Michigan. Written records remain scarce for this early period, but the beavers' activities around Chicago's continental divide left an enduring legacy on the waterways of the area. Building an extensive dam across a shallow marsh, the beavers constructed a barrier between the tributary streams of the Des Plaines River to the west and the Chicago River to the east. Given the size of the structure, the beavers must have begun building it long before the French arrived in the area during the last years of the century.

Beaver dams like the one at Chicago can last many decades as multiple generations of beavers maintain and add to them. The Chicago dam endured as late as 1682, when Robert Cavelier, Sieur de La Salle encountered it on his travels. As La Salle passed through, he noted the enormous dam as the central feature of the local waterscape. Along the main canoe route up the Chicago River, he reported, "There is a little lake divided into two by a causeway of the beavers, about a league and a half long." La Salle described how waters on one side of the beaver dam ran toward Lake Michigan while waters on the other emptied into the Des Plaines River, which ran south into the Illinois River and the Mississippi watershed.[1] La Salle was witnessing the drainage of the lowest, shortest continental divide between the watersheds of the Mississippi River and the Great Lakes Basin, which in 1682 ran along the top of the beavers' structure through the muddy lowlands of Chicago.

For most of the year, there existed a natural continental divide at Chicago, formed after the last Ice Age. This low glacial ridge, running between the beaver's marshland lake and the beginnings of the Chicago's southern branch, kept the Mississippi and St. Lawrence drainages separate by a thin strip of saturated earth. Wet prairies, comprising much of the terrain in the immediate area, tended toward seasonal flooding as the impermeable layer of topsoil and grasses became waterlogged from snowmelt and spring rains.[2] This was "contested ground," as one naturalist has described it, muddying the distinction between water and land and varying from one rainfall to the next. While times of low precipitation allowed the low-lying prairies to temporarily dry out, in seasons of flooding the waters between the Chicago and Des Plaines Rivers could totally submerge the low divide, connecting the watersheds periodically and inundating the entire area, despite even the beavers' best efforts to regulate water levels.[3]

A labyrinth of sluggish waterways coursed over the marshy terrain of Chicago, feeding a rich wetlands ecosystem between the glacial waters of Lake Michigan and the tallgrass prairie of the Illinois Country. Several rivers ran through this matrix of sloughs, marshes, oak groves, and submerged prairies, providing paths for drainage but also routes of travel. La Salle described the entire area as a "junction of several rivulets or prairie ditches" that came together to form a crossroads of waterborne movement.[4] At least two ancient portages served to connect the waterways, providing short overland pathways on which Indigenous peoples carried canoes and supplies between waters that led into the Great Lakes and waters that flowed south toward the Mississippi River. One of these carrying routes, called the *Portage des Chênes* by later French, ran overland between the south branch of the Chicago River and the Des Plaines, while the other portage, *Portage des Perches*, ran from a creek off the Calumet River into another tributary of the Des Plaines. Another route, up the northern branch of the Chicago River, may also have been used at times, though archaeological evidence remains scarce.[5] Given the varying water levels between seasons, and the intersecting backwaters, the lengths and locations of the portages varied from year to year, and even season to season, yet the multiple portage routes through Chicago's swamps provided travelers with a range of options to carry canoes and supplies only a short distance between the two watersheds.[6]

Nonhuman activity contributed directly to the efficacy of this watery highway across the continental divide. Beavers built dams like the one at Chicago to control variations in the water levels of the surrounding area. A dam across the wetlands ensured consistently more water surrounding the beavers' lodges in winter and supported the stability of the local water table. In turn, this maintained a suitable habitat for the beaver colony, providing ample aquatic vegetation for food and higher waters to protect the beaver lodges from predation.[7] While the beavers' designs for the dam revolved around local subsistence and survival, the repercussions of separating or combining the waters of the Chicago and Des Plaines Rivers held wider implications for the human geography of the continent's waterways. The presence of beaver dams in the watershed helped maintain higher water levels overall, which kept the river channels more navigable for Indigenous canoers moving between Lake

Michigan and the Illinois River valley. Taking their cue from the local fauna of Chicago's wetlands, over the next several hundred years, various peoples—both Indigenous and European—would struggle to control and connect the water-ways of Chicago as they came to envision this small patch of muddy ground as the key to power in the North American interior.

Harnessing Chicago's Portages

As the beavers stripped logs, layered mud, and built their dams across North America's lowest continental divide during the seventeenth century, a wide array of human traffic passed through the portages along these water-borne crossroads. Early in the seventeenth century, Oneota people, identified by their Huber-phase archaeological remains, exploited the wetland resources along Chicago's river routes. As more aggressive Algonquian peoples like the Illinois and Miamis displaced these earlier Oneota people, the portages came to serve as overland links between trading networks of the Great Lakes and Mississippi River valley during the latter half of the seventeenth century.

Such strategic geography soon attracted various groups, including Illinois slave traders, Anishinaabe middlemen, Haudenosaunee war parties, French fur traders, and Jesuit missionaries. These various groups came to envision Chicago's portages as strategically important routes even while local control of the area proved elusive. Chicago's accessibility as a waterborne thorough-fare cut both ways in the tumultuous latter decades of the seventeenth-century Great Lakes. While the ease of the route made it a thriving conduit of movement and exchange, its openness left local inhabitants and itinerant travelers vulnerable to invasion and attack. But by the end of the century, as Miami-speaking Weas, Catholic missionaries, and French officials began to establish collaborative footholds at the portages, Chicago's role began to change. No longer simply an opening for waterborne movement, a diverse group of Native peoples and Europeans began to hope for a more permanent presence along the confounding waterways of Chicago's wetland.

Examining Chicago as a transitional space, both ecologically and in human geography, offers a more fully fledged picture of Indigenous-European relations in the Great Lakes during the seventeenth century. Much of the ebbs and flows of power remained rooted in the particular geographies of encounter. The Chicago portage, serving as a transition zone between eastern woodlands and tall-grass prairies and a bridge between the waterways of the Great Lakes and the Mississippi watershed made for a natural conduit of movement both before

European contact and after French incursions into the west. As a corridor connecting the regions' waterways and one situated between powerful Native peoples, Chicago began the century as an ecological and geographic space *in between*.[8] As a diverse array of people became enamored with the potential of the portages and surrounding landscape, Chicago would transform into a thoroughfare for trade and, eventually, a borderland of contention as the century wore on and European intrigues triggered a cascade of Indigenous violence westward.[9] While no one group proved capable of claiming and controlling Chicago during this early era of contact, the route through the wetlands remained a strategically and commercially significant space of movement throughout the period.[10]

Native peoples near the portages understood this. As new motivations for trade and new threats of violence arose in the wake of European arrival, Native peoples would harness the portages as an access point and geopolitical negotiating tool to protect themselves from spreading warfare and increase their access to commerce. At the same time, French adventurers and missionaries projected their hopes for empire onto the localized space of Chicago, despite its geographic challenges. Tracing the Chicago portages' various stages as ecotone, trade lane, and Indigenous borderland reveal the violent disruptions that came about in the wake of European arrival in the Great Lakes. In analyzing Chicago's changing status throughout the seventeenth century, we can also appreciate the ongoing significance of geography in shaping relations between both French and Native peoples, and among the various Indigenous groups that saw Chicago as a vital nexus of mobility and power in the early borderlands of contact.

A Transitional Environment

From the seventeenth century forward, various peoples found Chicago's local environment and its potential for transcontinental connections appealing. While the waters of the area intersected sporadically across a continental divide, Chicago also lay along an ecological transition zone, called an ecotone, between the tallgrass prairie and the woodlands of the Great Lakes basin.[11] In this ecotonal borderland, marshy rivers ran through what later visitors called "Oak Openings." These oak savannas originally characterized much of the transitional meadowland running along the southern rim of the Great Lakes, from what would become western Ohio, southern Michigan, northern Indiana, northeastern Illinois, and southern Wisconsin. Alternatively described as prairie parkland, the oak openings featured groves of large burr or black oak, interspersed by open prairies and underbrush. The French named the portage at Chicago *Portage de Chênes*, "Portage of the Oaks," after the distinct oak groves lining the corridor between the lake and the prairies. This name for the vital portage pathway offers a fitting description of how the broken landscape of the transitional "Openings"—interspersed groves of oak trees amid wider prairies—would have looked in the early seventeenth century, when the beavers constructed their expansive dam across the waterways.

Ecologically, oak openings represented a rich border zone between the tallgrass prairies of the Illinois Country and the deciduous woodlands of the lower Great Lakes. The oak openings offered ideal habitat for quail, plovers, grouse, turkeys, and deer as an "edge environment," and their open patches of prairie provided ample grazing for herds of elk and bison. Louis Jolliet, who would be one of the first Europeans to travel through Chicago, described the region's oak openings as "equally divided into prairies and forests." He wrote that this broken terrain of savanna and oak groves provided "fine pastures for the great number of animals with which it abounds."[12] Father Claude Allouez, an early Jesuit missionary, described the openings on a voyage along Lake Michigan's shoreline in 1676, relating, "We advanced coasting always along vast prairies that stretched beyond our sight; from time to time we saw trees, but so ranged that they seemed planted by design to form alleys, more agreeable to the sight than those of orchards."[13]

Father Allouez's allusion to a cultivated orchard was not far off the mark. Native peoples of the region maintained the oak openings by actively setting fires across the prairies, which helped to maintain the oak openings as ideal hunting habitat. Such controlled burns eliminated woody plants and tree saplings while encouraging new growth of tender grasses. This welcoming forage in turn attracted herds of bison, elk, and deer. Thick-barked oaks, resistant to low-level fire, survived the blazes, and in this way, Indigenous peoples sustained the important transition zone ecology of the openings. Herds of large animals, attracted to new-growth grasses, further perpetuated the transitional nature of the environment, maintaining the savannas through intensive grazing.[14] Thus, at the same time Chicago's waterways offered a riverine crossroads between continental watersheds, the local landscape thrived as a transition zone between woodlands and prairie ecosystems within the oak openings of southern Lake Michigan. Chicago's locality bustled with life, as both a geographic and ecological borderland spanning waterways and bioregions as a space in between.

Chicago's position as the shortest and lowest continental divide along the southern rim of the Great Lakes was distinct, but its marshy terrain remained characteristic for much of the region west and south of Lake Michigan during the early modern period. Ranging from the extensive Limberlost and Kankakee wetlands in the east to Chicago's marshes and wet prairies, the region fostered a diversity of flora and fauna across a topography that blurred the lines between land and water.[15]

These rich marshlands provided a conduit of movement for a wide array of animals. Waterfowl, particularly, utilized the marshes of the region as a stopover on their yearly migrations through the Mississippi flyway. Teals, geese, swans, sandhill cranes, mallards, and wood ducks all flocked to Chicago's wetlands in the spring and fall. The marshes offered ample aquatic vegetation for food, and in the fall, stands of wild rice proved especially nourishing as the birds made their way south. The secluded and still waters of the marshlands and rivers also provided ideal spawning grounds for a host of fish species. The position of the inland marshes, adjacent to the open waters of Lake Michigan, offered a nursery for young fish and a productive hunting ground for ambush predator species like muskellunge, northern pike, longnose gar, and bowfin. Meanwhile the neighboring rivers teemed with catfish, sturgeon, bass, suckers, turtles, and freshwater mussels.[16] Just off the coast, in the waters of the Great Lake, still more aquatic species thrived. Whitefish and ciscoes schooled in the depths, while the rocky shoals of Julian's Reef, fourteen miles offshore, provided ideal spawning grounds for massive lake trout that grew to over twenty pounds.[17] Sturgeon, the lake's largest fish, cruised along Chicago's shoreline and could exceed five feet in length. Chicago's diversity of aquatic and terrestrial habitats, along an ecological borderland between lake and river, woodland and prairie, oak openings and wetlands, offered an ideal setting for a host of species that flourished along such a transition zone.

An Indigenous Corridor

Such a rich ecological area, endowed with oak savannas, abundant wetlands, and the vast resources of Lake Michigan, appealed to the region's Native peoples long before our first written descriptions of the area. These ecotonal transition habitats—oak savannas and wetlands—provide two of the most productive environments found in the Great Lakes region, and Native peoples relied on such spaces as valuable hunting and foraging grounds.[18] Archaeological remains dating to the fourteenth and fifteenth centuries from sites along the margins of Chicago's wetlands demonstrate that Native peoples exploited Chicago's faunal resources for centuries. Remains of wetland and aquatic species made up over two-thirds of the material found in ancient trash pits.[19] Among the most common refuse at such sites, archaeologists have found shell remains from mussels and turtles, along with bones from waterfowl, fish, and beaver. Scapula, antlers, and other remains from large ungulates such as elk, deer, and later, bison, are also present, demonstrating that Native peoples living along this ecological border zone utilized resources from both the oak openings and the surrounding waterways.[20] Discoveries at a site near the confluence of the Des Plaines and Kankakee Rivers—just downstream from Chicago's portages— contained remains of lake sturgeon, indicating that these early inhabitants were accessing resources directly from Lake Michigan as well. Archaeologists date these digs from the opening decade of the seventeenth century as late-Huber-phase sites of the Oneota people, identified by their distinct pottery, and prob-ably linked to later Siouan-speaking Ho-Chunk (Winnebago) people of southern Wisconsin. As more bison herds moved east into the prairies and oak openings of the upper Illinois River valley in the early seventeenth century, these sites be-came even more ecologically productive, providing Huber-phase Oneota people the chance to exploit various overlapping resource habitats within the immedi-ate area around Chicago.[21]

As appealing as the richness of these ecological borderlands proved to early Native peoples, additional value lay in the proximity and occasional intermin-gling of Chicago's waterways as routes of travel. During the early seventeenth century, Oneota communities gave way to more aggressive Algonquian-speakers from the south and east.[22] Local Huber-phase settlements began to disappear around Chicago as groups of Miami extended their influence westward from the St. Joseph River valley. Later oral traditions among the Miamis relayed how they became "so numerous at St. Joseph's as to render it necessary for a part of the tribe to migrate." Miami leaders set out along the riverways of the region, "each of the hunters" seeking "a suitable place for the formation of a new settle-ment." The Miami-speakers who first journeyed into the Chicago area may have been Wea, led by their progenitor, Wuyoakeetonwee, who established settle-ments from the Wabash and Tippecanoe Rivers northwest toward the western shoreline of Lake Michigan. Other Miami-speaking Piankeshaws ventured even farther west, onto the rivers leading toward the Mississippi, where they "found great numbers of water fowl" and learned to hunt "almost always upon the water." Notably, some of these Miami-speakers who ventured west toward Chicago and the Mississippi drainage adopted the regular use of canoes—a de-parture from their more landbound kin to the east—and so took on a new name, "Misoaleeaukee," designating them as a "Canoe people."[23] As Miami-speakers entered the crossroads of Chicago, their very identities took on a new significance within this waterborne geography.

At the same time these early Miamis began expanding along the waterways south and west of Lake Michigan, linguistically similar Illinois peoples estab-lished themselves in the river bottoms and prairies farther south, exploiting the bison herds of the grassy uplands and capturing slaves from neighboring Native groups across the Mississippi.[24] Both the Miami and Illinois spoke similar Algonquian dialects, and they designated the river route through Chicago as *šikaakwa siipiiwi*, or, the Leek River, named for the patches of wild alliums growing near the portage. Tellingly, these Native peoples did not distinguish between the waters that flowed south into the Illinois River (*nisiipi-iwinaani*) and those that flowed east into Lake Michigan (*kihčikami*), seeing both rivers (the Chicago and Des Plaines) and their adjacent wetlands as parts of the same body of intersecting waterways. As far as these savvy Indigenous observers were concerned, the watersheds were united through the muddy waters of šikaakwa siipiiwi whether flowing toward the Great Lakes or the Mississippi (see map 1.1).[25]

As the Illinois peoples gained strength along the Illinois River valley and Miami settlers began practicing intensive maize agriculture nearer the lake, the portages at Chicago offered one of the easiest conduits for trade linkages northward into the Great Lakes. As early as 1634, a wandering French adventurer, Jean Nicolet, encountered Illinois traders in his voyages of exploration along the western Great Lakes.[26] Such early Illinois incursions into the Great Lakes demonstrated the wide reach of these powerful people as they traveled north from the Illinois valley along the river routes leading to Lake Michigan.

Most significantly, as Illinois power grew along the Mississippi watershed to the south, Chicago provided an opening into the Great Lakes region for trading ventures in captives. By 1667, at least "fifteen cabins of Islinois" traders had established themselves among the Potawatomi settlements at Green Bay, while other Illinois merchants journeyed as far north as the St. Esprit Mission of the French Jesuits on Chequamequon Bay, in Lake Superior.[27] The Jesuit missionaries had come to the Great Lakes to establish missions, learn Indigenous languages, and convert Native populations into the Catholic faith, but groups like the expanding Illinois saw such newcomers as valuable trade partners and potential allies in the constantly changing political geography of the Great Lakes region.[28] Father Claude Dablon, writing from St. Esprit in 1670, reported that the Illinois "repair to this place from time to time in great numbers, as Merchants, to carry away hatchets and kettles, guns, and other articles that they need."[29] While the Illinois occasionally traded directly with the French newcomers, most of their trade northward centered on the Native peoples of the Great Lakes. Trading with Anishinaabe groups like the Potawatomis and Odawas, the Illinois offered Indigenous slaves in exchange for European goods that the Anishinaabeg had acquired in the fur trade. Father Jacques

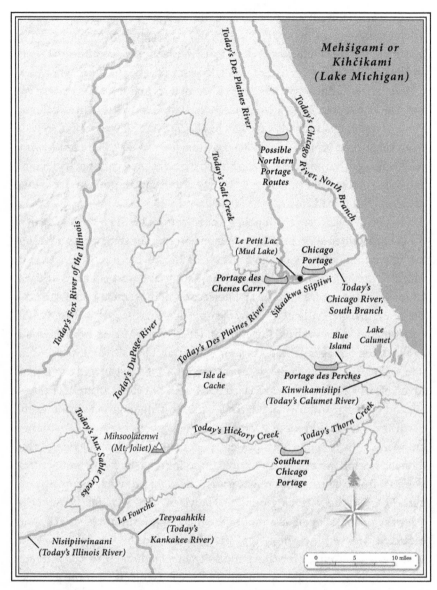

MAP 1.1 Probable portage routes and watercourses of Chicago, late seventeenth century, by Gerry Krieg.

Marquette, another Jesuit, relayed from his St. Ignace Mission near Michili-mackinac how "the Illinois are warriors and take a great many Slaves, whom they trade with the Outaouaks [Odawa] for muskets, powder, kettles, hatchets, and knives."[30] Such a slave trade appealed to Anishinaabeg in the Great Lakes as they recovered from waves of European epidemics and endured a costly war with the Haudenosaunee, or Iroquois, to the east. Captives sold by the Illinois to the Potawatomis and Odawas could be adopted into kin networks to replace fallen family members and maintain tribal strength.[31] In turn, these Anishinaabe middlemen proved content to trade French manufactured goods to the Illinois, and by the 1670s (if not earlier), they, too, were utilizing the portages at Chicago to access Illinois markets from the Great Lakes.[32]

Despite the wide variety of peoples passing through the portages, none of the earliest written sources from the region describe any settlements in the im-mediate vicinity of the Chicago routes.[33] As Miami and Illinois groups con-solidated their power along the river valleys south and west of Lake Michigan, Chicago came to serve more as an intersection of movement rather than as a site of permanent settlement. Traversed by Anishinaabe groups like the Po-tawatomis and Odawas from the north and Illinois and Miamis from the south and east, the portages offered a clear route of travel but also an exposed loca-tion, openly accessed via the water routes of Lake Michigan and the Illinois River. Nearby Kickapoos and Mascoutens used the Chicago area for travel-ing and hunting, and Meskwaki, Sauk, and Ho-Chunk groups living to the north and west shared some awareness of the portages' utility.[34] Positioned between so many Indigenous powers, perhaps the low-lying divide proved too tenuous for any one group to hold on to for long, or the potential threat of attack from the labyrinth of waterways discouraged would-be Native set-tlers. Possibly the various groups of Native peoples remained content to leave the portages open as a thoroughfare for trade and travel as exchange escalated between the Mississippi River valley and the Great Lakes. With so few written sources for this period, it remains difficult to say with cer-tainty, but long before Europeans arrived at Chicago, it is clear the portages bustled as an open thoroughfare of overlapping Indigenous activities, sea-sonal movements, and trading ventures with no single group asserting domi-nance in the local area.[35]

As Native Illinois merchants journeyed north into the Great Lakes and Anishinaabe trading parties voyaged south through Chicago to the Illinois Country, groups of travelers exchanged geographic knowledge about the best routes between the Great Lakes and the Mississippi. While much earlier ar-chaeological evidence demonstrates a thriving exchange between the lower

Illinois River valley and the upper Great Lakes in goods like pipestone, Lake Superior copper, and Cahokian chert, Native peoples began reconstituting these trade linkages during the seventeenth century.[36] Bridging the waters from the lower Great Lakes toward the Mississippi River valley, Indigenous travelers utilized several main portage routes to travel through the midcontinent. Finding the shortest overland divide between waterways, Native American mariners hauled their canoes and supplies from one waterbody to the next, continuing their travels afloat by these short overland carries. The best route could vary by season, water levels, or by the human geography of alliances and trade partners. Chicago functioned as one of these routes, but the portage between the St. Joseph and Kankakee Rivers to the east, and the portage between the Fox and Wisconsin Rivers to the northwest, both served similar purposes, allowing for a freshwater network of movement between the two great watersheds.[37] Well before the first French entered the region, various Indigenous groups utilized their geographic knowledge of these waterborne routes for trading, warfare, hunting expeditions, and seasonal migrations between homeland regions.

European Inroads

It was the Indigenous geographic knowledge of these routes that first brought the French to the portages at Chicago. In late 1672, the incoming governor-general of New France, the Comte de Frontenac, commissioned Louis Jolliet, a fur trader from Quebec, to explore rumors of a large river to the west, which might "provide a passage" to the Pacific Ocean, or "Sea of China."[38] On December 8, 1672, Jolliet arrived at the Jesuit Mission of St. Ignace, on the north shore of the Straits of Mackinac. There he recruited the Jesuit Father Jacques Marquette to accompany him on the venture. As the expedition members waited out the winter, they worked to glean geographic knowledge from Native people that had traveled the unfamiliar waters to the south and west. Marquette described how they "obtained all the information that we could" from Native people that had traveled the unknown regions. With this information, the French "even traced out from their reports a map of the whole of that new country; on it we indicated the rivers which we were to navigate, the names of the peoples and of the places through which we were to pass, the course of the great river, and the direction we were to follow when we reached it."[39] Sketching maps from the descriptions provided by local Odawas and other Native visitors, Jolliet and Marquette constructed a geographic picture of the lands that lay before them and the river routes they hoped to follow that spring.

Armed with these Native-informed maps, Jolliet and Father Marquette set out on May 17, 1673, to find a route from the Great Lakes into the Mississippi River, along with five other voyageurs—French adventurers employed to paddle the vessels. In two birchbark canoes, they made their way to Green Bay, on the western shore of Lake Michigan. Meeting Menominees, Kickapoos, Mascoutens, and Miamis along the route, the party members ascended the Fox River, then with the assistance of two Miami guides, carried their canoes into the Wisconsin River. While they had learned of this route from their Indigenous informants that winter and along the way, Marquette still found the route difficult. He complained, "The road is broken by so many swamps and small lakes that it is easy to lose one's way" en route to the portage. He was sure that without the two temporary Miami guides, they would have lost themselves on this leg of the journey. Once they entered the Wisconsin River, things failed to improve in Marquette's mind, as the sandy shoals of that channel "render[ed] its navigation very difficult."[40] Nevertheless, they persisted and followed the river down to the Mississippi, entering its broad course on June 17. From there, they contacted Illinois settlements as they descended the Mississippi River as far as the mouth of the Arkansas River. After ascertaining that the river emptied into the Gulf of Mexico, and not the Pacific Ocean, Jolliet determined to return north, having gotten to within "two or three days journey" of what they feared to be Spanish territory.[41] The party began to work their way back up the river, only gaining twelve or fifteen miles a day against the Mississippi's current, despite "paddling from morning to night."[42]

After ascending as far as the Illinois Country, Jolliet and Marquette's expedition visited a village of the Kaskaskia, where one of the settlement's leaders offered to show the French a route that "greatly shortens our road" as they sought to return to Green Bay and Michilimackinac.[43] This Kaskaskia headman and several of his people guided the French party up the Illinois River in September 1673, into the tributary that the French would eventually call the Des Plaines River. From the Des Plaines, Marquette recalled, "there is only one portage of half a league" into the waters of the Chicago River. Here, Jolliet, Marquette, and their five voyageurs carried their canoes "with but little effort" back to the Great Lakes watershed.[44] Crossing the muddy ground separating the Des Plaines and Chicago Rivers, Marquette and Jolliet found themselves back along the shore of Lake Michigan. Somewhere along the portage, the Frenchmen parted with their Illinois escorts and headed north to Green Bay, arriving in late November.[45] None of their discoveries, however, would have been possible without the steady flow of geographic information provided by their Indigenous hosts and guides along the way. It was

Native knowledge of the river routes that allowed Jolliet and Marquette to reach the Mississippi, and the Kaskaskias' awareness of the quickest portage back to the Great Lakes, up the Illinois and Des Plaines Rivers, that brought Jolliet and Marquette through Chicago at all.

Marquette stayed to recuperate at the St. Francis Xavier Mission, in Green Bay, through the winter, content in having baptized at least one dying child during the expedition, near Lake Pimiteoui. He believed he had laid a foundation for future missions among the Illinois towns and had promised them he would return to instruct them. Jolliet rushed on to Sault Ste. Marie and then Quebec, eager to report his discovery of the Mississippi to Frontenac. On the final leg of Jolliet's journey, however, his canoe capsized in the rapids of the St. Lawrence River. Several of the crew and an enslaved Indigenous boy that Jolliet had brought from the Illinois Country drowned in the turbulent water. Jolliet lost all his notes and mapping from the expedition.[46]

But the shipwreck failed to douse Jolliet's enthusiasm for the recent discoveries. He arrived unscathed to make his oral report to officials in New France during the summer of 1674. He had found the Mississippi River, confirming rumors that French Jesuits and officials had been fielding from Native informants in the Great Lakes country for decades. Though he had not reached the ocean, he had gotten within days of it, he believed. Given his gleanings from Indigenous informants, and the general flow of the river, he had furthermore determined that it must empty into the Gulf of Mexico, rather than the Western Ocean.

As significant as these discoveries proved, Jolliet also remained eager to highlight the importance of the overland passage his Kaskaskia guides had shown to him on the return journey. The portage at Chicago ran only a short way between the two watersheds and lay over very low prairie. When Jolliet and Marquette had gone through, water levels had been high, easing their journey. They reported only carrying their canoes for half a league—a little under two miles. Jolliet ventured to suggest that the French could dig a small canal there, which would eliminate even this short portage, and thus join the waters of the Great Lakes watershed with the upper reaches of the Illinois and Mississippi drainages.[47] At Chicago, Jolliet saw the potential for uniting France's claims across the continent, extending French mobility, and thus, French power, along the waterways of the interior.

Relaying Jolliet's ambitious vision for Chicago, the Comte de Frontenac, governor-general of New France, wrote to Jean-Baptiste Colbert, France's minister of state, in November 1674. Brushing over the details that the expedition had discovered the Mississippi River, he reported that Jolliet had found "a navigation so easy through the beautiful rivers . . . that a person can go from

Lake Ontario and Fort Frontenac in a bark [boat] to the Gulf of Mexico." The only thing impeding such waterborne movement, according to Frontenac, lay at a portage around Niagara Falls. In his mind, and from Jolliet's description, the marshland of Chicago posed no barrier, especially if Jolliet proved correct that the French could easily dig a canal there. Frontenac remained hopeful that "these are projects which it will be possible to effect" as soon as he succeeded in establishing a general peace between the Haudenosaunee and other Native nations of the Great Lakes. Once Frontenac established order in the region, New France would be ready to pursue a water route to the Mississippi drainage through Chicago "whenever it will please the King to prosecute these discoveries." In the meantime, Frontenac forwarded Jolliet's redrawn map of the route on to Versailles (see figure 1.1).[48] With Jolliet's report and Frontenac's endorsement, New France's colonists seemed poised to take control of Chicago's route and thus open an internal system of waterborne connections that would both bolster French imperial claims to the continent and facilitate swelling trade and communication between the Great Lakes and Mississippi River valley. As Frontenac and Jolliet implied to the Crown, the only requirement would be commitment from the French imperial state toward a canal project through Chicago's marshland.[49]

The Jesuit fathers stationed in Quebec likewise heard Jolliet's message. Framing his discoveries within their wider efforts for proselytizing in the North American interior, they even included his firsthand account in their annual report after he lost his notes in the rapids of the St. Lawrence River. With the Jesuit missionary efforts in mind, Father Claude Dablon wrote encouragingly that Jolliet's exploration "opens up to us a great field for the preaching of the Faith, and gives us entrance to very numerous peoples, who are very docile, and well-disposed to receive it." Father Dablon rejoiced that "all these countries abound in lakes and are intersected by rivers, which offer wonderful communications between these countries."[50] Most importantly, Jolliet and Marquette had shown "a very great and important advantage, which perhaps will hardly be believed"—that a water route could be established between New France and the Gulf of Mexico. Dablon speculated that missionaries "could go with facility to Florida in a bark [boat], and by very easy navigation." According to what Jolliet had relayed, "it would only be necessary to make a canal, by cutting through but half a league of prairie, to pass from the foot of the lake of the Illinois [Lake Michigan] to the river Saint Louis [Illinois River]."[51] For Father Dablon and his Jesuit colleagues, the existence of a water route, or even the potential for one, linking the Great Lakes and the Mississippi held as much importance for the expansion of the heavenly kingdom as any earthly

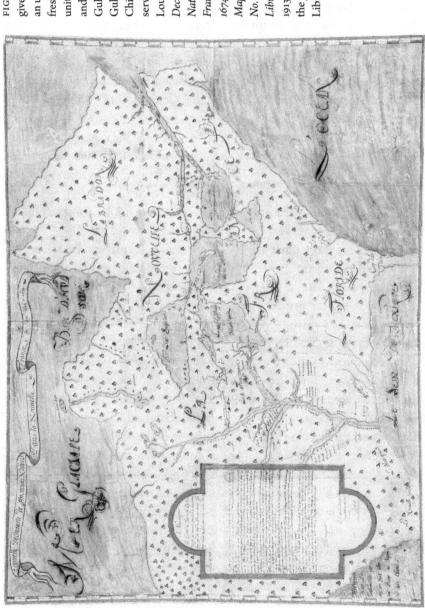

FIGURE 1.1 Joliet's map gives the impression of an unbroken route of freshwater channels uniting the French Empire and stretching from the Gulf of St. Lawrence to the Gulf of Mexico, with Chicago's passageway serving as the key linchpin. Louis Joliet, *Nouvelle Decouverte de plusieurs Nations Dans la Nouvelle France en l'année 1673 et 1674*, in K. Kaufman, *Mapping of the Great Lakes*, No. 8; John Carter Brown Library, *Annual Reports*, 1913, 21–22. Courtesy of the John Carter Brown Library.

venture. Waterways and the portages that connected them, it seemed, would unite all in this new, amphibious Eden.

While Jolliet preached Chicago's potential in Quebec, Marquette returned to the portage to continue his ministry in the late fall of 1674. Marquette brought two voyageurs with him, Pierre Porteret and Jacques Largillier, at least one of whom had ventured to the Mississippi during the first expedition. Initially, Marquette hoped to push on toward the Kaskaskias' settlements downriver on the Illinois, but on the voyage south from Green Bay, he began experiencing severe abdominal issues—possibly symptoms of dysentery or a chronic illness. Traveling up the Chicago River with accompanying parties of Potawatomi and Illinois traders, Marquette and his men decided to winter over at the trailhead of the portage rather than traveling on to the Illinois Country. On December 14, 1674, Marquette recorded in his journal that "it was impossible to go farther, since we were too much hindered and my ailment did not permit me to give myself much fatigue."[52] His Indigenous companions journeyed on without him, but not before trading tobacco in exchange for three buffalo robes, which would serve to keep the Frenchmen warm over the winter at the portage.

Waylaid by illness, Marquette struggled against his symptoms but continued to keep a journal of events at Chicago during the winter of 1675. From this, we get some of our first written glimpses of the environmental and human activities around the portage route. Marquette's companions hunted while Marquette convalesced in the cabin at the trailhead. The trio lived well on bison, deer, grouse, and turkeys throughout the winter from the surrounding oak savannas. Dried corn provided a staple as well. Even in the midst of winter, when visiting Illinois Indians expressed concern that the French squatters might starve, Marquette assured them "we feared not hunger."[53] When spring came, the French feasted on waterfowl and pigeons, and Marquette described how the wetlands teemed with "bustards, geese, ducks, cranes, and other game unknown to us."[54] These migratory birds moved through the Chicago marshes on their way north along one of the busiest flyways of the continent.

The portages, acting as a conduit of both human and animal movement, meant Marquette encountered many Native parties moving through the hub during his stay. The Jesuit and his companions had initially traveled from Green Bay to Chicago in the wake of Potawatomi and Illinois canoes, carrying trading parties south. Throughout the winter, Marquette met with Illinois visitors from downstream on the Illinois River who came to Chicago for a variety of reasons. Some passed through on hunts, and at least one party encamped in preparation for a longer-term trading expedition to the Native towns at Green Bay. Marquette relied on some Native travelers to act as couriers, taking letters on his behalf north to fellow missionaries in the Great Lakes. Others came intentionally to visit Marquette and his companions. Diplomatic and trade envoys exchanged gifts of beaver pelts and meat with the French while asking for tobacco, ammunition, knives, and other European goods. Some of these Illinois inquired after Marquette's health and requested that he come to their settlements soon. Marquette did his best to assure them he would come among them in the spring. Marquette's journal makes clear that even in the dead of winter, the portage remained a lively place of movement and interaction.

By 1675, Chicago's access as a route of trade had enriched and empowered an array of ambitious Indigenous merchants among the Illinois people. Some individuals had grown so distinguished as transregional traders that Marquette took special notice of them in his account from the portage that winter. Chachag8essi8, for instance, had preceded Marquette's voyage to Chicago, leading his own trade flotilla from the Green Bay area back to the Illinois towns for the winter. Marquette described him as "an Illinois greatly esteemed among his nation, partly because he engages in the fur trade."[55] Another Illinois, Nawaskingwe, meanwhile coordinated shipments of furs back and forth across the portage that winter. As Marquette put it, these shrewd Indigenous merchants "act like the traders, and give hardly any more than do the French."[56] By this period, as the exchange of furs, European goods, bison robes, and enslaved captives intensified between the Great Lakes and Illinois Country, Native individuals like Chachag8essi8 and Nawaskingwe staked their commercial fortunes and personal prestige to the mobility Chicago's geography allowed. Though residing far downriver, they and other Illinois traveled and traded widely, aided by the accessibility of Chicago's route.

These Illinois merchants were not the only visitors Marquette entertained at Chicago in 1675. Early in the winter, Largillier, one of Marquette's voyageurs, ventured to the nearest Illinois town several days' journey away, and met two French traders living there. Marquette called one of them "le chirurgien," or, the surgeon, and the other "la Toupine," whose given name was Pierre Moreau. These two coureurs de bois, or "woods' runners," had established themselves among the Illinois about seventy miles down the river at "a fine place for hunting cattle [bison], deer, and turkeys."[57] Such coureurs de bois—unlicensed traders— ranged far from the reach of New France to obtain furs without the oversight of colonial officials. From Marquette's account, it seems clear that some were already hunting and trading in the upper Illinois River valley by 1675.[58]

The geography of the Chicago region proved ideal for such unlicensed French traders, just as it did for Native merchants. While the neighboring prairies and oak openings provided ample hunting for bison, a source of both meat and hides, the river valleys and local marshes around the portages offered prime habitat for beavers and other wetland furbearers like mink, otter, raccoon, and muskrat. That spring, Marquette's party crossed paths with "the Surgeon" as he journeyed north to stow a canoe-load of beaver pelts at his cache along the Chicago portage route. The accessibility of riverways made the nexus of portages an optimal location for storing supplies and other valuables, and even Marquette made sure to stash a sack of corn at the trailhead for later use.[59] Moreover, the water routes connected by Chicago's portages meant open mobility for clandestine traders who traveled by canoe. From Chicago, they could access hunting grounds and trading towns downriver and still return east to sell their furs and hides at the end of the season, all by water.

In his journal, Marquette noted the seasonal variations that made Chicago's waterborne movement feasible, albeit unpredictable. That spring, while Marquette and his companions remained encamped at the trailhead, the ice of the surrounding rivers and marshlands began to thaw after a long winter. The waters of the Chicago and Des Plaines Rivers rose so quickly on March 29 that Marquette claimed, "We had barely time to decamp." The men sought shelter on high ground while they secured what goods they could in the branches of nearby trees. Marquette reported that the "very high lands alone are not flooded" and estimated the rivers must have risen twelve feet in a short time. Despite this initial panic, the rising waters proved more beneficial than not. The next day, the party decided to embark toward the Kaskaskia towns southward along the Illinois River. They paddled toward the portage path, only to find most of it submerged in navigable water. The men ended up having to portage their boats and supplies "about half an arpent"—only a third of a mile.[60] The variable waters of spring's snowmelts meant less work at the portage and more extensive navigable travel for the canoes, easing the party's voyage to the Illinois considerably.

Marquette and his companions traveled on to the Grand Village of the Kaskaskia, where Marquette said mass and spent Easter. Even so, his health continued to deteriorate. Soon, he and his companions determined it best to return to the Mission of St. Ignace at the Straits of Mackinac. The party turned north, retracing its route up the rivers and portaging over yet another pathway between the marshlands and Lake Michigan, before coasting the eastern shoreline of the lake, following the favorable currents.[61] But the waterborne retreat could not save Marquette from his illness. The Jesuit missionary died on the journey and was buried in the dunes overlooking Lake Michigan. Despite his death, Marquette and his two voyageurs had laid the groundwork for both future ecclesiastical and entrepreneurial adventurers at Chicago as more French learned of the area, its waterways, and the geographic potential the site seemed to offer.

Despite the activities of Marquette and the various French coureurs de bois in the area, the Chicago portages continued to act more as a channel of movement than as a permanent site of settlement in the years following the Jolliet expedition. This remained true for both the region's Native peoples as well as the incoming French. None of the earliest French travelers through Chicago mention local Indigenous settlements near the portages, though by the 1670s, the powerful Illinois downriver held enormous influence over the area.[62] Early observers constantly encountered parties of Native peoples on the move through the route, many of them Illinois or their allies. Marquette met trading and hunting parties composed of both Potawatomis and Illinois during the winter of 1675. In April 1677, when Father Claude Allouez arrived at the Chicago River to continue the evangelizing work of the Jesuits in the Illinois Country, a party of over eighty Illinois welcomed him and escorted him downriver to the Grand Village of the Kaskaskia.[63] Allouez found this Illinois settlement "largely increased," having attracted as many as eight different Illinois-speaking peoples and containing a population of at least five or six thousand inhabitants.[64] By 1677, Father Allouez encountered a burgeoning regional power downriver as the Illinois peoples positioned themselves to exploit both the bison resources of the tallgrass prairies and the water routes northward that linked their trade to the Great Lakes by way of the Chicago portages.

The Perils of No-Man's-Land

While the collective Illinois peoples worked to consolidate power along the river routes south of Chicago, the portages to the north also posed a threatening entryway for aggressors from the east. During the latter half of the seventeenth century, the Great Lakes erupted into a theater of intra-Indigenous warfare as the Haudenosaunees, members of the Iroquois League south of Lake Ontario, launched a series of offensives against the various Indigenous peoples of the upper Great Lakes. Armed with guns from the Dutch and, later, English traders in New York, Haudenosaunee raiding parties campaigned increasingly westward in efforts to gain captives, acquire individual prestige, and possibly expand their hunting districts.[65] The crossroads at Chicago's portages became an epicenter of conflict as the violence of these Mourning Wars spilled toward the Illinois Country. As early as the 1650s, a large force of Haudenosaunees, campaigning near Green Bay, turned south toward the Illinois River valley. Somewhere between Lake Michigan and the Mississippi River, they attacked "a small Illinoet Village," and while the Haudenosaunees killed the settlement's women and children, the Illinois men rallied reinforcements from neighboring towns. Pursuing the Haudenosaunees, the Illinois "utterly defeated them," somewhere near the shoreline of Lake Michigan.[66] These early bursts of bloodshed between the Haudenosaunees and Illinois pitted two potent confederacies against one another, and intermittent warfare between the two powers would continue throughout the rest of the century. The portages at Chicago, an established route from the lakes to the Illinois River valley, presented a beachhead for Haudenosaunee aggression as these campaigns raged back and forth along the southern coast of Lake Michigan.

Historians have often characterized Haudenosaunee campaigns to the west as a zero-sum conflict that pitted the Iroquois Confederacy against the various Algonquian peoples of the Great Lakes.[67] Along Chicago's dangerous crossroads, such delineations were far more complicated. Native people living and traveling along this gateway to the Illinois Country often found themselves caught between Haudenosaunee aggression from the east and Illinoian hegemony to the south. Father Claude Allouez, traveling with a group of Meskwakis during their winter hunt in 1674, reported that these people found themselves in "constant danger" from raiding enemies on all sides, including the Dakotas to the west, Illinois to the south, and Haudenosaunees to the east.[68] Several years earlier, "at the foot of the Lake of the Ilinioues [Lake Michigan]," eighteen Haudenosaunee raiders had attacked "six large cabins of these poor people" who they "put to rout," taking thirty women captive.[69] While the Meskwakis contended with Haudenosaunee attacks from the east, Allouez also reported that "the hilinois [Illinois] made raids upon them, and Carried off others into captivity."[70] Miami groups, despite linguistic and cultural similarities to the Illinois, soon found themselves in a similar position. When a party of Illinois visited Marquette at the Chicago portage in 1674, the Jesuit missionary had to plead with them not to "war with the Muiamis [Miami]" in the area.[71] By the 1680s, these Miamis complained to the French that they were "involved in disputes" with the numerous and powerful Illinois and "feared them hardly less than the Iroquois [Haudenosaunees]."[72] Living, hunting, and even traveling through the Chicago area was becoming treacherous business for other Native peoples as both Illinois and Haudenosaunee war parties expanded their raids and captive taking along this thoroughfare.

Caught between the violent incursions of these two captive-seeking confederacies, Miami, Meskwaki, Mascouten, and Kickapoo groups all vied to keep their powerful neighbors at bay, attempting to play the Illinois and Haudenosaunees off one another via the strategic geography of the portages. At times, this meant hosting Haudenosaunee diplomats and entertaining plots to join in their campaigns as they raged through the Chicago area. At other times, it meant forming alliances with the Illinois and even relocating to their thriving cities, such as the Grand Village of the Kaskaskia, downriver.[73] Surviving along the watery crossroads of various Indigenous powers at Chicago required tremendous diplomatic flexibility.

In some cases, these local intrigues worked effectively, while at other times, the captive taking continued, forcing groups to flee. In 1673, for example, the Meskwakis living northwest of Chicago sent an envoy to the Haudenosaunees "to solicit aid in the war which that tribe was then waging against the Illinois." Seeking an alliance with the Haudenosaunee against the Illinois, the plan backfired for the Meskwakis. The Haudenosaunees took the delegate prisoner and "carried away five families" with them back to Iroquoia as captives.[74] The Mascoutens, as well as some Miamis, chose flight over diplomacy. Groups of Miami refugees sought safety to the west along the Mississippi and north among the Jesuit missions of Green Bay, while some Mascoutens relocated their settlement to the headwaters of the Des Plaines River, "eight leagues," or about twenty miles, northwest of the Chicago portages and well clear of the main routes of travel.[75]

Buying off the aggressors worked too. Sometime in the 1670s, a war party of over eight hundred Haudenosaunee combatants came against the intertribal towns of the Miamis, Mascoutens, and Kickapoos south of Green Bay. Rather than resist such a force, the collected Native peoples received a Haudenosaunee diplomat into their midst with "the honors of the calumet" and "presented to him peltries, and two large canoes for transporting the presents which he received on every side."[76] Placated, the Haudenosaunee force divided, sending two hundred warriors down the Chicago River where they instead attacked a trading party of Illinois and Odawas, capturing or killing nineteen of them. Miami diplomats, in turn, journeyed east to negotiate a more permanent peace with the Iroquois Confederacy of the Haudenosaunees.[77] Around this same time, the Meskwakis successfully negotiated with the Haudenosaunees to establish that "the river Chigagon [Chicago] should be the limit of their raids."[78] This agreement offered advantages for both sides. The Iroquois agreed not to raid Miami, Meskwaki, Mascouten, and Kickapoo settlements west of the Chicago River, but also secured their access to the Chicago portages as a main route south in their campaigns against the Illinois. In this way, the Chicago portage route operated as both a barrier and a bridge for violence during the tumultuous decades of the Mourning Wars, with various Native peoples recognizing its importance as a dividing line as well as an entry point of waterborne war making in the region.

While the Miamis, Meskwakis, Mascoutens, and Kickapoos negotiated for security around Chicago, the Illinois brokered their own alliances to the south. In the winter of 1679, Robert Cavelier, Sieur de La Salle, had arrived, along with thirty men, at the Kaskaskias' winter encampment on Lake Pimiteoui. Offering these French adventurers corn and shelter, the Kaskaskias incorporated them as allies along the Illinois River. La Salle had entered the Illinois Country with visions similar to Jolliet's, of uniting the waterways between the Mississippi River and the Great Lakes to create a fur trading network through the heart of the continent. He had built a forty-five-ton vessel, *Le Griffon*, on the Great Lakes the year before, with plans to build a second ship on the Mississippi to relay furs and supplies throughout North America. The short continental divides running south along Lake Michigan from the St. Joseph-Kankakee and Chicago portages provide the only overland connections in La Salle's vision for a maritime empire deep within the interior. By 1680, however, he and his second-in-command, Henri de Tonti, found themselves enmeshed in the Native politics and violence along these once-promising water routes.

In September 1680, the Haudenosaunees launched their most destructive campaign yet into the Illinois River valley, sending more than five hundred Haudenosaunee warriors, along with one hundred Shawnee allies, against the large Illinois town of Kaskaskia. Henri de Tonti remained with the Illinois at their main settlement while La Salle had returned east for supplies. Tonti urged the Kaskaskia "not to scatter themselves at a time when they were expecting a pitiless war with the Iroquois."[79] Despite Tonti's pleading, many of the Illinois dispersed to hunt bison or raid for slaves, leaving the settlement along the Illinois River weakened when the Haudenosaunee arrived on September 18. The Illinois dispatched the women, children, and elderly downriver by boats under the cover of darkness while their warriors "passed the night after their manner in feasting, singing, and dancing," in preparation for their coming battle with the Haudenosaunees.[80] The following day, Tonti attempted to negotiate a peace, without success, and over the next "eight or ten days" the Haudenosaunee and Illinois forces skirmished back and forth on the eastern bank of the Illinois River before the Illinois warriors fled downriver in retreat. Tonti was wounded during the fighting, but he and his small group of French survivors—two Recollet priests and three voyageurs—fled northward by canoe. Passing over the Chicago portage and into Lake Michigan, Tonti's band eventually reached the safety of the Potawatomi settlements near Green Bay, while the Haudenosaunees continued to wreak havoc and take captives along the upper Illinois River valley.[81]

The Wea and other Native peoples near Chicago had colluded with the Haudenosaunees to allow for such a large and destructive campaign to take place. According to La Salle, the Miamis had entertained Haudenosaunee envoys, "listened to the proposals," and "came to an agreement with them." Several Haudenosaunee diplomats had even taken up residence among the Miami settlements in the St. Joseph River valley on the eastern shore of Lake Michigan.[82] Writing in 1681 about events in the Illinois Country, Jacques Du Chesneau, the intendant of New France, described how "the Miamis were engaged" in the Haudenosaunee campaign against the Illinois the previous year. The Miami, "being in great dread of [them]" worked with the Haudenosaunees to "seek an accommodation" exchanging diplomatic gifts and entering into "an amicable arrangement."[83] As the local gatekeepers of the St. Joseph–Kankakee portage route, the Miamis opted to cooperate with Haudenosaunees and allow the war parties an open door to the Illinois settlements downriver.

The Miamis were not the only group who attempted to steer—and even benefit from—the carnage running through the region's portage routes. On the heels of the Haudenosaunee attack against the Kaskaskias' settlement, a party of two hundred Kickapoo from west of Chicago had been bold enough to descend the Illinois River, looking to plunder what remained of the Kaskaskias' settlements. Father Membre, fleeing north with Tonti, described the opportunism of these Kickapoos. The Kickapoo "sent some of their youth in war-parties against the Haudenosaunees, but learning that the latter were attacking the Illinois, the war-party came after them" instead.[84] Scouts from this group even killed Father Gabriel de La Ribourde, one of Tonti's party, as he fled north from the battle.[85] Positioned along the water routes between the powerful Illinois towns and Haudenosaunee invasions, groups like the Miami and Kickapoo had little choice but to remain flexible and, when possible, profit along this thoroughfare of violence.

Realignments in the Illinois Country

In response to the Haudenosaunee campaign, La Salle returned that winter and redoubled his efforts to fortify the Illinois River valley against further attacks. Such plans hinged on successfully brokering an alliance between the Illinois and the Miami groups that lived along the two main portage routes into the Illinois River valley—those at the St. Joseph–Kankakee portage and the Wea of the Chicago region. Sending two Indigenous diplomats, Nanangoucy and Oui-ouilamet, to negotiate with the Miamis, La Salle went among the Illinois himself, where he "urged them to make peace with the Miamis" along the portage routes. La Salle argued that "as long as they [the Illinois and Miami] remained divided, the Iroquois would despise them and would defeat them separately."[86] Meanwhile, as parties of Illinois returned to the Illinois River valley and began launching counterraids against the Haudenosaunees, the Miamis, Kickapoos, and other groups near Chicago began realizing that their precarious position between two powerful confederacies remained intact. The Haudenosaunee offensive of 1680 had not removed the threat of the Illinois downriver, and in the face of reconstituted Illinois power, the Miamis began "fearing that the Illinois would take vengeance upon those who had taken the part of their enemies."[87] Well aware of their collusion with the Haudenosaunees the previous fall, and their ongoing vulnerability along the portage routes, the peoples in between again weighed their options.

Such apprehensions lent La Salle's negotiations some traction. By the end of 1681, not only the Illinois but also various groups of Miamis and Shawnees began to resettle near the Grand Village of the Kaskaskia downriver as they repositioned themselves both diplomatically and geographically closer to the Illinois heartland. In the interim, La Salle led an expedition down the Mississippi to explore. When he returned in the fall of 1682, he established Fort Saint-Louis at Starved Rock, near the main Kaskaskia settlement. The following spring, as rumors circulated of another Haudenosaunee invasion of the area, La Salle convinced even more Miamis that "they would be safer near my fort, by joining with the Islinois and the others who were settled there."[88] With more Native people resettling the area around the Grand Village of the Kaskaskia, La Salle placed Tonti in command of his new fort at Starved Rock and journeyed north once again to the Chicago portages in the summer of 1683.

La Salle had originally entered the Illinois Country with aspirations to unite the water routes between the Mississippi River and the Great Lakes, but by the time he departed Chicago in 1683 to return to France, he had realized both the political and geographic complications involved with such ambitions. His ship, *Le Griffon*, had sunk on its maiden voyage on Lake Michigan in 1679, unfit for the Great Lakes' sudden and violent gales and rocky shorelines. The bloodshed of the Haudenosaunee campaign in 1680 had further challenged La Salle's geopolitical hopes for the interior. By 1683, the political landscape around Chicago had somewhat stabilized with newly forged alliances between the Miami and Illinois, but the local geography of Chicago's wetland remained discouraging. Strong currents and wind from Lake Michigan piled up a sandbar at the mouth of the river, preventing any real harbor for larger vessels to navigate the river. The spring floods that made travel easier for Indigenous canoes complicated any European-style shipping, and La Salle complained that "the prairies by the lake over which travel is necessary are flooded by a great volume of water flowing down from the hills whenever it rains." Such flooding would also complicate the canal Jolliet had originally proposed along the portage, since between the Great Lake currents and the wetland flooding, "it is very difficult to make and maintain a canal that does not immediately fill up with sand and gravel." La Salle was convinced that a canal could only be dug "with a great deal of expense," and even then, it might be useless because the Des Plaines and Illinois Rivers ran too shallow for large boats during most of the summer.[89] La Salle's and Jolliet's visions for an interior maritime network proved unfeasible from a European perspective. Indigenous-style canoes and the overland portages would continue to serve as the only practical means of connecting the routes of travel between the two watersheds.

Before his final departure, La Salle traveled back and forth from Chicago to Fort Saint-Louis and its neighboring Indigenous settlements several times as he consolidated his Indigenous alliances and trade relations. During his sojourns between Chicago and the Illinois River settlements, he wrote how "all these tribes are uniting and coming to settle" near the Grand Village of the Kaskaskia. Notably from the Chicago area, La Salle shared "the Ouiatenon [Wea], to the number of a hundred and twenty huts are there now, having come away from their villages with me."[90] The same year, La Salle's men constructed a small post—possibly just a cabin—near the portage at Chicago, to serve as a depot along the route between Lake Michigan and the Illinois River. While not a garrisoned fort like Tonti's position downriver at Saint-Louis, La Salle intended the depot at Chicago as an auxiliary site for trading and caching furs, while also straddling the main waterborne route for canoe travel between his Illinois colony and French posts in the Great Lakes. La Salle's movements to and from the portages during the summer of 1683 and his construction of the depot there demonstrated that Chicago remained a major corridor between Lake Michigan and the Illinois River even amid the Indigenous populations shifts across the region.[91]

The Persistence of the Portages

New hopes of stability along the route led to an expansion of French trade through the Chicago portages. Brigades of voyageurs, or licensed fur traders, began entering the Illinois Country along its river routes in the wake of Indigenous resettlements and newly formed alliances. Drawing up contracts for trade in Montreal and Quebec, French merchants dispatched crews of canoe men into the western Great Lakes with goods in hopes of returning cargoes of furs the following year. A surviving contract between Henri de Tonti and two voyageurs from September 1684 demonstrates the typical terms of such trade voyages through Chicago's portages. Tonti signed as the venture's outfitter, providing all the trade goods, supplies, food, and ammunition needed for the voyage from Quebec to Tonti's post at Fort Saint-Louis on the Illinois River, where the crew was to deliver the shipment of trade goods. Tonti even supplied the canoe that the voyageurs would paddle and portage through the Great Lakes and into the Illinois River. In exchange for their labors, Jacques Neveau and Antoine Duquet Madri, the two voyageurs, received part of the profits from trade as well as two cloaks, three shirts, and a blanket each to trade for furs. The contract also guaranteed the two men 150 livres' worth of payment in the form of peltries and a signing bonus of ten beaver skins apiece.[92] Voyages from Quebec, through the treacherous waters of the Great Lakes and over portages like Chicago, posed many dangers and significant exertions. Men like Neveau and Madri received generous compensation for their efforts.

The voyageurs' pay was well earned in the face of continued danger in the Great Lakes. The threat of Haudenosaunee hostilities directed the course of the trade routes, as fur brigades from Montreal canoed north up the Ottawa River, across the Mattawa, down the French River through Lake Huron to the Straits of Mackinac, and then along the west coast of Lake Michigan to avoid Haudenosaunee attacks. Even with detours, French traders sometimes ran into trouble and Haudenosaunee war parties continued to threaten trade as far west as Chicago. The same year Tonti commissioned Neveau and Madri's canoe brigade to come west into the Illinois Country, a party of Haudenosaunee caught a French trade flotilla on the upper reaches of the Illinois River. Capturing fourteen men and confiscating seven canoe-loads of goods, ammunition, and muskets in late February, the Seneca and Cayuga Haudenosaunees stripped the French and carried them naked to the mouth of the Chicago River before releasing them.[93]

Around Chicago, Haudenosaunee warriors were not the only threat. Sometime in the 1680s, a party of Mascoutens, nominally allied with the French, "encountered at the river of Chikagon," two French traders, "who were returning from the Islinois." Disgruntled over an unproductive raid against the Haudenosaunees, the Mascoutens murdered the Frenchmen instead, and subsequently killed three Miami women to eliminate any witnesses. Taking the scalps, the Mascoutens then tried to pass them off as trophies from a victory over the Haudenosaunees.[94] The fate of such traders demonstrated the ongoing dangers of movement through the political vacuum around Chicago's wetlands. While Illinois and French power coalesced downriver around Starved Rock, the portages through Chicago often proved a no-man's-land where any traveler remained vulnerable.

Even when fur brigades avoided hostile encounters, the route through Chicago remained contingent on local environmental factors and Indigenous techniques well into the 1680s. Father Hennepin, a Recollet priest in La Salle's entourage, lamented that the local geography continued to stymie French ambitions for Chicago as a center of European power in the region. While Jolliet and La Salle had both projected Chicago's importance as a geographic linchpin for a waterborne empire, such hopes failed to manifest themselves in the face of environmental realities. Hennepin complained about his journey along "the Creek through which we went from the Lake of the Illinois, into the Divine [Des Plaines] River," which weaved through the wetlands, "that country being nothing but a Morass." Portaging through the swamps of Chicago proved difficult, but travelers also faced challenges in both directions along the route. Journeying east into the Great Lakes watershed, travelers faced a voyage "expos'd to the Storms" of Lake Michigan, and going southwest, the route followed the Des Plaines and Illinois Rivers along a course only navigable "for Canows."[95] Despite initial ambitions, French commerce and travel through Chicago remained curtailed by local geography and reliant on Indigenous craft and navigation. Chicago's portages remained a far cry from a European entrepôt of power.

While French priests and adventurers complained of such limitations, they had no choice but to continue to harness the Indigenous portages for travel between the Great Lakes and Illinois River valley. From 1683 forward, Chicago served as one of the main routes of supply for the growing French presence downriver at Starved Rock and the Jesuit missions among the Illinois. The Chicago corridor, connecting French outposts in the Great Lakes with the growing

colonial ventures of the Illinois Country remained crucial to the continued stability of the region. La Salle's efforts to strengthen the Illinois Country against
further Haudenosaunee invasions had already begun paying off, and when a
large expedition of Senecas and Cayugas attacked the Illinois River valley in the
spring of 1684, the collected alliance of Illinois, Shawnees, Miamis, and French
around Starved Rock repulsed the Haudenosaunee assault. The water route
through Chicago even allowed for rapid French reinforcements from the lakes.
Led by Olivier Morel, Sieur de la Durantaye, sixty French soldiers from the post
at Michilimackinac, at the northern end of Lake Michigan, portaged through
Chicago to further bolster the defenses at Fort Saint-Louis.[96] While French control over Chicago still proved impractical, the portages between the lake and the
rivers remained vital to their efforts in the Illinois Country.

La Salle himself did not return to the region after 1683, dying on an unsuc-
cessful mission to find the mouth of the Mississippi River in 1687. A remnant
party of French led by Henri Joutel fled northward following the collapse of
La Salle's expedition in the Gulf of Mexico, and from Joutel's journal, we
glimpse the continued openness of the route through Chicago. Arriving at
Chicago too late in the season for safe travel on Lake Michigan, Joutel and his
party returned to Fort Saint-Louis to wait for spring. While wintering at the
fort and recovering from their ordeal, Joutel described how resupplies to the
French and Illinois bastion at Starved Rock continued to flow south from Chicago. Late in 1687, for example, Tonti received notice "that three canoes, laden
with merchandize, powder, ball and other things, were arrived at Chicagou."
Freezing temperatures and ice, however, meant insufficient open water for the
voyage south from Chicago, so Tonti sent a party of forty French and Shawnees
to help carry the goods down from the portage.[97] Throughout seasonal cycles
of ice, floods, or low water, the Chicago portages continued to serve as a route
of travel and resupply for the French of the Illinois Country.

Joutel's small band waited until the fortuitous spring floods made the portages through Chicago shorter. Arriving in Chicago in late March, "the bad
weather obliged us to stay in that place till April" as gales continued to rage on
the open lake. Joutel's party subsisted on maple sugar, wild rice, and wild on-
ions growing near the portage. Eventually, the party carried "the canoe or
boat . . . as well as clothing" and other supplies over the portage, which Joutel
described as a task "there is no way of escaping." They collected furs they had
cached the previous autumn and "imbarked again and entered upon the lake
on April 8, keeping to the North [west] Side to shun the Iroquois." After enduring more storms on their voyage and "swelling waves like those of the sea,"
the party safely reached the French post of Michilimackinac.[98] Despite the

seasonal delays and inclement weather, Joutel's party successfully harnessed the Chicago portages to pass from the lower Mississippi River to the upper Great Lakes, and Joutel's description of the journey emphasized the enduring importance of Chicago's waterborne mobility.

New Hopes at Chicago's Portages

From the opening years of the seventeenth century forward to when Joutel's group of survivors passed through the portages in the spring of 1688, the portages at Chicago operated as a conduit for human movement. First navigated by Oneota people in the decades before written records, the portages soon fostered Illinois and Miami expansion along Lake Michigan's shorelines and undergirded trade links between these Illinois Country Algonquians and their Great Lakes neighbors to the north. Brought through the portage by Kaskaskia hosts, French explorers and Jesuit priests soon began to envision the strategic potential of this small patch of ground that bridged the continental divide between the Great Lakes and Mississippi drainage. Violence and ecological realities prevented individuals like Jolliet, Marquette, and La Salle from realizing their initial expectations for Chicago as a hub of European strength in the interior. Even so, the portages provided an important pathway between river channels that enabled both Indigenous and French mobility across the interior waters of a continent.

Sometime soon after Joutel's party passed through the portage, Chicago's role as simply a crossroads for movement began to change. The Wea, a Miami-speaking people, began moving into the area around the portages on a more permanent basis. It remains unclear what initially motivated the Wea to expand into the Chicago area other than the strategic geography and abundant ecology, but by 1690, Tonti's cousin, Pierre-Charles DeLiette, reported that the Wea "went to form a village at the river Grand Calumet, which also empties into this lake [Lake Michigan] twelve leagues from the Chicago toward the south." They also formed a second settlement at the forks of the Des Plaines and Kankakee, near the Chicago-area wetlands and directly along the main river routes.[99] Over the next several years, the Wea established more villages along the Chicago River, making the area the center of their largest settlements. Repositioning themselves along the main trading lanes, straddling the rich transitional ecosystems of wetlands and oak openings, these Miamis seemed intent to profit from their new homeland.

French encouragement might have directed these new Wea settlements to the Chicago area. As early as 1684, French traders had sent presents and invitations to the Miami groups like the Wea living north toward Green Bay to "settle at Chekagou" as they hoped to establish commercial links closer to their main route of travel. French traders did not hunt their own furbearers but relied on Indigenous expertise to collect pelts for them. The wetlands and low prairies around the Chicago portages provided ideal habitat for furbearing mammals, and these traders implored the Miami to "go hunting there."[100] As late as the 1680s, when La Salle passed through Chicago, beavers still dominated the waterways. Along with these apex engineers, muskrats, minks, otters, and raccoons all made use of the wetlands around Chicago's portages.[101] DeLiette, who eventually lived among the Wea at Chicago, still described the area as having "an abundance of animals."[102] All these semiaquatic mammals provided valuable pelts, and as the Wea became more engaged in the fur trade, Chicago's wetlands offered a new source of riches.

The intensive hunting of furbearers for trade signaled a shift from earlier Miami practices and helps explain the Weas' relocation to Chicago, even amid continued threats of Haudenosaunee attacks. When Nicolas Perrot had first made contact with the Wea north of Chicago in 1668, they had apologized and made "excuse for not having any beaver-skins" to trade, since "they had until then roasted those animals" for food rather than processed their hides for trade.[103] By 1690, the Weas' approach to furbearers had changed, and with it, their location. By the time they settled in the Chicago region, the Wea hunted beavers efficiently, and for commercial purposes. DeLiette made a distinction between the Wea of Chicago and other Native people of the Illinois River valley when he noted that "they are better beaver hunters than the Illinois, and they esteem the beaver more highly."[104] Prompted by their new focus on hunting beaver and other wetlands mammals for the fur trade, the Wea moved to Chicago to settle permanently for the first time in 1690. Whether responding to direct overtures from French traders or to the economic incentives of a fur-laden area, the Wea drew close to Chicago to exploit the portage routes and wetlands resources in a new way.

Just as Chicago's ecology and geography had influenced Indigenous and European aspirations and actions during the span of the 1600s, its environmental abundance prompted a shift in its human history in the last decade of the century. The Wea settlements at Chicago, straddling the watersheds of the Mississippi and Great Lakes, signaled a change in Chicago's geographic function. Long a conduit for human movement, the Wea settlers transformed the space around the portages into sites of occupation, intent to profit from fur hunting and the trade routes that ran through the portages. Attracted by Indigenous settlers like the Wea, French missionaries and imperial officials would follow in the years to come as Native peoples and Europeans began to envision Chicago in a new way—not just as a space to be traversed but as a place to establish permanent strength and settlements. Jesuits led by Fathers Pierre François Pinet and Julien Bineteau even founded the Mission of the Guardian Angel "among the Miami at Chicagoua" in 1696.[105] After decades of frustrated ambitions for the strategic geography of the portages, collaboration between Native and French peoples around Chicago's immediate area would define the coming years along North America's narrowest continental divide.

CHAPTER TWO

Barriers

Imperial Ambitions and Failures at Chicago's Portages

Wet and cold, a trio of missionaries trudged inland from the shores of Lake Michigan. Henri de Tonti had guided Reverend François de Montigny, Father Antoine Davion, and Father Jean François Buisson de St. Cosme to Chicago in late October 1698. Accompanied by several other lay brothers, missionaries, and voyageurs, the party beached their boats just north of the Chicago River. St. Cosme later reported that he and his companions had "considerable difficulty in landing and in saving our canoes." As winds rose and waves became more violent on the Great Lake, the missionaries from Canada "had to jump into the water" and wade ashore. Leaving the rest of the party behind with the canoes and supplies, the three priests made their way inland, about a mile and a half, to the recently established Mission of the Guardian Angel. The Jesuit father, François Pinet, had built the small mission among the Weas—Native Miami-speakers—two years earlier. When Montigny's party arrived, Father Pinet took the travel-weary Quebec seminarians in with "cordiality and manifestations of friendship," despite being from a rival missionary order. The waterlogged expedition was bound for the Arkansas Country via the Chicago portages. They took the next week to rest and wait out the stormy weather before setting out for the portage into the Des Plaines River.[1]

What the missionaries saw at Chicago encouraged them. St. Cosme described its position favorably, "built on the bank of a small river, with the lake on one side and a fine and vast prairie on the other." St. Cosme wrote enthusiastically about the mission's prospects, seated promisingly between land and water, along one of the region's major canoe routes between the Great Lakes and Mississippi drainages. The possibility that Chicago might become a center of Catholic influence and Indigenous alliance seemed a burgeoning reality when St. Cosme passed through in 1698.

Both Native peoples and Europeans had begun to imagine Chicago as an auspicious place for settlement by the end of the seventeenth century. The wetland environment fostered abundant populations of beaver and other fur-bearers. As the Weas became more invested in the fur trade, the Chicago area offered both ample hunting grounds and access to water routes for exchange, communication, and travel. Settling at Chicago placed them at the crossroads of a major trade route between the Great Lakes and Illinois Country. The newly established Wea towns around the portages attracted Jesuit missionaries led by Father Pinet who saw opportunity for evangelization. In the local vicinity, a settlement of Weas contained "over a hundred and fifty cabins," and further up the river, another village bustled with even more souls to save. St. Cosme believed that Pinet, aided by the blessings of God and his own zeal—along with his mission's promising geography—sat poised to convert "a great number of good and fervent Christians."[2] Indeed, Pinet shared with his guests that "several girls of a certain age and also many young boys have already been and are being instructed, so that we may hope that when the old generation dies off, they will be a new and entirely Christian people."[3] During most of the previous century, the fraught political landscape coupled with the variances of the local wetland environment had left Chicago unsettled by both Native peoples and Europeans. But now, as the Weas settled along Chicago's portages and the Jesuits progressed in their evangelism, French and Indigenous coexistence seemed at hand.

French officers in the Illinois Country, also eager to maintain a presence at the strategic junction and influence among the Weas, placed Pierre-Charles DeLiette at Chicago in 1698. Then, French administrators managed to negotiate a region-wide peace between the Haudenosaunees and France's Algonquian allies in 1701, appearing to secure a new level of stability and prosperity for both Chicago and the Great Lakes as a whole. Geographically, Chicago's portages had operated as a physical pathway connecting the waterways of the continent throughout the previous century. Now, as Wea settlers, Jesuit missionaries, and French officials came together upon the muddy ground of Chicago, the local area seemed poised to become a space of cross-cultural collaboration and connections as well.

But Chicago never became the nexus of such multicultural cooperation. Nor did it become a center of French power in the region. Just as permanent Indigenous-French settlement along the portages seemed to promise stability and successful alliance in the western Great Lakes, new complications arose. Renewed threats of Indigenous warfare scrapped the Weas' efforts to occupy the portages in any permanent way. At the same time, new administrative divisions created a jurisdictional border through Chicago, nullifying French attempts to quell the bloodshed or bring control to the area. Chicago's geography as both a passageway of violence and an unenforceable border of French dominion directly exacerbated tensions in the region.

In the decades following the establishment of the Wea settlements and the Jesuit mission, any hopes of a stable French-Indigenous alliance in the west unraveled. Chicago's portages would become a launching point of French aggression and Indigenous resistance, rather than a site of settlement and collaboration. And while the portages would continue to foster movement between the Illinois Country and the Great Lakes during the following decades, most of the traders portaging through Chicago did so clandestinely and in defiance of French authority. Chicago's promising geography continued to thwart both French and Indigenous ambitions well into the eighteenth century as complications over a new set of borders disrupted efforts to control the portages.

What unraveled at Chicago during the early decades of the eighteenth century demonstrated two important points about French empire in the western Great Lakes. First, that France projected a maritime vision for its Great Lakes empire, which meshed with its existing imperial, Atlantic frameworks. The navigable waterways of the interior allowed for such a vision, as French officials, missionaries, and traders learned Indigenous techniques for canoeing and portaging the interior lakes and rivers of the continent. French newcomers acquired specific Native knowledge of the environment and geography that enabled waterborne movement across North America. Along with Indigenous knowledge, bottlenecks like the Chicago portages proved crucial to such maritime ambitions, functioning as the connective geographies for an interior waterborne network of mobility. This ambitious vision was Atlantic in its approach, yet also continental in its scope, and it necessarily relied on Indigenous knowledge, technologies, and alliances to even be contemplated. Understanding French imperial aspirations for Chicago and the wider, waterborne interior offers insights into France's ambitions for North America's freshwater geography. French officials' focus on seemingly peripheral places in the Great Lakes and Mississippi River watershed, rather than being provincial anomalies or self-serving projects, fit within wider imperial approaches to French maritime empire, albeit in an inland, freshwater context.[4]

Second, the view from Chicago calls into question the reach and strength of the alliance system France maintained in the so-called middle ground of the *Pays d'en Haut*, or upper country, even at the height of French power in the region. This perspective joins a generation of new histories that challenge just how much control French officials were ever able to assert at strategic sites like Indigenous Chicago. Early hopes of French and Native settlement and collaboration at Chicago's portage withered under threats of continued warfare following the Great Peace of 1701. Local French representatives seemed unable to address the situation and, in many cases, exacerbated tensions between Native groups. By 1712, violence had erupted full force across the region in what the

French called the Fox Wars, a series of conflicts pitting the Meskwakis and their allies against French, Anishinaabeg, and Illinois peoples. Chicago would become a route of war, a point of rendezvous, and a site of ambush as the conflicts escalated.[5]

Historians have shown how Anishinaabeg and the Illinois peoples orchestrated the Fox Wars for their own agendas and used the Native slave trade to undermine French attempts to broker peace.[6] Alongside such Indigenous actions in perpetuating the conflict, the view from Chicago demonstrated how colonial borders further compounded French struggles to forge a peaceful alliance in the region. With the creation of the Louisiana colony, the French drew a jurisdictional borderline through Chicago's continental divide, separating New France's Great Lakes from the Illinois Country, which became part of Louisiana. This administrative division between the Great Lakes and Mississippi drainage complicated French efforts at coordinated policy toward the Meskwakis and other Native peoples who straddled the watersheds. Ambiguous colonial borders weakened any comprehensive Indian policy for France in the Upper Country, demonstrating how Chicago's complex local geography directly influenced the course of empire. Chicago's portages remained beyond the control of both New France and Louisiana throughout the eighteenth century, persisting as a route for illegal traders and defiant Native people rather than ever becoming a space of imperial power.

A Collaborative Space

Yet, in the 1690s when St. Cosme's party passed over Chicago's portage, things seemed promising. Their host, Father Pinet, had enjoyed success both in his conversion efforts and linguistic work along the Illinois River. Sometime in 1698, he baptized Peoria, a recalcitrant Illinois leader who had resisted earlier Jesuit overtures. While at Chicago, Pinet also began his magnum opus, a Miami-Illinois dictionary aimed to help future missionaries in the Illinois Country as they grappled with the related dialects of the region. Drawing from his experience among the Weas, he compiled 672 pages of text in the languages, providing future missionaries with a primer for translation and a tool for cross-cultural communication. Bridging both religious and cultural divides along the connecting pathways of Chicago's portages, Pinet's efforts at the Mission of the Guardian Angel gave hope for French and Indigenous cooperation and stability in the midcontinent.[7]

While the Jesuits toiled to learn Native languages and convert the local Weas, Pierre-Charles DeLiette pursued French imperial and commercial agendas at

Chicago. A relative of Henri de Tonti, DeLiette had first come to the Illinois Country in 1687. He spent time among the Illinois at Fort Saint-Louis, or Starved Rock, where he learned their language and lifeways. In 1698, Tonti sent him north to live among the Weas at Chicago, where he stayed until 1702. Accustomed to Illinois culture and language, he proved equipped to deal with the Miami-speaking Wea as well. He wrote that he "found no difference between their customs and language and those of the Illinois."[8] His account of his time at Chicago offered detailed descriptions of Wea and Illinois cultures and emphasized how these Native people prospered amid a favorable environment and strategic geography.

In his memoir, DeLiette emphasized Chicago's dual role as a thriving place of settlement and a bustling space of movement. The specific geography and related ecology of the region allowed the Weas and French to utilize the Chicago area in both capacities. Writing around 1700, he estimated that Weas had settled there ten or twelve years earlier, establishing "a village at the Grand Calumet"—near the southern portage route—and another "at the fork of the Kankakee River." These towns, plus the one located closer to Pinet's mission near Chicago's northern portage, accounted for a total Indigenous population of around 4,500 people in the immediate area, according to the Jesuit estimates.[9] For the western Great Lakes at the time, Chicago had become a significant population center.

These Wea settlements prospered largely on the labors of their women, who produced "a great store of food" through diligent agriculture.[10] Later French observed the Weas' prowess as farmers, describing them as "very industrious." Specifically, French commentators noted the Weas for their maize agriculture, raising a higher quality strain of corn "with much finer husks and much whiter flour."[11] DeLiette, at Chicago, reported that along with corn, Native women reaped harvests of "many fine watermelons" in the summer, and an "abundant supply of fine pumpkins" every fall.[12] The Wea women harnessed the rich soils of the Chicago lowlands to enjoy plentiful crops and consistent food supplies for their large settlements.

The wetlands that predominated the local ecosystem offered still more abundance. According to DeLiette, Indigenous women gathered *macopine*, a "big root which they get in the marshes," as well as other edible roots and tubers.[13] "In autumn and spring," these same marshes were "full of bustards, swans, ducks, cranes, and teals"—plenty of waterfowl for hunting, while fish schooled in the rivers.[14] Aside from food resources, the wetlands vegetation also provided the reeds necessary for making typical Wea and Illinois dwellings. DeLiette described how the women "procure bundles" of these cattails, which he called *apacoya*, from the marshland, drying them and using them to make roofs for their cabins.[15] Furbearing animals who made the wetlands their home also became increasingly important as a resource for hunting and trade. The teeming populations of beavers, muskrats, and raccoons had initially motivated the Weas to relocate to the Chicago area.[16] In the following years of settlement, this abundance in pelts allowed the Weas to capitalize on the growing fur trade. Many French commentators might have viewed the marshes and wet prairies of Chicago as an unenviable homeland, but the resources of its wetland environment sustained the Weas' way of life and helped their large settlements prosper for over a decade.

While Chicago had clearly become a center for Wea settlement by this time, DeLiette portrayed his own station on the Chicago River, "a little river only two leagues long," as positioned along a thoroughfare of travel as well. DeLiette described the "Portage of the Oaks," one of Chicago's two main portages, where "a short portage simply of one's baggage is made." This pathway, connecting the Chicago River with the Des Plaines, remained "the route usually taken to go to this country [the Illinois]."[17] The Wea towns benefited from occupying the space connecting these trade lanes. Miami-speaking peoples such as the Wea strategically settled along portage routes throughout the lower Great Lakes during the seventeenth and eighteenth centuries, exploiting such geographies' access to trade, selling agricultural products to passing travelers, and sometimes even exacting tolls from parties of trade canoes.[18]

As the Weas exploited their location along a major trade lane, French officials like DeLiette and his superiors also calculated ways to harness Chicago's nexus of movement. At the time DeLiette relocated to Chicago's portage, Tonti and François Dauphin de la Forest, the two French commanders in the Illinois Country, had a trade warehouse near one of Chicago's portages.[19] Part of DeLiette's assignment at the strategic junction likely involved maintaining the building as a supply depot for French trading parties crossing from the Great Lakes into the headwaters of the Illinois River, while also living among and maintaining good relations with the Wea villagers to ensure safe passage for these brigades of French canoes en route to the Illinois Country with supplies and fur trade wares.[20]

Imperial successes farther afield improved prospects locally for both DeLiette and the Weas. In 1700, after decades of intermittent warfare, the Haudenosaunee had agreed to peace talks with the French governor in Canada. Augustin le Gardeur de Courtemanche, an emissary from Quebec, arrived among the Wea at Chicago the following spring where "he obliged them to lay down their arms" and made them "promise to send deputies to Montreal."[21]

Inviting diplomats from all the western Nations who had fought against the Haudenosaunee, French governor Louis-Hector de Callières brokered a collective peace in Montreal in August 1701. The Wea delegation from Chicago participated and signed the treaty documents on August 4 with a pictographic image of a furrow, or row of corn, signifying their agricultural prowess.[22] With their mark, the Chicago Weas ensured an end to the violent Haudenosaunee forays through Chicago and aimed to consolidate their interests with those of the French. With a general armistice secured, the Weas could hope for increased traffic and trade along the portage routes. Returning from the Montreal peace talks, the Weas looked forward to a new era of collaboration with French officials, traders, and missionaries at Chicago, free from the threats of continued warfare in the region.

Envisioning a Maritime Interior

The French likewise looked to shore up their own influence along Chicago's route. DeLiette's four years living with the Weas at Chicago represented the local manifestation of France's imperial hopes for the portages at the dawn of the eighteenth century. Even as commitment to western investment wavered in the French ministry in Europe, officials throughout New France continued to tout the importance of Chicago in their schemes for French empire in the American interior. Antoine de la Mothe, Sieur de Cadillac, one of the most outspoken of these champions, had commanded at Fort de Buade, along the Straits of Mackinac, from 1694 to 1696. He wrote extensively about the geographic potential of both the Great Lakes generally and Chicago's portages specifically, offering a reimagining of the visions first expressed by Jolliet and La Salle in earlier decades. Cadillac argued that the Great Lakes promised one vast arena of waterborne mobility, each lake and river "communicating and flowing from one to the other." All French travelers had to do to unite the continent was "to make a portage, that is, to carry baggage" occasionally between adjacent bodies of water.[23] With the exception of these few short overland junctions, Cadillac observed, "one can navigate the interior of the country on fresh water in a ship."[24] Such an arena of waterborne movement, reaching well into the interior, offered the potential for French maritime empire extending deep within the continent.

Cadillac believed Chicago loomed large in such visions. He explicitly associated Chicago's importance with its geographic position along a continental divide, describing "a certain height of land near Chicago sloping north and south which divides the water route between the Southern Sea and Quebec."[25] Along this continental divide, DeLiette's one-man "post at Chicago" offered the possibility to "navigate through 2,300 leagues of country." In Cadillac's reckoning, Chicago stood as one of only three strategic portages that could unite French interests across the waterways stretching from the Gulf of St. Lawrence in the north to the Gulf of Mexico in the south.[26] This small patch of ground remained key to French hopes for empire in the interior.

As European officials grappled to understand the elusive interior geography of North America, they looked for strategic bottlenecks like the portages at Chicago. Such spaces, on maps, seemed to offer localized centers of control for vast swaths of territory. Focusing on junctions of waterways, French officials like Cadillac believed they could re-create their more familiar maritime empire by controlling strategic sites along the freshwater routes of the continent. These French officials applied the same imperial, maritime logic that had benefited them across their wider Atlantic and Caribbean empire to make the geography of North America's Great Lakes legible. Chicago, as a portage that connected navigable waterways, allowed French officials to imagine an empire in the interior even as they struggled to make it a reality.[27]

European mapmakers also projected an imagined geography onto spaces like Chicago. Their efforts reflected similar imperial ambitions. Around the turn into the eighteenth century, a "Fort Checagou" appeared on many European maps along the southwestern shoreline of Lake Michigan. Jean Baptiste Louis Franquelin first included a "Fort Checagou" at the portages on his 1688 map of North America (see figure 2.1).[28] While he likely based his map's fort on scant references to fur trading posts, or depots, built by La Salle and his men, Franquelin embellished the reports to show a full-fledged military fortress at Chicago.

Franquelin's invented fort projected French power at the crucial portages and fit within the wider message and style of the map. Along with the fictionalized "Fort Checagou," Franquelin also included images of tall-masted, European-style sailing ships upon Lake Michigan's waters. While no such ships sailed the Great Lakes at the time, Franquelin's map made a clear statement about the projection of French maritime power into the North American interior. Later mapmakers and publishers replicated the embellishment and perpetuated European visions for Chicago as a fortified site of French power along the edge of the maritime Great Lakes. A wide array of European cartographers continued to include a fort at Chicago, and the embellishment appeared on maps through the 1750s.[29] While never realized, the fictional "Fort Checagou" demonstrated French designs for the site as a key junction in the region.

FIGURE 2.1 While the tall-masted ship and the fictionalized fort at Chicago were meant to project "imagined" French power in the Great Lakes, the dotted line accurately portrayed the main routes of travel and trade through Chicago's portages. Jean Baptiste Louis Franquelin, *Carte de l'Amerique Septentrionnale: depuis le 25, jusqu'au 65° deg. de latt. & environ 140, & 235 deg. de longitude* [detail] (1688). Courtesy of the Library of Congress.

As Cadillac penned his treatise and European cartographers drew their maps, both highlighted the Great Lakes as a burgeoning maritime arena for French empire. At the time, such predictions seemed well on their way to fruition. DeLiette and the Jesuits made inroads among the Weas, and trade continued to flourish across Chicago's portages after the Weas settled the area. With continual exchanges between Indigenous groups and the expanding French fur trade, canoes and their cargoes poured through Chicago. And it was not just French voyageurs portaging through Chicago. DeLiette reported on the still robust trade that the Great Lakes Anishinaabeg carried on with the Illinois peoples farther south. Among other things, the Illinois eagerly obtained porcupine quills from the coniferous forests of the northern Great

Lakes to decorate their moccasins.[30] In addition to materials from the Northwoods, Native Illinois and Weas also acquired European goods passed south through Chicago by Anishinaabe middlemen. Illinois traders, in turn, shipped bison meat and hides north. They also maintained an expanding trade in Indigenous slaves captured from beyond the Mississippi River and sold to the French and other Native peoples of the Great Lakes.[31] All such trade relied on Chicago's portages as they connected the navigable river routes of the region.

The proliferation of the birchbark canoe in the Great Lakes enhanced the potential for such waterborne mobility across these portages. Northern Algonquian peoples such as the Wendat, Wabanaki, Ojibwe, Odawa, and Potawatomi had perfected birchbark canoe construction over centuries of riverine travel. Native builders crafted the outer hull of the vessels out of waterproof sheets of bark stripped from white birch trees, which grew abundantly throughout the Great Lakes region. Boatbuilders lashed these sheets to cedar ribbing, gunwales, and thwarts using the fibrous roots of black spruce and jack pine. As a final touch, Indigenous builders applied a caulking of pitch made from the sap of pine or spruce trees, which sealed seams and prevented leaks. The result was a lightweight vessel, easily carried between waterways but still sturdy enough to transport goods and people across both the shallow waters of the region's rivers and the wider expanses of the open lakes—a technological marvel of Native American artisanship.[32]

Most Europeans remained content to defer to Native skill in canoe building. In 1705, writing from New France, Denis Riverin noted, "the Indians, and particularly their wives, are excellent canoe makers" but lamented that "few French are successful."[33] A canoe could take anywhere from several days to several weeks to build, depending on the size and design, and during the early eighteenth century, these vessels cost around 150 French livres. While Europeans eagerly embraced canoes to traverse the region, they continued to rely on Indigenous ability and ecological innovation to supply their preferred watercraft. Father Joseph-François Lafitau, a Jesuit missionary living in the eastern Great Lakes, described birchbark canoes as "masterpieces of native art," relating that "nothing is prettier and more admirable than these fragile craft in which people can carry heavy loads and go everywhere very rapidly."[34] Lightweight and constructed from plant materials gathered from the surrounding woodland, these canoes could be easily carried between waterways. Such canoe technology, coupled with Native knowledge of the best travel routes, fostered a highly mobile network of Indigenous waterborne movement across the interior waterways of the continent, hinging on portages like those at Chicago.[35]

French traders, missionaries, and officials who entered the Great Lakes necessarily learned the particulars of such travel, and Native-built canoes remained the standard mode of transportation throughout the French period in the western lakes. Europeans' disinclination to abandon Indigenous methods of travel on the Great Lakes came from a pragmatic realization that such movement worked through a balance of environmental factors and Native ingenuity. The French willingly embraced such a system of mobility as the most functional and affordable option for traveling and transporting goods through the lakes.[36]

Cadillac, who had watched convoys of Native mariners and French voyageurs pass through Michilimackinac for years, understood the labor involved in such canoe travel. He particularly stressed the hard work of connecting navigable waterways at portages like Chicago. His memoir remains an important early account for understanding the travel and labor of the fur trade, since many of the men engaged as voyageurs remained illiterate and left few records.[37] Trading parties left Montreal in the spring, braving rapids along rivers and full-blown gales on the Great Lakes, reaching Michilimackinac "by this continual and laborious work." The most arduous tasks lay at portages, as crews left Michilimackinac for places like Chicago and other tributaries in the western lakes. At such carrying places, crews disembarked and carried "the canoes on the men's shoulders through the woods, along with all the merchandise or beaverskins."[38] Using their own human labors, these voyageurs knit the region's waterways together over connective portages to create maritime networks of trade and travel throughout the interior.[39] In so doing, they followed the lead of Indigenous mariners who had navigated the waterways of the continent in a similar fashion for centuries.

Chicago's portages served as a key link in this wider network of makeshift maritime travel. One voyageur passing through Chicago during the early years of the eighteenth century recorded a typical trade voyage from Montreal to the Illinois Country, highlighting Chicago as a significant nexus between watersheds. Captured as a boy from Deerfield, Massachusetts, Joseph Kellogg grew up in New France and entered the fur trade in 1710, signing onto a trading voyage along with six other voyageurs to carry goods to the Mississippi River "in two Cannoos made of birch bark." The voyaging brigade traveled the river routes from the Ottawa River into Lake Huron, linking disparate water routes and avoiding rapids through a series of portages. In the language of the fur trade, the brigade of two canoes then "coasted" the shorelines of Lake Huron and Lake Michigan, paddling within sight of land lest a storm should arise, until they reached the southwest end of the lake. Here Kellogg's party "carried their canoos over land a full league to a Branch of the River Ilinios [the Des Plaines] and this was their biggest carrying place of the whole Voyage and is called Chigaquea." As the party passed from the waters of the Great Lakes and entered the Mississippi drainage, Kellogg described it as "a New World, compared with the Rivers [of] Canada."[40] Crossing the Chicago portage proved a profound experience for voyageurs like Kellogg as they carried their canoes and supplies, bridging the distinct watersheds through their physical efforts.

Despite the labor-intensive nature of portaging, voyageur crews continued to use the Chicago route as a lucrative entrepôt for bartering European goods for pelts. Already by the 1690s, fur trade documents explicitly named Chicago as a destination for such commerce. In 1692, for instance, Marie Magdeleine St.-Jean, acting on behalf of her merchant husband, hired four voyageurs in distant Ville-Marie, Québec, to take two canoes and "travel to Missilmakina [Mackinac] and Chicagou to transport goods and purchase pelts of beaver."[41] These voyageurs, en route to St.-Jean's husband in the Illinois Country, would have necessarily paused as they portaged through Chicago. The Weas, settled along the main route of travel, could expect to exchange their pelts, acquired in Chicago's wetlands, for a variety of European goods that such travelers brought along for trade.

Chicago also remained key for the convoys of canoes transporting goods between nascent French settlements downriver, in the Illinois Country, and the Great Lakes posts. St.-Jean's husband, François Francoeur, and traders like him, awaited such shipments of fresh supplies to continue trade downriver from Chicago in the Illinois settlements.[42] Other examples of such shipments abound. In 1698, for instance, la Forest contracted a canoe to carry supplies from Montreal to his post at Starved Rock on the Illinois River. As la Forest and Tonti had already established a depot at Chicago's portages and placed DeLiette there in 1698, it seems likely this shipment passed through Chicago. Transporting "powder, ball, and other supplies" necessary to maintain their post downriver, la Forest and Tonti relied on Chicago's portage as their geographic link to the French colonial centers in the Great Lakes and lower Canada. When Kellogg and his companions passed over Chicago in 1710, they were en route to trade at French settlements at the mouth of the Illinois River before returning home to Quebec. As in the previous century, Chicago's portage route remained one of several main thoroughfares the French relied upon to resupply and maintain communications with their Illinois settlements downriver.

As more French traveled the interior water routes, learned the particulars of canoe navigation, and participated in the expanding fur trade, they confirmed the geographic potential of the Great Lakes as a region of maritime mobility. The natural wonders of the lakes and the waterborne movement they fostered caused some Europeans to wax romantic about their time spent traversing the Upper Country. The Jesuit Pierre François Xavier de Charlevoix wrote of his trip through the lakes in 1720, "If one always travelled, as I did then, with a clear Sky, and a charming climate, on a water as clear as the finest fountain . . . one would be tempted to travel all one's life." Charlevoix described in detail the ease of navigation between waterways, the accessibility of affordable food, fish, and game, and the stunning lakeshore scenery around him.[43] Charlevoix cast the region as a freshwater Eden, poised and ready for French mastery.

On a promotional map from 1718, Nicolas de Fer depicted an interior geography ideally suited for waterborne travel through French North America. Like many mapmakers of the time, de Fer included images in his cartography. He filled his map with beavers, bison, and game birds and peppered the North American interior with depictions of Native hunting parties. The effect portrayed a region bursting with natural resources. If Native Americans currently tapped into such bounty, so too could the French who now claimed all the riverine interior from Canada's Great Lakes to the Gulf of Mexico. In addition to the rich flora and fauna, de Fer's map was striking in its depictions of interior waterborne movement. De Fer included imagery of Native Americans traveling the Great Lakes waters. Off the coast of "Les Checagou," for instance, on the "Lac des Illinois [Lake Michigan]," de Fer showed a canoe loaded down with Indigenous mariners. Elsewhere on the map, he also portrayed Native people actively portaging canoes between waterways. With such detailed illustrations, de Fer conveyed the aquatic geography and the Indigenous methods that came together to create an interior world of waterborne mobility and gave hope to French plans for maritime empire in North America's interior (see figure 2.2).[44]

Partitions of Beaver and Bison

Explorers like Charlevoix and cartographers like de Fer portrayed the waterborne interior as geographically suited for French empire and in this way, destined for greatness. Chicago's portages fit into French visions of empire as they promised to knit the waters of the Great Lakes with the Mississippi's tributaries. But Chicago was not somehow *geographically determined* to benefit French designs in the interior. Despite such positive interpretations by French observers, there remained other factors that portended continued difficulty for French ambitions along Chicago's portage routes. The same

FIGURE 2.2 The French cartographer Nicolas de Fer drew on firsthand travel narratives to portray a continent of freshwater connections, hinting at the maritime potential of French North America while simultaneously depicting the centrality of Indigenous canoes and portages to mobility in the interior. Nicolas de Fer, *Le Cours de Missisipi* (Paris: Chez J. F. Benard, 1718) [detail]. Courtesy of the Newberry Library.

waterborne travel that allowed some to imagine a French maritime empire in the interior of North America provided mobility to illegal and insubordinate traders, from an official perspective. New imperial boundaries left Chicago ambiguously between colonial jurisdictions and weakened French control over the portages while further warfare between Native peoples again threatened movement along the route. Wea settlements wavered in the face of new dangers, while French travelers continued to contend with the environmental hazards of Chicago's maritime reality. Even as optimistic officials, cartographers, and travelers held out hope for the portages as a site of French power and Indigenous collaboration, a diverse array of complications rendered Chicago's route a dangerous conduit beyond the control of imperial dominion.

The activities of coureurs de bois, or illegal French traders, at Chicago's portages posed a continual problem for imperial officials. French fur trade policy required traders to obtain a license, called a *conge*, to legally exchange goods for pelts with the region's Native peoples. Officials capped these licenses to a limited quantity each year in efforts to control the supply of furs entering Europe and prevent too many traders from flooding the peltries market and driving down prices. Individual traders, unable to obtain licenses or unwilling to leave the Great Lakes country at the end of a trade season, chose to operate beyond the official sanctions of the conge system, instead smuggling goods into the interior and dodging French regulations. Chicago's portages, which offered an unfortified route between the Great Lakes and Illinois Country, enabled coureurs de bois activity and perpetuated flows of illicit goods across the continental divide.[45]

In the early eighteenth century, distant French officials redrew colonial borders across Chicago's continental divide, inadvertently strengthening illegal traders' position along the western routes. In 1699, Pierre Le Moyne, Sieur d'Iberville, founded Fort Maurepas on the Gulf Coast near the mouth of the Mississippi River on behalf of France. Soon, Iberville's colony spread, with a new capital at Mobile Bay by 1702. Connected by the course of the Mississippi River, the French and Native settlements and missions in the lower Illinois Country formed new trade linkages with these Gulf settlements in what became the nexus of France's Louisiana colony. Given these riverine connections, Iberville laid claim to the entire Mississippi drainage, asserting that the Louisiana colony held authority over the entire watershed. In the wake of successful colonization along the Gulf Coast and lower Mississippi, proposals for new jurisdictional boundaries in French North America began to muddle Chicago's position within French schemes of empire.[46]

Iberville's rival colony worried French officials in the Great Lakes. In 1700, Louisiana had sanctioned new schemes for processing bison hides in the Ohio River valley and sent envoys up the Mississippi to the Illinois settlements and beyond.[47] To Canadian officials, this looked like overreach into territory previously ruled by New France. Worse yet, Iberville's settlements began siphoning away furs from the interior and encouraging illegal traders to travel south to avoid official trade sanctions in Canada. Officials in New France complained to the king that "all the libertines and coureurs de bois . . . have carried all their beaver and other pelts to M. de Iverville," along the Gulf Coast, rather than back to Canada.[48] Ambitious Native traders also tapped into this new commercial outlet. The governor of New France, Louis-Hector de Callières, reported that thirty Odawas from Michilimackinac had taken their beaver pelts down to Iberville—an incredible journey of more than 1,200 nautical miles. As he penned his complaints, he shared that "10 more canoes have gone down" from the Great Lakes to the Gulf and "other coureurs de bois are preparing to do in like manner."[49] Callières accused Iberville of welcoming these illicit traders with open arms in Louisiana, "where they will take their furs" from the Great Lakes into the new colony. Callières equated this southern flow of furs to "wholesale robbery" and worried over the economic consequences Louisiana's founding held for Canadian merchants and his own colony.[50] As coureurs de bois and Native traders made their journeys southward, the portages that had promised to unite French imperial interests in the interior facilitated illicit shipments of pelts between France's two competing colonies instead.

With furs flowing between the Great Lakes and the Mississippi, Iberville pressed his colony's claims to the waters of the upper Mississippi drainage. Iberville worried that New France held designs for all lands north of the Ohio River and specifically fretted over his Canadian rivals' attempts to build a permanent post at Chicago. He feared that "if they establish a warehouse there, it is evident that all the Indians on the Illinois river would go there," drawing away trade with the Mississippi.[51] To curb this escalating competition between the two colonies, Iberville proposed a formal geographical boundary along the continental divide that separated the Great Lakes and Mississippi watersheds. Iberville argued that "all the streams falling into the Mississippi, up to their sources, and the tribes upon them, should belong to the Mississippi" while those rivers flowing into the Great Lakes and the people living on them, "such as the Miamis of Chicago," should fall under Canada's authority.[52] Iberville insisted, "The Illinois should be informed that they must not expect to trade with Canada in future, but must do their trade with the posts on the Mississippi."[53] With the founding of the new colony to the south, French officials like Iberville pushed for clear geographic boundaries to separate the peoples and commerce of the midcontinent.

Iberville also believed the French could delineate colonial jurisdictions based on specific ecological resources to alleviate trade rivalries. He proposed that traders and Native groups under Louisiana's authority remain limited to trading bison hides while those under Canadian jurisdiction enjoy a monopoly on beaver pelts through the fur trade.[54] He would urge all Native peoples along the Mississippi's tributaries to "devote themselves to hunting oxen and roebucks"—bison and deer—rather than furbearers.[55] King Louis XIV approved of such clear-cut proposals. He assured officials in New France that he would allow traders in the new colony to deal "only in the hides of buffaloes" and promised to "prohibit them also hunting Beaver."[56] With the king's endorsement, the French looked to neatly partition the regions into separate territories of beaver and bison. Far-flung officials failed to account for places like Chicago, where people like the Weas hunted both bison and beavers locally in the transitional landscape of oak openings, marshes, and prairies, trading pelts and hides at will across the two watersheds. Ecological resources could not be so easily divided by those who understood Chicago's complex environs.

To French heads of state, Iberville's proposals seemed determined by clear geographic and environmental divisions. French territorial claims in the Americas often adhered to the boundaries of watersheds, and in this, Iberville's proposals benefited from imperial precedent. French state-builders relied on distinctions between river drainages to make their claims of sovereignty against European rivals.[57] But Chicago's wetlands fit uneasily into such neat understandings of nature and empire. For one, the continental divide at Chicago varied by season, and in high water, ceased to exist at all. The marshy terrain of Chicago left no clear border between the Mississippi and Great Lakes watersheds. Additionally, groups like the Wea exploited both bison and beaver resources in their local surroundings. The Wea had intentionally positioned themselves at Chicago's important junction between prairies, wetlands, and oak savannas to enjoy the benefits of overlapping environments and waterways. The notion that they would now comply to new boundaries distinguishing which ways they could transport and trade specific commodities surely seemed absurd. While French officials in Versailles, Quebec, and the Gulf could rest easy with these new borders, the ambiguity of such boundaries at a place like Chicago demonstrated the limits of imperial comprehension and authority along the western portage routes. While Chicago's portages had existed as a

liminal space between ecosystems and watersheds for centuries, French officials added a new administrative layer to Chicago's borderland complexity that
would further complicate their efforts to govern the area.

Barriers to Entry

While fictive borders and ecological ignorance complicated French control over
Chicago, tangible environmental factors troubled movement along the canoe
route as well. Lake Michigan remained treacherous for canoes as they coasted
along its shores between Michilimackinac and Chicago. Foul weather on the
lake forced Father St. Cosme's party ashore just north of the Chicago River in
1698, soaking the travelers and threatening to destroy their canoes. As he related,
"One must be very careful along the lakes, and especially Lake Mixcigan" where
"the rollers [waves] become so high in so short a time that one runs the risk of
breaking his canoe and losing all it contains."[58] Father Charlevoix, who would
coast these same waters several decades later, related that "the ocean, in its greatest fury, is not more agitated" than the Great Lakes in a gale.[59] This was no idle
observation. The lakes, being comparatively shallower than the ocean, become
rougher much faster as winds instigate wave action. The offshore sandbars and
shoals of Lake Michigan led to fast-moving squalls and quick-rising swells.
As Charlevoix asserted, "When the wind comes from the open lake," conditions
could deteriorate and "render the navigation more dangerous."[60] When storms
rose off Chicago's coastline, canoe crews needed to make for shore immediately
or run the risk of shipwreck.

Jean Baptiste Perrault, a voyageur who passed through Chicago's portage
later in the century, witnessed the deadly results of such a wreck farther north
on Lake Michigan. Perrault's own canoe brigade had been driven ashore by the
rising of a gale, the lake being "as white as a sheet," but another party of traders
had continued despite the danger. While Perrault watched, the second party's
canoe foundered, and eleven men disappeared beneath the waves. The following day, Perrault's brigade found the wreckage "here and there along the beach"
and recovered the bodies. Some they found still "rolling in the waves" while
others lay "half buried in the sand."[61] Such maritime tragedies occurred with
frequency. Father St. Cosme, writing about Chicago's coast, noted that "many
travelers have already been wrecked there."[62] Charlevoix reckoned that "there
are few places in Canada where there are more wrecks" than Lake Michigan.[63]
The realities of wind and wave complicated the nautical aspirations that French
officials projected onto the Great Lakes waters and made voyages off Chicago's
coast harrowing business.

Apart from natural perils, plenty of human-induced hazards continued to jeopardize travelers along the portage route. Portages like the ones through Chicago posed opportune points for Indigenous gatekeepers to demand goods from outside travelers. When the missionary Dominic Thaumur de La Source passed north through the Chicago portage in 1699, he reported that the Wea themselves sometimes exploited travelers along the route. Though his party came well armed and "not in the humour of letting ourselves be plundered," the Weas "boasted they would rob us" as La Source's crew traveled up the Illinois River to Chicago. When they arrived at the portages, La Source complained that "the Miamis are trying to pick a quarrel with us."[64] While La Source and his party were able to force a passage through Chicago, other parties of less brazen Europeans were not so lucky. Cadillac, in his memoir, emphasized the dangers of attacks along portages, where it remained easiest "to prepare ambushes in narrow passes along the route."[65] Weighed down as they carried boats and supplies across the muddy pathway between navigable waters, canoe crews amid portaging remained most susceptible to such opportunistic plundering or demands for payments of goods. Part of the Miamis' strategies across the region lay in exploiting the bottlenecks of portages to either exact tolls from vulnerable brigades of voyageurs or to threaten pillaging those who did not comply or acknowledge the Indigenous control over these passageways.[66]

Weas profiting on their portage geography occasionally troubled French parties crossing the Chicago route, but rumblings of more sinister violence also plagued the area. In July 1700, Father Pierre-Gabriel Marest, a Jesuit living among the Kaskaskias downriver, reported that groups of Miamis had "been routed by the Sioux [Dakota]," raiding down from Wisconsin. The Dakota raids into northern Illinois proved threatening enough to unite the Kickapoos, Mascoutens, Meskwakis, and Miami groups into a temporary alliance in 1700.[67] The raiding even extended down the Mississippi and Illinois River valleys, and Father Marc Bergier, a priest living with more southerly Illinois Indians shared that his village experienced "frequent alarms" from war parties of Dakotas.[68] Even after his success in coordinating the Great Peace at Montreal in 1701, Governor Callières worried that "the affairs of our allies in the Upper Country do not appear to be entirely in as good a condition" as the French would have hoped. Trouble in the Great Lakes region persisted, according to Callières, because hostility between the Dakotas and other Native peoples "increases every day." The governor "feared that these Upper Nations will be drawn into a general war."[69] Just as the French and their allies began shoring up a peace in the east with the Haudenosaunee, new incursions from the west seemed on the rise along the borderland between the Mississippi River and Lake Michigan.[70]

Ongoing rumors of war threatened the tentative security of Chicago's recent settlements and the rest of the western Great Lakes. Squabbles between various Indigenous allies of France cast doubts on the general peace the French believed they had established after 1701, and Dakota incursions from the west continued to trouble the upper Illinois River valley. At the same time, Miami-speaking groups like the Wea clashed with Anishinaabeg from Green Bay, Michilimackinac, and Sault Ste. Marie as they forged new relations with the Haudenosaunees to the east. The Sault Ste. Marie Ojibwe had already raided Meskwaki settlements along the Fox-Wisconsin portage and the collective Anishinaabeg posed a formidable threat. Potawatomis and Odawas had utilized the Chicago portages for decades in trading with the Illinois downriver, and as experts in waterborne mobility, these powerful Anishinaabeg could easily strike the Wea settlements there. Growing tensions and threats of attack along the portages escalated until the Weas became wary of their own safety at Chicago.[71]

Faced with the risks of exposure posed by the open route through Chicago, groups of Wea migrants retreated from their promising marshes sometime after 1700. Resettling along the Wabash River, these Weas sought a secure location away from Lake Michigan and the potential incursions of hostile peoples. As Jacques-Charles de Sabrevois, commander at Detroit, later explained, "The Ouyatanons [Wea] were also at Chicagou, but they feared the canoe people, and consequently left the place. It would be difficult to make them return." While Sabrevois's explanation of "canoe people" remained vague, he likely meant Anishinaabe groups like the Odawa, Ojibwe, and Potawatomi, but it seems clear that Dakota raids in the Illinois Country might also have contributed to Wea insecurities at Chicago.[72]

With Weas moving inland, the Jesuits' position along Chicago's coast became redundant. Early in 1700, Father Pinet left the mission, traveling south along the Illinois River to continue his mission work among the Tamaroas of the Mississippi River. Father Jean Mermet, another Jesuit who had worked closely with the Weas and contributed to Pinet's dictionary, stayed on briefly. By then, with Miami-speaking groups relocating to the Wabash and St. Joseph Rivers, Chicago offered few opportunities for sustained evangelism. Mermet abandoned the area in favor of the St. Joseph River and the mission closed by 1702.[73] Around the same time, DeLiette left the portages, returning to the larger Illinois settlements downriver near Lake Pimiteoui. In the wake of new threats and the Weas' retreat, local French influence at the Chicago portages ebbed.

At the same time as the Weas fled from threats of Anishinaabe and Dakota attacks, French traders and missionaries traveled the river routes in fear of Meskwakis living to the north and west of Chicago. The Meskwakis called themselves the Red Earth People, but other Algonquians of the Great Lakes referred to them as the Outagamis, and the French called them the Renards, or Foxes. They had migrated to the western shores of Lake Michigan sometime before the French arrived in the region, but they had carried old feuds with them. During the seventeenth century, they had formed a close alliance with other groups in the Chicago region, including the Mascoutens and Kickapoos, through bonds of kinship and trade. Their main settlements lay along the upper Fox River of Wisconsin, but they were active in the Chicago area as well. And throughout the region, they remained concerned about French trade voyages to the Mississippi watershed, where they feared voyageur brigades would arm their longtime enemies, the Dakotas, with firearms and metal tools. Facing the potential of supply routes to their enemies running directly through their home country, they impeded French trade through the portages and river routes west of Lake Michigan. By the early eighteenth century, the Meskwakis' hostility toward incoming French threatened to curb travel along the western portage routes and foil French mobility in the region.

As early as 1698, during Father St. Cosme's missionary voyage from Michilimackinac to the Illinois Country, Tonti had steered the party through Chicago's portage to avoid trouble with the Meskwakis along the Fox-Wisconsin portage, an alternative route to the Illinois Country. As St. Cosme related, "We should have liked very much to pass by the bottom of that bay [Green Bay]," to ascend the Fox River, "but the Renards [Meskwaki] . . . will not allow any persons to pass." The Meskwakis were rumored to "have already pillaged several Frenchmen who tried to go that way." Instead, the seminary priests were "compelled to take the route by way of Chikagou," avoiding the Meskwakis' blockade to the north.[74] Rumors of ambushes and potential violence along the river routes demonstrated continued French vulnerability while traveling in the west.[75]

Roads of War

By 1712, forebodings of Meskwaki predations along the western routes had erupted into full-scale warfare across the region. At the newly established French and Native settlement of Detroit, Potawatomis, Wendats, and Odawas pushed the French commander there to besiege a town of resettled Meskwakis and Mascoutens after old rivalries flared. The resulting skirmishing, retreats, and attacks left over a thousand Meskwakis and Mascoutens dead and many more enslaved by the French and their allies.[76] Enraged by the betrayal, Mascoutens, Kickapoos, and Meskwakis to the north and west of Chicago launched retaliatory raids against various Native groups and French alike in an expanding theater of war.

The Meskwakis and their allies remained entrenched inland from Michigan's western coast, and France's Native allies warned that the Meskwaki would use the portages strategically as potential points of ambush.[77] As Father Marest wrote from Michilimackinac, the Meskwakis and their allies "would indeed prove truly formidable," especially for small parties of French in canoes. The Meskwakis' coalition included many "boatmen" and given their position along the portage routes of the west, Marest feared they would prove a danger both "by land and water," now that hostilities had begun.[78] After the bloodshed of 1712, the route through Chicago became openly dangerous for French and their Indigenous allies.

As Marest had predicted, reports soon filtered back to Canadian officials about French flotillas waylaid by war parties along Chicago's route. In 1712, the Kickapoos—in alliance with the Meskwakis—captured a French messenger named Langlois en route from the Illinois Country and intercepted "many letters from the Rev. Fathers, the Jesuits of the Illinois villages" as well as important correspondence from the Louisiana colony. The Kickapoos destroyed the mail and robbed the man of all his trade goods.[79] By the fall of 1714, Claude de Ramezay, acting governor, and Michel Bégon de La Picardière, the intendant of New France, wrote to Paris to share that "five boats, in which were twelve Frenchman, including a jesuit brother . . . had been destroyed by a party of Renards near chicagou."[80] Though the following year, the pair of high-ranking officials amended their initial report—the attack at Chicago had never been verified—they shared new intelligence that "a French boat, in which there were five men, carrying corn to Michilimakinak, was surprised by a boat of 22 Renards [Meskwakis], who killed these Frenchmen" on the open water.[81] Even while rumors swirled and reports of violence circulated, such stories from Chicago's portages drove home how vulnerable French travel through the west had become with the renewal of Indigenous conflict along the region's water routes.

Throughout the escalating violence, Chicago provided a pathway for war parties from both sides of the conflict. In 1713 and 1714, Meskwaki campaigns followed the Illinois River south to attack an Indigenous settlement of the Illinois near Lake Pimiteoui.[82] In 1715, the French and their allies also used Chicago as a planned rendezvous to unite forces of Illinois and Miamis from the south and Odawa contingents from the north. Over 450 Illinois, along with Governor Ramezay's son, arrived at the portage in late August to await the other contingents and "the munitions of war," which were "to be taken by canoes to chicagou."[83] No reinforcements materialized, however, and the Illinois warriors eventually returned south. In late November that same year, French-allied Wendats and Potawatomis passed through the area en route to attack the Kickapoos and Mascoutens, killing over a hundred people on the prairies along the Rock and Fox Rivers. The Meskwaki, in turn, launched a counterattack and intercepted the Wendats and Potawatomis along the upper Illinois River in early December.[84] Again, as it had during the Haudenosaunee campaigns of the seventeenth century, the Chicago area became a crossroads of bloodshed.

Amid the fighting, Governor Ramezay of New France emphasized the strategic importance of Chicago. He called for a new post at the portages as the west descended into chaos. Reporting back to the ministry in France in 1715, Governor Ramezay and Bégon outlined three main threats in the west that control over Chicago could help remedy: English traders had made inroads among the Weas and other Native groups south of the Great Lakes, the conflict with the Meskwakis raged, and renegade French traded illegally in the Illinois Country.[85] To remedy western disorder, Ramezay and Bégon stressed that it would "be useful to establish a post at Chicagou." This would help secure the strategic route between the Illinois Country and New France, and "facilitate access to the Ilinois and Miamis, and to keep those nations in our interests."[86] As Ramezay put it in another letter to the minister, a garrison at Chicago remained necessary, "in order to keep in bounds both the French and the *sauvages*" of the Illinois Country.[87] Ramezay also proposed relocating the Wea back to Chicago to strengthen French authority there, advocating "they must, if possible, be persuaded to transfer their village to Chicagou."[88] This would draw the Weas away from British traders on the Wabash and Ohio Rivers and ensure local Native support along Chicago's portages.[89] While imperial rule in the west unraveled, officials like Ramezay asserted that local control over Chicago's strategic geography offered a solution to French woes.

The proposed French fort at Chicago could also serve as a forward outpost for continued French campaigns against the Meskwaki threat. By 1715, the Meskwakis had fortified their own settlements on the upper Fox River near the Fox-Wisconsin portage while their Mascouten and Kickapoo allies remained along the Fox and Rock Rivers west of Chicago. A French post at Chicago's portages would provide support for campaigns against these various enemy strongholds, which all lay within easy striking distance—a seven days' march from Chicago. Ramezay believed troops at "Chicagou—where there is abundant game of all sorts"—could also better sustain themselves. Chicago's ecological abundance contrasted sharply with French supporting posts at Michilimackinac or Green Bay. Ramezay assured officials back in France that Indigenous and French combatants launching raids from Chicago against the Meskwakis and their allies "can easily subsist" at the portages without incurring high costs for supplies.[90] As Ramezay and those in New France looked to bring a swift end to the conflict, garrisoning Chicago's portages seemed a logical step in forcing the Meskwakis to peace terms or supporting further military campaigns in the area.

Only two years after Ramezay's proposal for the new post at Chicago, however, officials back in France permanently undercut Canadian authority at the portages. In 1717, the ministry officially transferred the Illinois Country to Louisiana. This effectively confirmed commercial divisions already approved by the king back in 1702, but now meant that French officers on the Wabash, Illinois, and upper Mississippi Rivers would answer to Louisiana's governor, rather than administrators in Canada. The transfer also served to ostracize the concerns of French Illinois from Canadian officials, who resented the loss of territory.[91] This administrative transfer perpetuated the conundrum over where Chicago fit within France's American empire and left the portages ambiguously between Louisiana and Canadian control.

Historians have shown how Indigenous groups orchestrated the so-called Fox Wars in efforts to stymie a pan-Algonquian alliance under the French and halt the inclusion of the Meskwakis into their diplomatic networks. Anishinaabeg had pushed France into the conflict at Detroit, while the Illinois Confederacy, bent on continued captive taking, exacerbated the conflict south and west of Lake Michigan by enslaving Meskwakis, Mascoutens, and Kickapoos and selling them to the French.[92] In both the Anishinaabeg and Illinois cases, hostility against the Meskwakis went back decades and served the Native groups' own interests while undermining French hopes for a pan-Indigenous alliance in the west. While aggressive Native peoples and their strategic slave trading played an essential role in escalating the war against the Meskwakis, ambiguous jurisdictional boundaries drawn by the French posed further barriers to any mediation.[93]

Such bureaucratic divisions left French officials in both the Illinois Country and New France unable, or unwilling, to effectively broker a lasting peace between the Meskwakis and other Native groups.[94] In the Great Lakes, powerful Anishinaabeg at Michilimackinac, Detroit, and St. Joseph pressured French commanders to maintain a belligerent stance against the Meskwakis west of Lake Michigan. In the Illinois Country, incessant warfare between the Illinois and the Meskwakis' allies, the Mascoutens and Kickapoos, barred French officials downriver from entertaining peace as an option. Meskwakis, Mascoutens, and Kickapoos living in the lands between the Mississippi and Great Lakes watersheds found themselves flanked by French-allied Natives on both sides. That the main theater of war straddled the administrative divide between the two French colonies only exacerbated the conflict as French officials, reluctant to extend themselves beyond their own jurisdictions and often ambivalent about the Meskwakis' fate, failed to stem the violence that raged back and forth across the continental divide.

The ambiguities of these colonial boundaries allowed commanders in both Illinois and the Great Lakes to shift blame for the ongoing violence to French counterparts in their neighboring colony. Officials in New France complained that those in Illinois, under the authority of Louisiana, would not control the belligerent Illinois' captive taking. Governor Vaudreuil, in Quebec, complained "all would be peaceful on this continent" if only the Illinois did not continue to attack the Mascoutens, Kickapoos, and Meskwakis. Indeed, Vaudreuil argued that the grievances that the Meskwaki coalition held against the Illinois "seemed reasonable" and believed the Kickapoos when they claimed, "It was not they who had begun the war, but that the Ilinois had attacked them." While Vaudreuil wrung his hands over the injustice of Illinois aggression against the allied tribes to the north, there seemed little he could do. He wrote that "it will prove impossible to arrange this peace, unless the officer in command among the Ilinois is able to induce that nation to make overtures to obtain it."[95] The Illinois people, after all, remained under the care of commanders in the Illinois Country who answered to Louisiana, and not Vaudreuil's authority in Canada.

Vaudreuil and other officials in New France remained content to turn a blind eye to the ongoing violence along the Illinois River. In 1716, Louis de Louvigny, an officer from New France, had negotiated a limited peace with the Meskwakis in Wisconsin. By 1718, Meskwaki leaders were working with Governor Vaudreuil to broker further peace agreements between Canada and the Kickapoos and Mascoutens. All the while, Illinois Natives warred back and forth against Meskwakis, Mascoutens, and Kickapoos in the borderlands of the upper Illinois River from Starved Rock to Chicago.[96] Meskwaki retaliatory raids into the lower Illinois valley left both French and Illinois casualties, and yet Canadian officials continued to believe they had solidified a peace.[97]

When another Canadian officer, François-Marie Le Marchand de Lignery, secured a second armistice with the Meskwakis in the Great Lakes region in 1724, both French and Illinois Natives downriver from Chicago felt betrayed.

Claude-Charles duTisné, commanding in the Illinois Country, expressed surprise "that those Gentleman at la Baye [Green Bay] should have concluded peace so soon," especially since hostile Kickapoos had just killed five more French traders on the Wabash River.[98] Missionaries such as La Source, living among the Illinois, became equally incensed by Canadian negligence in such negotiations, decrying "the peace that has been made" as "hurtful to this province."[99] While officials in New France continued to attempt a peaceful resolution with the Meskwakis, they conveniently ignored the plight of fellow French colonists and their Indigenous allies across the continental divide.

Worse than official indifference across the imperial border, French colonists in Illinois accused Canadian traders of actively supplying and encouraging Meskwaki raids against their Illinois valley settlements. After Louvigny's peace in 1716, unsanctioned coureurs de bois from Canada had returned to trade with the Meskwakis, providing them with guns, ammunition, and other European goods that the Meskwakis, in turn, used against the French and Natives of Illinois.[100] As duTisné complained from the Illinois Country, "We are killed everywhere by the renards [Meskwakis], to whom Canada supplies weapons and powder." These illegal Canadian traders actively instigated the Meskwakis to attack French in the Illinois Country, who they viewed as trade rivals. According to duTisné, the Canadian coureurs de bois assured the Meskwakis that the French of Illinois were "other white men" and that the governor of New France would never punish the Meskwakis for attacking them.[101] The French Jesuits similarly feared that coureurs de bois had misled the Meskwakis to "believe that we are not Frenchmen of the same nation as those of Canada."[102] Given official Canadian indifference at the time, such boasts seemed vindicated, and the Meskwakis and their allies continued to terrorize the French and Indigenous towns of the lower Illinois River unchallenged from the north.

The divisions in French policy became so stark as to perplex their Indigenous allies in the borderland between the colonies. In 1725, in response to Lignery's armistice with the Meskwakis, several Illinois leaders visited Fort Chartres on the Mississippi River to address the inconsistencies of French policy. The Illinois headmen named thirty-nine individuals, both French and Native, who had succumbed to Meskwaki attacks along the Illinois River corridor from "Kikagwa [Chicago]" to Starved Rock, Lake Pimiteoui, and as far down as Cahokia. Jouachin, speaking for the Metchigamea group of the Illinois peoples, likened the French of Canada and Illinois to two brothers who had gone separate ways. He questioned the way one could supply the enemy of the other. He asked how the Canadians could make peace when "the allies of the renards [Meskwakis] dance the scalp-dance around their scalps before the traders from Canada, who are of the same blood, and who supply them with powder in exchange for their beaver and marten skins."[103] To Illinois leaders like Jouachin and his compatriots, it seemed incomprehensible that jurisdictional boundaries trumped blood ties between the French and that Canadian traders and officials could so easily ignore violence downriver for the chance to trade pelts with the Meskwakis, Kickapoos, and Mascoutens in the Great Lakes.[104]

Campaigning across the Divide

By 1727, it had become clear to French officials in both Canada and Illinois that unless they coordinated policy across colonial borders, the Meskwaki wars would continue indefinitely. A new governor in New France, Charles de la Boische, Marquis de Beauharnois, had assumed command in Canada in 1726. With direct orders from Versailles to end the conflict, he proved much more willing to coordinate with imperial counterparts in the Illinois Country than his predecessors, embracing a brutal strategy of eradication to end the war against the Meskwakis and their allies.[105] He wrote to Charles Henri Joseph DeLiette, commanding at Fort Chartres in Illinois, and requested him to rendezvous once again at Chicago to coordinate a joint attack with Canadian forces against the Meskwakis.[106] In July 1728, DeLiette marched on Chicago with twenty French troops and several hundred Illinois allies, only to run headlong into a large Meskwaki hunting camp. In the ensuing battle, the Illinois killed twenty Meskwakis and captured another fifteen, but the fight proved costly for the Illinois and French allies as well. Discouraged by their own heavy casualties, the Illinois turned back south. Meskwaki survivors fleeing north, however, reported the unexpected advance on Chicago. That same summer, Meskwakis watched anxiously as a large force of French and Indigenous allies from the Great Lakes assembled at Green Bay. These coordinated campaigns convinced Meskwaki leaders in Wisconsin that French from both Canada and Illinois had reunited in their efforts to wipe out Meskwaki opposition.[107]

As French policy against the Meskwakis coalesced, the situation between the watersheds became more tenuous. Kickapoos and Mascoutens wavered in their support of the Meskwakis and opened talks with French representatives in 1728. The following year, once-sympathetic Ho-Chunks in Green Bay had also turned against the Meskwakis. For years, straddling the colonial border had allowed the Meskwakis to play French officials, priests, and traders against one another while they launched raids south along the Illinois River corridor and Mississippi valley. After 1728, however, the Red Earth People found themselves on shakier ground. Caught between French colonies and Indigenous enemies that appeared more willing to coordinate campaigns across boundaries, by 1730, over a thousand Meskwakis made plans to abandon the region entirely.[108]

The Meskwakis fled east, hoping to find sanctuary among the Senecas in the land of the Haudenosaunee south of Lake Ontario. Crossing the upper Illinois River in late July 1730, the Meskwakis made their way across the savannas south of Lake Michigan, skirmishing with parties of Cahokia Illinois as they went. These Cahokias, in turn, alerted French commanders in both the Great Lakes and Illinois, as well as a host of Native allies who descended upon the Meskwakis as they trekked east. Unlike operations in the past, however, this time French and Indigenous forces from the two colonies effectively coordinated efforts to intercept the Meskwakis, "about 60 leagues to the south of the Extremity or foot of Lake Michigan, to the east south east of le Rocher [Starved Rock] in the Illinois Country."[109] A party of Potawatomis, Kickapoos, Mascoutens, and Cahokias waylaid the Meskwakis, forcing them to halt and fortify their camp. Soon after, Robert Groston de St. Ange arrived with a hundred French soldiers from the Illinois settlements and Nicolas-Antoine de Villiers joined from Fort St. Joseph, under the jurisdiction of New France. Though hailing from different colonies, St. Ange and de Villiers worked with their Native allies to manage a sustained siege of the Meskwaki camp for the next twenty-three days.

When Governor Beauharnois of New France sent orders to annihilate the Meskwakis without negotiation, the French commanders and their Native allies coordinated their efforts with grim efficiency. The Meskwakis attempted to break the siege and escape in early September but failed. Caught on the open prairies, the Meskwaki warriors turned to face their pursuers in one final stand as their women, children, and elders tried to flee. In the battle and rout that followed, the French and their allies slaughtered over five hundred Meskwakis and captured four to five hundred women and children.[110] Reporting the carnage from Canada, Beauharnois boasted that only 450 Meskwakis remained alive, mostly in captivity as slaves of the French or their Indigenous allies.[111]

Yet even in their ruthless victory, the French bickered over the ambiguity of their colonial border. Despite their cooperation during the fighting, the French officers disagreed over the location of the siege following the battle, contesting whose district had hosted the carnage. Meanwhile, the governors of the two colonies raced to send dispatches back to France to report the victory. When Louisiana's messages arrived first, officials in New France "were greatly mortified . . . at not being the first to convey information of this happy success."[112]

Both de Villiers, the officer from New France, and St. Ange, the commander from Illinois, claimed the battle had occurred within their jurisdictions in the lands south of Lake Michigan. Since the battlefield lay on the open prairie between waterways, it remained impossible to verify the location with any certainty.[113] While the individual French officers quarreled once more over geographical borders, however, Canadian officials exalted over the victory's geopolitical implications. Reports from Quebec rejoiced that "the routes to the Mississippi . . . will now be open," and hopes ran high among the French that the licensed fur trade could again flourish in the west.[114] By eliminating the Meskwaki threat along the western routes, Beauharnois believed he had "reestablished tranquility in Canada and communication with Louisiana."[115] For the first time in decades, portages like the ones at Chicago stood open to flows of legal French commerce and promised renewed connections between the Great Lakes and Illinois Country, albeit at an incredible cost in human life.

Despite Beauharnois's confidence for renewed connections, trouble continued to plague French activities along the western routes. Some Meskwakis still held out along the tributaries of the Fox and Illinois Rivers to the west and north of Chicago. The following years saw still more investment in French campaigns against the surviving Meskwakis with few tangible results. Soon, the French and their Illinois and Anishinaabe allies found themselves pulled into more intricate conflicts with not only the remaining Meskwakis but Sauks and other Native peoples who had harbored survivors after the devastation of 1730.[116] Uneasy over these new conflicts, Anishinaabe leaders pushed for peace, and Beauharnois finally relented, granting a general pardon to the Meskwakis and their allies in 1738. The wars waged against the Meskwakis, Mascoutens, Kickapoos, and eventually Sauks devasted the Indigenous populations of Lake Michigan's western shore, but in the end, French control over the region's river routes remained as feeble as it had ever been.[117]

The conflicts undercut any hopes of future French rule at Chicago's portages. Administrative wrangling about which colony—New France or Louisiana—held authority over Chicago's continental divide had exacerbated the violence and prevented any uniform response from French officials for years. Even after officers in Illinois and the Great Lakes began cooperating, the conflict had continued to disrupt trade and travel along the portage routes that connected the two watersheds. The wars had thwarted fur trade profits and sapped money from the French treasury. Even as France finally acquiesced to peace in the Great Lakes in the mid-1730s, new conflicts to the south with the Chickasaws siphoned money and resources away from any efforts to rehabilitate French imperial authority along the western routes.[118] Despite earlier French aspirations for a fortified post at Chicago, and a wider imperial vision for the portages as a key junction connecting the Great Lakes and Mississippi, the route through Chicago lay beyond official French dominion.

An Appealing Periphery

By 1740, Chicago remained the only major portage route in the Great Lakes region outside of imperial French control. It also persisted as the only unmonitored route connecting the Mississippi and Great Lakes watersheds. While French garrisons guarded movement through other portages and French officers mediated relations with local Indigenous people at strategic sites like Green Bay, Michilimackinac, Fort St. Joseph, and Ouiatenon, Chicago's portages remained conspicuously open and unfortified.[119]

But lack of official presence along the portages did not signal an end to the geographic significance of the space. Following the wars against the Meskwakis, trade continued to flow through Chicago, albeit illegally. As Governor Ramezay had warned decades earlier regarding Chicago, without "troops to restrain them" along the western routes, the coureurs de bois had "become even more lawless."[120] After the wars against the Meskwakis ended and routes became safer, new waves of coureurs de bois harnessed the portages at Chicago to illegally trade for furs with the area's remaining Native people and smuggle goods between the French colonies. As the only route without an official French presence connecting the watersheds, the Chicago portages offered the primary channel for clandestine movement and trade between the Great Lakes and Illinois Country.

Often, such illicit traders directly undermined French authority in the west, as they had when they continued to trade with the Meskwakis during the late conflicts. Even after the wars ended, "the French who came to Chikagon" as illegal traders continued to stir up Indigenous bitterness against the French Crown. De Villiers, commanding at Fort St. Joseph, reported that coureurs de bois "pass the winter at Chicagou and Meolaki [Milwaukee]," where they continued to instigate Meskwakis against French authority. These illegal traders "advised them Secretly," warning the Chicago Natives against returning to the French fold. This had, according to the Meskwakis, alarmed "their Warriors" and "their old men" who remained convinced of official French plots to destroy them.[121] Whipping up resentment among the area's remaining Native holdouts, the coureurs de bois of Chicago explicitly defied French officials and further weakened imperial influence along the portages.

With encouragement from these subversive traders, Native people, wary of relations with France, also found refuge in the Chicago area. By 1741, for instance, some Sauks and Meskwakis had migrated to Chicago's portages, away from their main settlements at Green Bay and beyond the reach of the French Fort La Baye. According to French commanders at Green Bay and the St. Joseph, coureurs de bois living and passing through Chicago's portages kept these Indigenous refugees well supplied. Outraged by the coureurs de bois' relations with the Meskwakis at Chicago, Beauharnois sent orders to his western commanders to prevent any French traders from going to Chicago to trade. He also forbade "all voyageurs who have set out for the Illinois to carry on any trade" with the Meskwakis or Sauks still living along Chicago's route, "under penalty of being declared coureurs de bois."[122] To French officials, Chicago's portages now loomed as a threatening space of insubordinate French traders and unconquered Native people.

Governor Beauharnois, still distrustful of the Meskwakis' activities, insisted that these "renegades" return to Green Bay. Rumors circulated that these hostile Meskwakis and Sauks, now allied with the powerful Dakotas to the west, still warred against remnant Illinois people in the upper Illinois River valley. Reports from the Illinois Country likewise confirmed that "the Ilinois had been warring with the Renards [Meskwakis]," once again taking prisoners. Reports circulated of a Frenchman named Gatineau "cut to pieces" by the Meskwakis near Lake Pimiteoui and of Sauk incursions against the Illinois downriver.[123] Though peace waxed and waned in the Chicago area, French officials seemed unsure of the details and incapable of preventing them. Events along the river route lay beyond their imperial purview.[124]

The situation at Chicago posed an insurmountable challenge to French officials by the 1740s. After two years of steady pleading with the Meskwakis to leave that place, Beauharnois reported that "ten cabins" of Sauks and Meskwakis still held out at Chicago's portage and there remained little he could do about it.[125] Chicago's space persisted as a route beyond French control both on account of its political geography and its wetlands ecology. Historians have increasingly begun to pay attention to the value of swamps as sites of refuge and resistance in early America—especially for Native peoples who knew how to tap into the waterborne mobility and aquatic resources of such impenetrable spaces.[126] Chicago seems to have become such a place for refugee populations of Sauks and Meskwakis desperate to escape the persecutions of French empire after the Fox Wars. So, too, with the clandestine French traders who supplied these Indigenous holdouts and lived among them despite Beauharnois's prohibitions. At Chicago's portages, a defiant array of Native people and coureurs de bois had found a wetlands refuge in which to thrive between colonial jurisdictions and beyond the reach of French authority.

A Stumbling Block of Empire

Despite renewed hopes at the beginning of the eighteenth century, Chicago's portages never became the site of imperial power that French officials had predicted. Even with the promising connective geography and the maritime mobility of a waterborne interior, the French struggled to harness the potential of Chicago's portages. With French failures and renewed warfare, the possibilities of stable French and Indigenous alliance and settlement at Chicago also faltered. In the years following 1712, violent raids by and against the Meskwakis continued to weaken French control over the western portage routes, as did jurisdictional wrangling about where Chicago's continental divide fit within French colonial geography. Following the conflicts, Chicago's route remained an open thoroughfare for illegal traders and unconquered Native peoples as French authority failed to penetrate this elusive space. Throughout these decades of uncertainty, Chicago's local geography continued to play a critical role in both French efforts at empire and alliance in North America, first as a boon of imperial aspirations and later as a thorn in the side of French control.

After the early 1740s, imperial priorities shifted elsewhere, and Chicago's route fell out of official correspondence. For a site that had garnered so much attention from the seventeenth century forward, Chicago's waning significance to French officials is noteworthy. An array of French adventurers, missionaries, and administrators had hung their hopes on Chicago's geography in their schemes for empire in the interior. Yet from early exploration in the 1670s forward, Europeans had proven unable to harness Chicago's portages for imperial purposes. French efforts at Chicago, and French inability to master its local terrain, stood out as a stark failure of European colonization. As official efforts to control Chicago's route yielded to the activities of clandestine traders and defiant Meskwakis, new opportunities would arise along Lake Michigan's southwestern shoreline. While the French had proven incapable of exploiting Chicago's geography, in the 1740s, new and better-equipped contenders would stake their fortunes on the portages' muddy ground.

CHAPTER THREE

Crossroads

Indigenous Resurgence at the Portages

In 1752, an assembly of Potawatomi and Mascouten diplomats made their way down the Illinois River to Lake Pimiteoui. They came to the town of the Peorias, a subgroup of the Illinois people, to "sing the Calumet." This was an ancient diplomatic ritual of the Illinois and Mississippi watersheds, practiced by Native groups across the region since before European contact. These Potawatomi envoys had journeyed downriver to ensure that the Peorias, who had suffered an attack from the Sauks and Meskwakis, would not seek vengeance against the recently settled Anishinaabe towns to the north. The Potawatomis came from their new towns south of Lake Michigan, representing a collection of Anishinaabe migrants that had moved into the Chicago area over the past decade. Accompanied by Mascoutens and implicated with the hostile Sauks and Meskwakis, these Potawatomis spoke for the wider array of Native peoples settled along the route connecting southwestern Lake Michigan with the upper reaches of the Illinois River. Their wielding of the calumet ceremony indicated their ambitions to become new brokers of both war and peace in the borderlands. Presenting the Peorias with the calumet—a ceremonial smoking pipe meant to signify respect and peace—these Potawatomis conformed to the region's diplomacy while also presenting themselves as the ascendant Native authority in the area between the Illinois Country and Lake Michigan.[1]

The Potawatomis' diplomatic mission downriver represented an unfolding debut of Anishinaabe influence along the Chicago portage route. Already by the early 1750s, French officials had begun referring to these Potawatomis "of the River Chikagou" as a distinct group to be reckoned with.[2] In the coming decades, the recently settled Potawatomis, Odawas, and Ojibwes that made up the Anishinaabe settlements at Chicago, Milwaukee, and the upper Illinois River would work to consolidate their powerful position along Lake Michigan's southwestern shoreline.[3] A variety of factors had drawn the Anishinaabeg to Chicago's portages, including kinship ties to local Indians, the promising diversity and abundance of the area's ecology, and its strategic geography along a route of travel and trade. Anishinaabeg had thrived elsewhere along the water routes of the Great Lakes region for centuries, and they came well prepared to prosper in the local waterscape of Chicago's crossroads.

Amid continental upheaval elsewhere, Anishinaabeg relocated to the Chicago portages to expand their influence in the western Great Lakes. Chicago's peripheral location, beyond French imperial reach and on the fringes of Anishinaabewaki—the homeland of the Anishinaabeg—proved a valuable setting in which Odawas, Ojibwes, and Potawatomis consolidated power in the decades to come. The new Anishinaabe settlements around Chicago prospered from the ecological abundance of a vacated quarter following the violence of the Fox Wars and the continual importance of the portages as overland connections between the Mississippi and Great Lakes watersheds. Equipped with the environmental knowledge necessary to thrive in Chicago's transition ecosystem and endowed with the expertise of waterborne transport as a "canoe people," these southwestern Anishinaabeg would carve out a sovereign space at Chicago in the years to come.[4]

In the mid-eighteenth century, Chicago's remote position proved a boon to Anishinaabe ascendancy southwest of Lake Michigan. Chicago's neglected space could become a valuable place in Indians' efforts to reconstitute power during a period of turmoil and change. By the 1740s, Chicago had become peripheral to imperial French interests in the region. But it was the portages' marginality in imperial schemes that made it so critical to Anishinaabeg moving westward. As Anishinaabe migrants concentrated around the portages, Chicago's distance from European centers of control in the Great Lakes and Illinois Country proved increasingly appealing—and crucial to its trajectory as a center of Indigenous strength. From the 1740s through the Revolutionary era, Chicago's Anishinaabeg would demonstrate a profound resilience undergirded by the power of their new location at this periphery-turned-homeland.

The area's ecological abundance and local geography continued to bolster Native power along Chicago's strategic portage route. Indigenous geographic and environmental knowledge allowed the Anishinaabeg to prosper in a space that had eluded earlier French efforts to project control over Chicago's portages. The collapse of European empires, the emergence of new settler regimes, and Nativist revivals all led to turmoil elsewhere across the Great Lakes in the years to come. Yet along Chicago's portages, southwestern Anishinaabeg found a new geography on which to reaffirm their autonomy. From these strategic new settlements at Chicago, Anishinaabe leaders exerted revitalized influence throughout the lower Great Lakes. In the following decades, Anishinaabeg from Chicago would defy a series of European empires while distinguishing themselves diplomatically from their Anishinaabe kin to the east, laboring to assert their own power in the region and maintain control over the portages.

Incorporating Chicago into Anishinaabewaki

The Anishinaabeg of the Great Lakes had known the potential power of Chicago's portages for more than a century when seasonal hunting parties traveled to the area during the early 1740s. Trading parties of Odawas and Potawatomis had journeyed over Chicago's portages as early as the seventeenth century in commercial enterprises among the Illinois Confederacy downriver.[5] Along with trading voyages, each winter, smaller groups of Anishinaabeg dispersed to hunt in the Kankakee wetlands southwest of Lake Michigan. But such forays through Chicago had remained temporary, until now. This time, as spring thaws flooded the wetlands between the Chicago and Des Plaines, these families of hunters stayed on. Refugee settlements of Meskwakis and Sauks at Chicago offered an appealing new residence for the Anishinaabeg who found their homeland resources stretched thin east of Lake Michigan. Some of these Anishinaabeg had kin among Sauks still living around the portages, as well as the Mascoutens and Kickapoos dwelling along the adjacent waterways of the Kankakee and Des Plaines Rivers. Potawatomis from the St. Joseph River had maintained blood ties with the Sauks and Mascoutens for decades. They likewise facilitated new bonds between their fellow Anishinaabeg—Odawas and Ojibwes—and local Indians around Chicago as they lingered near the portages.[6] Though records remain sparse so far beyond French colonial centers, later Anishinaabeg relayed that by 1743, Potawatomis had entered the area around Chicago's portages in force, along with groups of Ojibwes and Odawas, their fellow Anishinaabe relatives and allies.[7] Well versed in the waterborne travel of their Great Lakes homeland, the portage route through Chicago made strategic sense to trade-oriented Anishinaabeg looking to extend their influence and commerce in an ever-expanding geography of kinship and alliance.[8]

Another appealing reason for the Anishinaabeg to settle along Chicago's route in the 1740s remained rooted in the region's environmental abundance. The geopolitical realities of prior decades had directly contributed to this ecological richness, and after the devastation of the Fox Wars in the area, the ecological incentives to migrate into the Chicago area grew stronger for expansionist Anishinaabeg. Conflict zones like the contested route through Chicago had remained inhospitable places for hunting parties throughout the early eighteenth century. After the Weas vacated the portages in the early eighteenth century, Chicago's wildlife populations had ample time to recover. As war parties attacked and counterattacked up and down the Illinois River corridor during the conflict with the Meskwakis, few Indians or French dared linger long in the Chicago area to exploit its natural resources. By the 1740s, Chicago's wetlands hosted one of the few untapped hunting grounds for fur-bearers in the lower Great Lakes. In addition to otter, muskrat, raccoon, and beaver along the waterways, bison, elk, and deer still grazed on the surrounding oak savannas and prairies.[9] This environmental richness proved a compelling draw for an expanding Anishinaabe population from other parts of Lake Michigan seeking new hunting grounds for both furs and meat by the mid-eighteenth century.

The Anishinaabeg came prepared to prosper in Chicago's transition environment of wetlands and prairies on the edge of Lake Michigan. For centuries, Anishinaabeg had flourished in the Great Lakes through seasonal cycles of aggregation and dispersal that balanced agricultural production, foraging, hunting, and fishing.[10] According to Jacques-Charles de Sabrevois, a well-informed French observer in the early eighteenth century, the Anishinaabeg remained "very industrious, both in hunting and agriculture."[11] In summer, Anishinaabe clans, or *nindoodemag*, congregated while women tended to fields of maize and other crops in a political economy divided by gender. Men were responsible for steady hauls of fish caught from the lakes and their adjacent rivers. According to Cadillac, who lived near Odawas around Michilimackinac, Anishinaabeg relied on fish as "a daily manna which never fails; there is no family which does not catch sufficient fish in the course of the year for its subsistence." Cadillac described the methods Anishinaabe fishermen used to net "as many as a hundred white fish" in handwoven nets. During the winter months, Anishinaabe anglers also proved adept at hooking large lake trout through holes in the ice. Both whitefish and lake trout abounded in the deep waters off Chicago's shoreline. The area's inland waterways and marshes likewise held other species of catfish, suckers, sturgeon, bass, perch, and walleye that the Anishinaabeg could exploit.[12] Endowed with teeming waters and fertile lands, Chicago's local landscape offered a promising setting for the Anishinaabe way of life.

While Anishinaabeg cultivated crops and caught fish to sustain themselves in the summers, they transitioned to reliance on wetlands resources during the autumn months when millions of migrating waterfowl passed through the Great Lakes. This is when Chicago's waterscape would have proven most appealing for immigrating Anishinaabeg. The surrounding marshes and wet prairies played host to large flocks of sandhill cranes, teals, swans, ducks, passenger pigeons, and geese. The birds lingered along this geographic corridor on their transcontinental journeys, in part, because of the wild rice that grew in thick stands throughout Chicago's waterways. While Anishinaabe men captured rafts of ducks and geese in nets upon the water, Anishinaabe women harvested the ripened rice, beating its loose grains into the hulls of their canoes as they paddled through the marshes each autumn. Fall proved a plentiful time in Chicago's wetlands, especially for Anishinaabe settlers who came to the portages with a prior knowledge of how to harvest both the wild rice of its waterways and the waterfowl passing overhead.[13]

Anishinaabeg dispersed into smaller family groups over the winter to hunt furs and larger game such as deer, elk, and around Chicago, bison.[14] In addition to hunting for meat, Anishinaabe men had crafted elaborate methods for capturing Chicago's furbearing species such as raccoons, muskrats, mink, otter, and beaver. Lahontan, a French adventurer, described how Anishinaabe hunters made staked traps for otters, baited with fish, which could capture these otherwise wary animals.[15] For beavers, which would not fall for baited traps as herbivores, Anishinaabe hunters cut holes in their dams to lower water levels and drive them into nets. French observers like Lahontan noted that the Anishinaabeg were careful not to kill too many furbearers at any one time—they would purposely use these methods to catch beavers to control the numbers killed from each pond. They also took care to vary the areas they hunted from year to year to avoid exhausting the game populations.[16] Bringing a collection of relevant environmental knowledge and skills with them, the Anishinaabe people came ready to flourish along Chicago's marshland, prairies, and lakeshore in all seasons.

The Anishinaabeg who sought out Chicago's crossroads in the eighteenth century also remained well versed in the cultural and spiritual potency produced at places where land and water met. Sacred Anishinaabe *aadizookaanag*, or creation stories, told of a flood and a raft (in some versions, a canoe), where Nanabush, the Great Hare, gathered all the living animals for survival. To end the deluge and create the world, Nanabush employed the diving mammals—beavers, otters, and muskrats—to bring up sand and mud from below. When the muskrat retrieved a single grain of sand, Nanabush multiplied it into a pile of dirt. He scattered the soil upon the water, creating mud. He continued this process until this life-giving mud produced enough land that he and the other animals could disembark and venture out, finding their proper places upon the wet earth.[17] The Anishinaabeg who journeyed to Chicago's waterlogged marshes and pathways understood the significance of such an amphibious place. In the years to come, they would work to create their own home upon Chicago's muddy ground where land and water merged.

The selection of the Chicago portages as a townsite fit the geographic logic of the Anishinaabeg as well. For a century before coming to settle the Chicago area, the Anishinaabeg had prospered by controlling strategic bottle-necks and facilitating movement along the water routes of the Great Lakes region. Positioning themselves at key locations like the Straits of Mackinac, Sault Ste. Marie, and the waterborne trade routes entering Lake Huron from the east, Odawas, Ojibwes, and Potawatomis had become some of the most powerful and commercially successful Indigenous groups of the interior.[18] Along these routes, Anishinaabeg extracted tolls from French and other Na-tive people moving along their waterways and portages.[19] They also pros-pered in trade with waterborne travelers who relied on Anishinaabe suppliers for canoes, meat, fish, and corn. Aware of the similar role Chicago's portages played in linking Lake Michigan with the water routes to the Mississippi, An-ishinaabe settlers sought out the Chicago area as promising new territory for their growing populations during the 1740s.

The Anishinaabe who migrated to Chicago's portages had thrived in the waterborne world of the Great Lakes for centuries. French observers across the Great Lakes considered the Odawas, Potawatomis, and Ojibwes as first-rate mariners, marveling at the ways these Anishinaabeg traversed the expansive lakes in lightweight birch canoes that they fashioned from the woodlands around them. Some French officials simply called these peoples *canoe Indians*, a kind of shorthand meant to designate their ability to both navigate and build the vital watercraft. And Anishinaabe boatbuilders were the masters of their trade, constructing canoes for both their own use and for sale within the French fur trade at commercial depots like Michilimackinac.[20] Chicago's wetlands sat at the southern extremity of the range of the white, or paper, birch tree—a key botanical resource in Anishinaabe lifeways.[21] In the Great Lakes basin, crews of traveling European voyageurs or Indigenous traders relied on the availability of birchbark along their route to repair their fragile craft often. At Chicago, these travelers would take the opportunity to exchange their lightweight birch canoes for sturdier dugout pirogues if they were traveling south into the waters of the Illinois and Mississippi Rivers, where birch trees became scarce and made ca-noe repair difficult. When these crews returned north to enter the Great Lakes watershed once again, they could, in turn, purchase birch canoes from the ex-pert boatbuilders among the Anishinaabeg at Chicago. While slowed down at Chicago making the portage or acquiring vessels, such travelers also took the opportunity to resupply with Anishinaabe corn and other foodstuffs, and likely traded for locally acquired furs. By establishing themselves at the crossover point between birchbark canoe travel and the heavier pirogue navigation of the Mississippi watershed, the Anishinaabeg of Chicago could prosper on travelers' needs as they passed through the portages in both directions.[22] Recognizing the strategic significance of the Illinois River corridor on a number of ecological and geographic levels, the Anishinaabe migrants came to Chicago's shoreline with the skills and knowledge to make it an amphibious center of Native power in the decades to come.

The Anishinaabe migration into the Chicago area was part and parcel of the greater westward expansion of the Odawas, Ojibwes, and Potawatomis during the mid-eighteenth century. Extending out from their traditional centers of power such as Sault Ste. Marie, Michilimackinac, Detroit, and the St. Joseph River valley, Anishinaabe clans forged kinship ties with Sauks, Menominees, and Ho-Chunks and established new settlements all along the western shores of the Great Lakes.[23] The French were aware of these movements, as the collective Anishinaabeg remained some of their most important regional allies. In July 1741, for instance, the French governor Beauharnois addressed Odawas and Ojibwes from near Michilimackinac. He had heard stirrings that they wished to relocate "owing to the dearth of food that had prevailed the previous winter" near the Straits. The Anishinaabeg feared this might prove a recurring problem and had been searching for a more prosperous home for their growing population.[24] Beauharnois urged them to remain near the French, reminding them "of the advantages you have enjoyed" from such proximity. Beauharnois stressed that if the Anishinaabeg left, they "would not enjoy those advantages if you were far away from them."[25] Many in his audience took heed and only moved as far as L'Arbre Croche, down Lake Michigan's eastern shoreline.[26] But others went farther, seeking out lands far removed from French imperial influence and more prosperous in natural resources.

Indeed, the French remained powerless to prevent Anishinaabe relocations as part of this overall expansion. But officials in the interior held out hope in the Potawatomis and Odawas as some of their most loyal allies. As Jacques Legardeur de Saint-Pierre, a French officer stationed at Michilimackinac explained, the Potawatomis were the "only Nation to be relied on." Paul-Joseph Le Moyne de Longueuil, the commander at Detroit, concurred, considering the Anishinaabeg "the best disposed" and the only Native allies he could "confide in." Along this logic, the Chevalier de Bertel, commanding at Fort Chartres in the Illinois Country, "was using every appliance at his disposal to attract" Potawatomi migrations into the upper Illinois River south of

Lake Michigan.[27] French officers in the Illinois Country, like Bertel, encouraged Anishinaabe migrations to Chicago, or at least viewed the incoming Potawatomis as a better alternative to the Meskwakis and Sauks who lived along the portages at the time. But the Anishinaabeg who sought out Chicago's crossroads as their new home in 1743 were well aware of the site's advantages and needed little encouragement from French officials downriver. These Anishinaabe migrants came to Chicago's portages on their own terms and planned to harness Chicago's strategic waterways and adjacent lands as they saw fit.

While the French reassured themselves of Potawatomi dependability, the Anishinaabeg who resettled Chicago in the 1740s remained concerned with their own agenda of establishing power in the borderland. Removing themselves from French imperial centers, Chicago's Anishinaabeg showed little regard for French policies or programs. Indeed, the migrants to the area seem to have been a self-selecting lot of those Anishinaabeg less inclined to French allegiance. Soon after resettling along Chicago's route, the Anishinaabeg began actively undermining French alliance structures in the region by sparring with the Illinois Indians downstream. The Peoria Illinois, living along Lake Pimiteoui, had traditionally hunted bison as far north as Starved Rock during winter months—only several days downstream from Chicago by canoe. Beginning in the late 1740s, friction mounted as resettled Anishinaabeg from Chicago and the Kankakee began encroaching on the Peorias' prized bison hunting territory. In 1751, the Chicago River Potawatomis invited their Ojibwe kin from Michilimackinac to join them in a campaign against the Illinois. The allied Anishinaabe warriors advanced as far as Starved Rock in their search for Peoria victims, but only found one hapless French trader named Jean Brossac, who they killed "on the spot." French officials reported the bloodshed in Canada, and authorities worried as increasing numbers of aggressive Anishinaabeg "rejoined the Potawatomi at Chicago" to form additional war parties and stake their claim south of Lake Michigan.[28] The Anishinaabeg who gathered along Chicago's portages demonstrated their continued independence from French agendas in a new space, far beyond the control of imperial officials.

Simultaneously, the resettled Anishinaabeg began strengthening alliances with local Indians around Chicago to shore up their role as the region's new powerbrokers. Mascoutens, Sauks, and Kickapoos, now emboldened by their association with the Anishinaabeg, renewed attacks on the Peoria and other Illinois Indians downriver.[29] In 1752, Jean Jacques Macarty Mactigue, French commandant in the Illinois Country, expressed his sympathy to his Illinois allies as "the Foxes, Sauk, Potawatomi, Sioux, and many others ask to eat you up." By that summer, things seemed so dangerous for the Illinois living at Lake Pimiteoui that they dared not venture out of their palisade for the annual bison hunts, opting instead to go hungry.[30] In 1754, on the eve of the Seven Years' War, officials in Canada reported that the southwestern Anishinaabeg and their allies had "assembled together to go and destroy the Peorias."[31] Retreating under the sustained pressure from the Anishinaabeg and their neighbors, the Peorias vacated the upper reaches of the Illinois River entirely in the coming years. By the time a trader named Hamburgh traveled the route a decade later, he reported that the Peorias "had been driven away from the River Illenois by other nations."[32] In the wake of Illinois withdrawal, the Anishinaabeg commanded the Chicago portage, its valuable trade route, and the adjacent hunting grounds of the upper Illinois River valley.

By midcentury, the Anishinaabeg had effectively carved out a thriving cluster of towns around Chicago's portages and surrounding riverways. Hamburgh the trader, traveling from the Straits of Mackinac to the Mississippi, described the Native settlement he found near the entrance to the "Chycacoo River" as a mixed settlement of Odawas and Potawatomis along with some Sauks. He also described the ecological abundance around Chicago, recalling, "Here the Country Begins again to be very Pleasant, good soil, and Hunting very Plenty: Such as Buffiloes, Deer, Bears." While the local terrain offered an ideal location for game and farming, the Anishinaabe town also lay near the strategic portage route. He reported, "Up this River, the Chykocoo, the french used to make a Carrying Place into an other River for about 3 miles which falls into the river Illinois and is deep Enough for Large Battoes to go up or down."[33] Positioned so close to such a crucial route of travel, the entire upper course of the Illinois River lay at the disposal of Chicago's resettled Anishinaabeg for both trade and hunting. Traveling south along the Illinois River, the trader deemed it "the chief hunting ground of the Battowaymes [Potawatomis]" for over two hundred miles down to Starved Rock, with "the greatest plenty" of raccoons, otters, beavers, elk, deer, and buffalo.[34] Listing animals essential for the continued fur trade as well as food, the trader noted the dual bounty the Anishinaabeg reaped in their newly won territory. Chicago's Anishinaabeg, now in uncontested control over the vital portages, stood poised to thrive in a new homeland far removed from Indigenous competitors and European meddling.

A Refuge for Resistance

While Chicago's Anishinaabeg pressed their advantage along the strategic portages, geopolitical events farther afield signaled significant changes on their horizon. In the Ohio valley, fellow Anishinaabeg attacked a settlement of Miamis that had forged commercial ties with British traders at Pickawillany in 1752. The violence triggered further involvement from both British colonial forces and French officials as the two European powers vied for influence south of the Great Lakes. The fighting escalated into the upper reaches of the Ohio River, eventually leading to war between the two empires and their Native allies. While Anishinaabeg from the St. Joseph River, Detroit, and Michilimackinac became caught up in the imperial turmoil of the Seven Years' War and subsequent French defeat, the Chicago Anishinaabeg remained far removed from the conflict and its aftermath.[35]

Anishinaabeg along Chicago's route stayed insulated from the geopolitical fallout of French defeat, but they nevertheless witnessed the final, chaotic end to France's claims over their region. During the winter of 1761, following the surrender of Montreal, two separate groups of French soldiers tried to avoid surrendering to the British by escaping through the river routes of northern Illinois. Louis Liénard de Beaujeu de Villemonde fled Michilimackinac with a force of 132 soldiers and militia in October 1760. British leadership later learned he had planned "to proceed to the Illinois Indians by way of Chicago," but early ice caused him to abandon his boats, traveling down the Fox River route to Sauk and Meskwaki villages along the Rock River of Illinois instead. He wintered there, exchanging all the merchandise he had brought from Michilimackinac to secure basic food and shelter from his Native hosts.[36]

At the same time, Pierre Passerat de la Chapelle, commanding a force of militia from Detroit, had fled overland to Lake Michigan. Here, his force built rafts, sailed to the southern shore of Lake Michigan, and marched overland to the upper Illinois River. Chapelle was fortunate that a force of 110 Odawa and métis auxiliaries had accompanied his command, bringing with them expertise for navigating the region. They proved their worth in guiding his force to the safety of Starved Rock, and in constructing a stockade before winter set in, which he named Fort Ottawa in their honor. His Odawa and métis militia also deployed their expertise in building him boats, which he planned to use for his retreat downriver come springtime. But after Chapelle made contact with Beaujeu's ill-supplied forces on the Rock River, quarrels and accusations broke out between the two officers over rank and command. By spring, when both forces made their way south to Fort Chartres and the Illinois Country, they had become more caught up in internal squabbles than in an orderly retreat. Many of the militia and Odawas stayed on at their makeshift fort and may well have linked up with settlements of fellow Anishinaabeg along the waterway.[37] With Chapelle's and Beaujeu's embittered retreat down the Illinois

River that spring, what little official French influence had persisted along Chicago's route ceased altogether, leaving the Anishinaabeg in uncontested control of the region.

While the French fled down the Illinois River in the wake of British victory, the Anishinaabeg settled at Chicago would do no such thing. Rather, in the years following the British conquest of Canada and the occupation of the Great Lakes, the Anishinaabeg west of Lake Michigan proved persistent in their opposition to British incursions into the region. Their distance from Great Lakes forts, now occupied by the British, likewise made these southwestern intertribal towns appealing sites of refuge for other Anishinaabeg unwilling to yield to British authority. The waterways southwest of Lake Michigan became a stronghold for anti-British resentment in the years to come, with Chicago's towns offering harbor to Anishinaabe holdouts from across the Great Lakes.

Minweweh arose as one of the most determined opponents to British occupation, and one of many Indians who found support in the Chicago towns after 1761. Minweweh was an Ojibwe from northern Michigan whom the British called "The Grand Saulteur" for his commanding height. When the British first arrived at the Straits of Mackinac in 1761 to claim the Great Lakes after French defeat, Minweweh expressed his scorn. He admonished Alexander Henry, a newly arrived English trader: "Although you have conquered the French, you have not yet conquered us." Invoking Anishinaabe sovereignty in clear environmental terms, he reminded Henry that "these lakes, these woods and mountains were left us by our ancestors. They are our inheritance; and we will part with them to none." Dismissing dependence on European goods, Minweweh assured the British trader that he and his people could thrive on the ecological abundance of "these spacious lakes and woody mountains."[38] Unfazed by British claims to dominion, Minweweh stood prepared to defy imperial incursions. He and his kind would find allies and safe haven along Chicago's portages in the years to come.

Lake Michigan's southwestern basin offered seclusion for Anishinaabeg unwilling to live under British rule. Between Fort St. Joseph and La Baye, the French had left no forts for the British military to occupy. The shoreline from the Milwaukee River down to the Calumet River remained one of the few areas uncharted by the British after their occupation of the Great Lakes. In 1762, Thomas Hutchins, a royal engineer tasked with mapping the western lakes, had been forced to make a significant detour for fear of hostile Anishinaabeg in the area. Crossing the lake from Green Bay directly to the St. Joseph River, he sidestepped the entire southwestern basin, including Chicago, to avoid trouble.[39] Before the occupation of the Great Lakes, the

British had only a vague understanding of Lake Michigan's geography. With Hutchins's failure to penetrate its southwestern coast, British knowledge about the vital Chicago portage and the area's Anishinaabe settlements remained murky. Chicago's route and the adjacent river channels, lakeshore, and marshlands existed beyond the imperial eye of British officials. As such, the Anishinaabe towns along Chicago's route offered a sanctuary for warriors like Minweweh who preferred to distance themselves from European authorities in the wake of French defeat (see figure 3.1).[40]

It is unclear when Minweweh first came to the Chicago portages. He may have had kin ties to Anishinaabeg already settled at Chicago's portages by 1761 or maintained commercial relations with French merchants in the Illinois Country. Either way, by 1763, reports circulated among the British that "a tribe of the Chepwas, from Mitchilimacina, with an Indian called the Grand Soto [Minweweh] at their head, are much disaffected to the English interest." According to these reports, Minweweh and his warriors had traveled toward the Illinois Country and were threatening to resume hostilities against the British while they drew supplies from French *habitants* downriver. Minweweh returned to Michilimackinac during the summer of 1763, long enough to orchestrate a successful attack on the British garrison there as part of Pontiac's War—an Indigenous uprising against British rule that had erupted across the Great Lakes. He returned to the Anishinaabe settlements southwest of Lake Michigan soon after, becoming a leader of the composite Anishinaabe town at Chicago.[41] Over the next year of conflict, the British would stave off defeat in the Great Lakes by holding onto their garrisons at Detroit, Niagara, and Fort Pitt, eventually establishing peace with most of the region's Native peoples. But Minweweh, now representing both himself and "the people of Chicag8" held off on making amends with the British until September 1765, long after most of his fellow Anishinaabeg to the east had ceased fighting.[42] Well aware of their distance from any real imperial retribution, the Chicago Anishinaabeg under Minweweh offered a tardy, halfhearted promise of peace to the British at Detroit before returning to their southwestern reaches.

Minweweh and the "people of Chicago" had no intention of keeping peace with the British, at least not in the security of their own homeland. While the eastern Anishinaabeg in places like Detroit, Michilimackinac, and even the St. Joseph River valley began to forge closer ties to the British soldiers and traders living in their midst after Pontiac's War, the distance of the Chicago area from European posts allowed for little interaction between the southwestern Anishinaabeg and incoming British.[43] The absence of local European traders encouraging the southwestern Anishinaabeg toward reconciliation made the

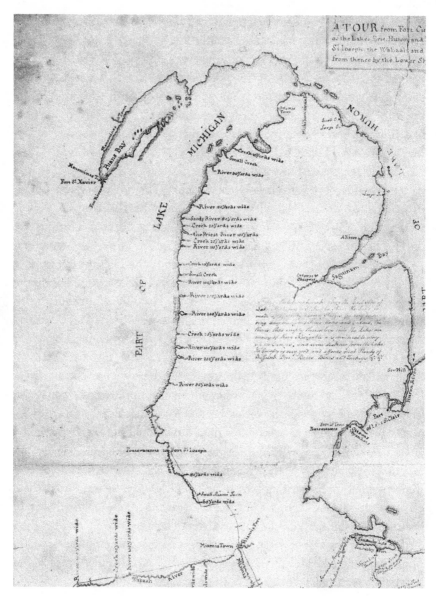

FIGURE 3.1 Thomas Hutchins's attempted map of Lake Michigan represents just how little grasp the British had over the lake's southwestern basin during the 1760s. Detail from Thomas Hutchins map (1762), original Map HM 1091. Courtesy of the Huntington Library.

towns southwest of Lake Michigan hotbeds of anti-British resentment even after Pontiac's War. Some Anishinaabeg from the eastern shores of the lake even migrated to join southwestern kin at this time, attracted by the remoteness of the settlements at Chicago, Milwaukee, and the upper Illinois River. As Hugh Crawford, a British trader, reported in 1765, "part of the Chippawa" as well as Potawatomis from the St. Joseph River, had migrated "to settle about the Illinois."[44] Further specifics from another trader, William Howard, confirmed that "fifty Canoes of Chippawas were gone to the Illinois" drawn by hopes to "get Amunition of the French the other side of the Lake" and obtaining the "great plenty of Rum at Shekago [sic] from French coureur de bois there."[45] British officers, now tasked with reoccupying the western Great Lakes as well as the lower Illinois Country, likewise reported the movement of unreconciled Anishinaabeg to the southwestern settlements.[46] As British control over the Great Lakes solidified after Pontiac's War, the Anishinaabe settlements along Chicago's route became a place for those who would not accept British intrusion into the region.

Chicago's remoteness from European centers proved part of its appeal for Minweweh and other Anishinaabeg who would not yield to British rule. As one British officer at Michilimackinac told it, Minweweh "never durst Venture near an English Garrison." Chicago's portages, far from any European posts, provided an ideal center for continued Anishinaabe resistance to British incursions. From Chicago, Minweweh and other Anishinaabeg could continue "to Plunder and kill the English" while remaining beyond the reach of reprisal.[47] In December 1765, only months after Minweweh's visit to Detroit, rumors again circulated that he and his southwestern brethren continued to encourage resistance to British pretenses. Sir William Johnson, superintendent of Indian Affairs for the British, received warnings that the southwestern Anishinaabeg "remained very ill disposed" and attempted to assuage them by sending specific presents earmarked for Minweweh and his Chicago followers.[48] But gifts proved ineffective, and the Odawa, Potawatomi, and Ojibwe settlers of the upper Illinois remained aloof to British overtures. Even as British troops occupied the lower Illinois River valley after 1765, Anishinaabeg positioned between the Illinois Country proper and the colonial centers of the Great Lakes maintained a hostile stance toward the British.[49] In 1766, Minweweh ventured as far east as St. Joseph with war belts, hoping to distribute them among his eastern kin in an ongoing effort to renew their resistance against the British.[50] Chicago and the other southwestern towns served as an ideal base from which Minweweh could continue to agitate for war against the British while remaining beyond the reach of imperial officials.

Chicago's portages also presented the perfect jumping-off point for Anishinaabe warriors looking to go on the offensive. The mobility provided by Chicago's portages, intersecting river channels, and trails allowed the waterborne Anishinaabeg to strike at hapless British merchants and soldiers across an extensive geography. Patrolling the river routes from the Kankakee marshlands as far west as the Mississippi, Chicago's warriors continued to threaten British traders in the years after Pontiac's War.[51] The route down the Illinois River from Chicago allowed Minweweh and the other southwesterners to raid against British outposts farther south, in the Illinois Country.[52] Anishinaabeg from the southwestern settlements likewise attacked British traders who risked traveling south of Lake Michigan. In 1767, Anishinaabe warriors along the Kankakee River confiscated the trade goods of a British merchant, then killed him. They sent news of his death east to the St. Joseph Potawatomis along with a message that they "would not suffer an English man to come near their Place."[53] In 1768, southwestern Anishinaabeg killed two more British traders who failed to heed their warning. Another English colonist fell to Illinois River Potawatomis in 1771.[54] Given the waterborne mobility provided by the portages and river routes, Chicago served as the geographic nexus for these acts of resistance. Minweweh found ideal geography along the riverways and ready support from the Anishinaabe towns around Chicago to continue the fight in the west well after Pontiac's uprising had ended.[55]

Minweweh's changing whereabouts during the late 1760s remain beyond the grasp of archival sources, but he met his end along the southwestern shoreline of Lake Michigan. A British trader, William Bruce, had lost merchandise and friends in Minweweh's attack on Fort Michilimackinac in 1763. By 1770, Bruce had married a Meskwaki woman and lived with her people west of Lake Michigan. When a misunderstanding resulted in violence between the Meskwakis and Anishinaabeg at the Milwaukee town, Bruce used the bloodshed as leverage to rally a war party. Surprising Minweweh's camp sometime that autumn, Bruce stabbed and killed the Anishinaabe headman in his tent. British officials received news of the attack shortly after, with enthusiasm. One Indian agent in Canada remarked that Minweweh's death was as beneficial to the British as the murder of Pontiac. The British commander at Michilimackinac also relayed the news, happy to be rid of "one of the most Mischievous Indians that ever was in this Country."[56]

In killing Minweweh, Bruce eliminated one potent enemy from Lake Michigan's southwestern shoreline. But many more Anishinaabeg, still hostile to the British, remained entrenched along Chicago's portages and the surrounding settlements that now stretched from Milwaukee in the north to the

Kankakee River in the east and Lake Pimiteoui to the south. Standing at the crossroads of this new expanse of Anishinaabe territory, Chicago's portages linked a collection of towns and people that continued to remain beyond British ken and control. In the decades to come, these Anishinaabeg living on the margins of both British dominion and at the edges of Anishinaabewaki would prove how potent their peripheral geography could be.

Local Liaisons

The Native Chicagoan's rejection of British commercial infiltration along the portage route proved especially perplexing for imperial officials. Given the continued hostility of the southwestern Anishinaabeg, few traders risked the voyage between Lake Michigan and the Illinois River. As late as 1771, the British commander in chief of North America, Thomas Gage, relayed to Sir William Johnson "that none of our Traders dare, even yet, [to] go amongst that Nation to Trade." While the rest of the Great Lakes peoples had come around to new alliances and commercial ties with the British, the southwestern Anishinaabeg continued to resist. Gage expressed his vexation to Johnson, pondering the "Strange Situation that no English Trader dare go into their Country."[57]

Faced with an impenetrable political geography along Chicago's route, British officials like Gage dismissed the region as strategically irrelevant to imperial security. Gage reported back to British heads of state that "the Ilinois River is of little Consequence, having no Powerfull Nations dwelling upon it; and leads only up to a Tribe of the Pouteatamies." Tracing these Anishinaabeg to their St. Joseph progenitors, Gage described them as "a Licentious People who have often done Mischief."[58] Southwestern Anishinaabeg continued to foment hostility against the British, but given their remote location, officials like Gage dismissed both the people and region as beyond his concern. Diplomatic estrangement and persistent low-level violence thus left the southwestern fringe of Anishinaabewaki beyond British influence even as Anglophone traders and agents forged ever closer ties with Anishinaabeg east of Lake Michigan.[59]

But even as the Chicago Anishinaabeg blockaded British commercial ventures through the portage route, they welcomed continued trade with Francophone smugglers operating in the area. These coureurs de bois had traded illegally along Chicago portages since the French era. After the British claimed the Great Lakes in 1761, they continued to supply the Chicago area with European goods in exchange for pelts and furs, beyond official jurisdiction. According to one British trader familiar with the region, in 1765 a French trader named Tullong "wintered at Shekago [sic] with Rum, powder, Ball &

other merchandise," meant for hostile Anishinaabeg there.[60] Similarly, in May 1770, Jean Orillat outfitted two canoes under the command of Jean Baptiste St. Cyr to travel from Montreal to "Chiquagoux" to establish trade with the local Anishinaabeg. Unlike concurrent British ventures, the Anishinaabeg seemed to have welcomed St. Cyr's commercial advances.[61] Sometime in the 1770s, another trader, Jean Baptiste Guillory, established a short-lived trading post and farm near the forks of the Chicago River.[62] Farther downriver, other French habitants had forged good relations with the Illinois Country Anishinaabeg and reestablished a trading base at Lake Pimiteoui by 1773—now called Peoria. The prominent Francophone merchant Jean Baptiste Maillet was active in farming ventures in the area and kept up commercial exchanges with the Anishinaabeg farther upstream.[63] These few unsanctioned French traders kept Chicago's Anishinaabeg supplied and allowed them to remain commercially, as well as diplomatically, independent from the British.

Chicago's Anishinaabeg also actively sought to exploit their geographic position for diplomatic gains. Living along the corridor of routes between the Great Lakes and the Mississippi River, the southwestern Anishinaabeg remained in a unique and advantageous position to play European empires off one another. Historians have often cast 1763 as a watershed date in Native North America because it permanently removed the French as a counterweight to British claims over the interior. With France gone, Native peoples in many parts of the continent lost their ability to pit European empires against each other in trade and diplomacy. But at Chicago, 1763 came and went without such geopolitical fallout. The British occupied Fort Chartres and the Illinois Country in 1765, but the newly founded trading town of St. Louis also lay downriver from Chicago. This commercial center, across the Mississippi River from British-held Illinois, belonged nominally to the Spanish after 1762. Many of its inhabitants, traders, and military officials were simply French who had relocated from the Illinois Country proper. While Britain claimed dominion over Chicago and the rest of the southwestern Anishinaabe towns, Spanish and French allies in St. Louis offered a counterweight to British authority southwest of Lake Michigan. The riverine connection to St. Louis offered an outlet to the Anishinaabeg at Chicago who planned to maintain their separation from British imperial centers in the Great Lakes region.[64]

Exploiting their position along the trade routes connecting the Great Lakes with the Mississippi, Chicago's Anishinaabeg found a vital substitute for trade and gifts in Spanish-held St. Louis, even while shunning relations with the British. As early as 1768, eastern Potawatomi leaders warned the British at Detroit that their southwestern compatriots had assembled at the forks of the Kankakee and Des Plaines Rivers to receive messages from the Spanish. This location—a longtime site of Indigenous settlement—lay only a day's journey downriver from Chicago's portages and provided a natural meeting point at the confluence of the Kankakee and Chicago routes.[65] French emissaries delivered the Spanish message, accompanied by eight strings of wampum commending the southwestern Anishinaabeg for keeping the British "at a distance." The Spanish message also encouraged further resistance to British rule along Lake Michigan's western shore.[66] While overtures from the Spanish perplexed the eastern Potawatomis, now allied with the British, their southwestern Anishinaabe kin from Chicago and Milwaukee proved receptive. According to the Spanish commander at St. Louis, Francisco Rui, in 1769, large numbers of Native people living between Lake Michigan and the Mississippi journeyed south to treat with the Spanish. Many of these visitors were Potawatomis and Odawas from the southwestern settlements. Rui also reported distributing gifts and meeting with Anishinaabe leaders from the Illinois River.[67] Well after the British had established a tentative hold over the Great Lakes, southwestern Anishinaabeg found alternative patrons in the Spanish at St. Louis.

The southwestern Anishinaabeg had learned to thrive along this space in between. Straddling the imperial fault line of British and Spanish patronage and relegated to the edges of Anishinaabewaki, the Chicago Anishinaabeg began to carve their own strategic path in the borderlands. Though many of the Anishinaabe people living in the southwestern settlements originated from Odawa, Ojibwe, and Potawatomi centers on the eastern side of Lake Michigan, by the 1760s, the geopolitical situation along the southwestern river routes differed strikingly from the diplomatic collaboration between British garrisons and Anishinaabe headmen east of the lake. As the terms of cross-cultural diplomacy changed, the actions and alliances of Chicago's Anishinaabeg diverged from the rest of Anishinaabewaki. Increasingly, southwestern Anishinaabeg pursued separate interests as they worked to establish the southwestern settlements' autonomy from their progenitors at Michilimackinac and the St. Joseph River. Given their geographic location, in between spheres of effective British control, the Chicago-area Anishinaabeg could continue to exploit their connections to the Spanish in St. Louis while avoiding closer collaboration with British officials and traders in the Great Lakes.

Toward Diplomatic Autonomy

Southwestern Anishinaabe headmen such as Sigenauk and Nakiowin began cultivating these Spanish connections in earnest during the mid-1770s.

Sigenauk acted as a spokesman for the Anishinaabe town of Le Petit Fort, north of the Chicago River. Nakiowin may have served as war captain in the Anishinaabe town at the confluence of the Kankakee and Des Plaines Rivers, or one of the other settlements in the vicinity of Aux Sable Creek. Both southwestern Anishinaabe leaders traced their ancestry to the Michilimackinac area but had come of age in the intertribal settlements west of Lake Michigan.[68] By the 1770s, they and their fellow Anishinaabeg from Chicago, Milwaukee, and the Illinois River towns began to harness their ties to Spanish St. Louis to solidify their separation from British dominion.

As Spanish officials in St. Louis reported in 1777, these southwestern portions of Anishinaabeg seemed well inclined to a Spanish alliance.[69] They named Sigenauk as the man presenting himself as the leader of these collective Chicago and Milwaukee Anishinaabeg. Despite Sigenauk's posturing, Spanish officials realized "this tribe being so large, and being divided into various districts," it remained impossible to know the differing allegiances of the collective Anishinaabeg. Indeed, the Spanish at St. Louis even seemed aware of differing allegiances within Anishinaabewaki, as they reported internal rupture among the St. Joseph Potawatomis who "are somewhat in revolt at present" over their mixed alliances with the British and Spanish. Regardless of the fissures forming between southwestern and eastern Anishinaabe diplomatic strategies by the 1770s, the Spanish in St. Louis proved happy to accept the Chicago-area Anishinaabeg as allies, recording that Sigenauk himself gave "signs of great affection" toward the Spanish at St. Louis.[70]

While the Spanish shared their enthusiasm for the Chicago Indians' divergent diplomacy southwest of Lake Michigan, such activities worried the British in the Great Lakes. In the spring of 1777, a trader in the British interest, Laurent Ducharme, wrote to the British commander at Fort Michilimackinac to report Sigenauk's actions. Ducharme warned Arendt Schuyler De Peyster "that the chief Siginakee or Letourneau has received a parole [order] from the Spanish Commandant to raise all the Indians between the Mississippi and the Little Detroit of La Baye [Green Bay]." Sigenauk had begun actively courting the Spanish in St. Louis for recognition—and even a military commission—according to Ducharme. In June 1777, De Peyster received further confirmation that the Spanish had reciprocated Sigenauk's overtures by circulating belts of wampum and even sending agents among the Native towns to the west of Lake Michigan, which only strengthened Sigenauk's influence among the Chicago-area Anishinaabeg.[71]

Such news could not have come at a worse time for British officials in the Great Lakes. As the outbreak of the American Revolution expanded from a colonial uprising in the east to a full-fledged war for the continent, the British feared a host of new security threats along the backcountry. Embracing the existing links between his people and the Spanish at St. Louis, Sigenauk raised the stakes of the diplomatic rivalry in the western Great Lakes at the very moment British officials faced an expanding revolutionary conflict and the threat of Native defection along their Great Lakes frontier.

When George Rogers Clark led a force of Virginians against British posts in southern Illinois, the Chicago-area Anishinaabeg pursued a new alliance with the American rebels as well. A delegation of southwestern Anishinaabeg led by Sigenauk were in St. Louis meeting with the Spanish when Captain Joseph Bowman captured the eastern shore on behalf of Virginia in July 1778. Initially fleeing north, Sigenauk later sent a wampum belt of peace and requested an audience with Clark directly. Clark followed up this liaison by making "strict enquirey about Black Bird" (Sigenauk's translated name) among the region's traders. He learned that Sigenauk "had great Influance" among "Considerable Bands" of Anishinaabeg south of Lake Michigan. Clark went so far as to pay a trader 200 dollars to contact the allegedly powerful Sigenauk and invite him to Kaskaskia for talks. Clark must have wondered if he had overpaid when Sigenauk and Nakiowin arrived that fall with a retinue of only eight warriors from the towns around Chicago and Milwaukee.[72]

Whatever Clark's doubts, Sigenauk impressed the American commander with his frankness and pragmatism, an accomplishment that represented a significant diplomatic coup for the southwestern Anishinaabeg, given Clark's well-known prejudices against Indigenous people. Unlike many of the other Native leaders Clark had met with in the Illinois Country, Sigenauk forsook the "Utial preparations" and "seremony" that remained a staple of borderlands diplomacy. Instead, Sigenauk got down to the "Business of Consequence that conserned boath our nations."[73] Hailing from the young towns of migrants around Chicago, Sigenauk presented himself as a new man of the borderlands, someone who, like Clark, was able to set aside diplomatic protocols and push for pragmatic alliances on behalf of his faction's agenda in the midst of a continental revolution.

Operating from the margins of Anishinaabewaki and feigning a powerful hold over the composite Indigenous settlements along southwestern Lake Michigan, Sigenauk proved convincing enough for Clark. Sigenauk asked Clark to explain the American cause to him, and Clark spent the day providing background to the colonial rebellion. By the end of the meeting, Sigenauk flattered the frontiersman with assurances "that he thought the Americans was perfectly Right" and confessed "he long suspected" the British of poor conduct. Sigenauk promised to return north to Lake Michigan and take "pains too tell the Indians of every denomination of what had passed between us," offering to bring an end to any alliances between the Anishinaabeg and the British. In turn, he requested that Clark send an emissary north with him, that they might continue their communication in the coming months. Clark was impressed, describing the Anishinaabe leader as "a Polite Gen[tleman]" and "inquisitive Indian," going so far as to mention him by name in dispatches back east to Patrick Henry and Virginia's congressional delegates. As it was relayed to Patrick Henry, Sigenauk, or "The Great Blackbird," was a chief worthy of special recognition from the Americans. To secure such a potent ally, Clark gave Sigenauk gifts valued at twenty or thirty pounds for himself and his family, several packhorses of goods for his people, and a promise of future correspondence.[74] Sigenauk turned home toward Chicago successful on all counts.

After Sigenauk's initial contact with Clark, the Americans kept up communication with the southwestern Anishinaabeg throughout the fall. Mech Kegie, a Chicago leader, wrote at least one letter to Major Joseph Bowman, the American commander at Cahokia. Clark confidently reported that meanwhile, Sigenauk had "not only stopt his own tribe" from supporting the British cause west of Lake Michigan but had persuaded "great numbers of Indians in that quarter" to abandon their alliance with the British as well.[75] The geographic position of the southwestern Anishinaabeg, far removed from British influence and eastern Anishinaabe leadership, allowed Sigenauk, Mech Kegie, Nakiowin, and other Chicago-area Anishinaabeg to begin crafting their own diplomatic course, distinct from their eastern kin and in direct opposition to British authority.

Despite these inroads among the southwestern Anishinaabeg, by spring the Americans grew impatient with the seeming divisions across wider Anishinaabewaki. Responding to Mech Kegie, Bowman urged the Chicago Anishinaabeg to "listen attentively" to the Americans and "stay quietly at home, to hunt, to support their wives and children," and to avoid becoming mixed up in the Revolutionary War one way or the other. Bowman threatened any Anishinaabeg "who wish to fight for the English," recommending they "sharpen well their tomahawks" to prepare for American retribution. Clark sent a similar letter to another Potawatomi, Nanaloibi, urging him to cast out any British agents or traders among his people.[76] Among Mech Kegie and the southwestern Anishinaabeg, the correspondence seemed to work. A group of the southwestern Anishinaabeg journeyed downriver to visit Bowman in May 1779. Their friendly meeting assured the American officer that "Peace and Quietness seems to be established Among the Indians" upriver.[77] But the Americans failed to understand the diverging strategies between the southwestern and eastern factions if they thought blanket threats could neutralize the various groups of Anishinaabeg living around the southern rim of Lake Michigan. While Sigenauk, Nakiowin, and Mech Kegie along the southwestern shore entertained communication and even alliance with the Americans and Spanish, their kin to the east remained more sympathetic to British interests.[78] By 1778, the Chicago Indians, operating from the edges of Anishinaabewaki, were forging their own diplomatic course in the borderlands.

We should not, however, conflate the Chicago towns' machinations against the British with a direct break from their eastern Anishinaabe kin. Rather, such multifaceted diplomacy could benefit the collective people of Anishinaabewaki. Chicago-area Anishinaabeg, who continued to thrive between Spanish and British spheres, and eastern Anishinaabeg, who had grown increasingly close to the British in the Great Lakes, were facing divergent geopolitical realities by the 1770s. As the terms of cross-cultural diplomacy changed, the southwestern Anishinaabeg built relations with the Spanish and Americans while eastern Anishinaabe leaders maintained good standing with the British. In this way, Sigenauk's efforts in the early years of the conflict reflected southwestern Anishinaabe impulses to exploit the alliances that offered the greatest advantage to them without severing their connections to the eastern Anishinaabeg aligned with the British. This was a delicate process, but one Sigenauk and his compatriots from Chicago proved adept at choreographing. In the years ahead, they would work to maintain good relations between eastern and southwestern factions of the Anishinaabeg even as they established their own authority and the diplomatic autonomy west of the lake (see map 3.1).[79]

Chicago's War of Independence

The defection of the southwestern Anishinaabeg toward the Spanish and Americans perplexed the British, given their reliance on other Anishinaabeg as Indigenous allies elsewhere in the Great Lakes. As the Anishinaabeg of the Chicago area continued to diverge from British wishes, disturbing reports began to float back toward British officials in the region. One British-allied Anishinaabe, Gorce, from east of Lake Michigan, warned the British that "a disaffected Chief of Milwaukee (named Sagenake) [Sigenauk]" actively consorted with the rebels around Chicago.[80] De Peyster, commanding at Michilimackinac, likewise heard rumors that the southwestern Anishinaabeg were actively building boats at Chicago, meant for an American invasion of Lake Michigan. Furthermore, he reported to General Haldimand in Canada that he "understood that the Virginians were at Chicagow."[81] While this was unlikely,

MAP 3.1 Chicago in Anishinaabewaki, mid-eighteenth century, by Gerry Krieg.

it was even rumored that if the Americans launched an offensive against Michilimackinac, Sigenauk and his Anishinaabe warriors would lead the troops.[82] Just
years before, General Gage and other British officials had scorned the Chicago
portages as an irrelevant periphery controlled by unruly Anishinaabeg on the
margins of their Great Lakes dominion. Now, Chicago's portages and the disaffected Anishinaabeg living along its route posed a grave threat to British security
in the west.

The British took such rumors about Chicago seriously. Since the 1760s,
the Chicago portages had remained a thorn in the side of the British, first as a
hotbed of ongoing Indigenous resistance after Pontiac's War and now as a potential open road for American rebels and their allied Anishinaabeg under
Sigenauk. Lieutenant Dietrich Brehm, one of the few British officers to have
reconnoitered the western Great Lakes by the 1770s, anticipated Michilimackinac lay vulnerable to attack "by way of the Illinois & Chikago Rivers."
He warned General Haldimand that the portages at Chicago, as an open
route from the Illinois Country, posed the greatest risk to British control over
the lakes.[83] Faced with the looming threat of Chicago's open geography, De
Peyster resolved to deploy his Indigenous allies to rein in their southwestern
kin and close off Chicago's route to the Americans.

During the winter of 1779, De Peyster ordered Charles Langlade, a prominent métis mediator for the Odawas around Michilimackinac, to travel south
to Green Bay and Milwaukee to neutralize the southwestern Anishinaabeg in
the area. But Langlade, usually influential among the region's Native peoples,
met with resistance. He learned that rival American agents had made inroads
with the Anishinaabeg of the Illinois River, who remained "much divided"
and uncommitted to the British cause. As spring came and De Peyster gained
more details about the situation southwest of Lake Michigan, he learned that
Maurice Godfroy Linctot, acting on behalf of the Americans, had journeyed
from Cahokia as far north as Peoria to buy horses and meet with the region's
Native people. De Peyster suspected that the southwestern Anishinaabeg of
Chicago and Milwaukee were complicit in Linctot's activities. To corral them,
he sent Lieutenant Thomas Bennett with twenty British soldiers, sixty trad-
ers, and some eastern Anishinaabe representatives to treat with the St. Joseph
Potawatomis and rally Britain's Native allies along lower Lake Michigan. De
Peyster also purchased the armed sloop *Welcome* and sent it to report on ac-
tivities and frighten any wavering individuals back into the British alliance.[84]
But this last-ditch effort by the British to incorporate or intimidate Chicago's
Natives failed to do much good. The southwestern Anishinaabeg seemed un-
fazed by British activities along their lakeshore.

At a loss, De Peyster turned to his Anishinaabe allies around Michilimacki-nac to rein in their errant kin from Lake Michigan's southwestern shoreline. Traveling to meet with the Odawas that summer at L'Arbre Croche, De Peyster expressed frustration at his allies' inability to curtail Anishinaabe defiance around Chicago. Incredibly, De Peyster later put his vexations into rhymed verse in an annotated rendering of his July 1779 speech at L'Arbre Croche. He referred to the southwestern Anishinaabeg as "those runagates at Milwakie," and "Eschickagou," describing "Sly Siggenaak" and his fellow defectors as "a horrid set of refractory Indians."[85] Faced with ambivalence among his own Anishinaabe allies, De Peyster castigated their southwestern kin as rogue actors in defiance of legitimate Anishinaabe leadership as well as British authority.

De Peyster understood Native politics sufficiently to realize that the Chicago-area Anishinaabeg originated from the same people as his own Odawa and Potawatomi allies, deeming them wayward members of a united Anishi-naabe confederacy. But De Peyster missed that these southwestern Anishinaa-beg, now far removed from eastern Anishinaabe centers, did not answer to eastern leadership any longer. A new generation of leaders to the west of Lake Michigan, including individuals like Sigenauk, had risen to prominence and had begun politicking in their fellow southwesterners' best interests amid the turmoil of revolutionary intrigues. The imperial dynamics that dominated An-ishinaabe considerations at Michilimackinac, Detroit, and St. Joseph remained less relevant in Milwaukee, Chicago, and the Illinois Country. Sigenauk's people acted accordingly and, in the process, moved ever further from both British de-mands and eastern Anishinaabe diplomatic strategies. True to the localized real-ity of Anishinaabe governing structures, De Peyster's Odawa and Potawatomi allies at L'Arbre Croche and St. Joseph remained disinclined to address such interregional disagreements about leadership and diplomacy. The southwest-erners, emboldened by their remote but vital geography, were operating on their own terms, welcoming Spanish and American overtures while using their control over Chicago's route to threaten the British hold over Lake Michigan. Chicago's Anishinaabeg found themselves masters over a strategic stretch of geography in a changing imperial borderland. They were now eager to leverage their position as they worked to expand their own power in the region.

Unsatisfied with his own Odawa allies' inability to curtail the southwest-ern Anishinaabeg, De Peyster resorted to force. In October 1779, De Peyster ordered the armed sloop *Felicity* to sail south along Lake Michigan's coastline to capture Sigenauk and intimidate the southwesterners. Samuel Roberts pi-loted the ship around Michigan's entire southern shore in search of the head-man, passing Chicago and Le Petit Fort in a show of force. On November 3,

Roberts sailed into Milwaukee harbor where he bribed "a chife named Chambolée who lives close by Saganac [Sigenauk] to attempt to fetch him in either by fair or forc'd method." Chambolée happily accepted gifts of food and tobacco from the British captain and apologized for following Sigenauk in the past. Whether Chambolée ever attempted to apprehend Sigenauk, however, remains doubtful. Given the lateness of the season and the stormy weather, the *Felicity* could not wait around indefinitely and returned north to Michilimackinac empty-handed.[86]

Despite their inability to capture Sigenauk, the British were planning a grand offensive that would drive his American and Spanish allies from the west, in the process undercutting southwestern Anishinaabe support. Spain had officially joined the war against Britain during the summer of 1779, and by the following winter, British officials began assembling a force to descend the Mississippi River to retake the Illinois Country and capture Spanish St. Louis. In May 1780, the British-allied fur trader Emanuel Hesse led several hundred Native warriors and European traders from Prairie du Chien against Spanish St. Louis while Langlade led a smaller contingent of Odawa warriors through Chicago's portage and down the Illinois River in support. Though the British forces outnumbered the defenders of St. Louis and Cahokia, the attacks failed after a day of skirmishing, and the Indigenous and British forces retreated.[87]

While it is unclear whether Chicago's Anishinaabeg played an active role in the British failures at St. Louis, they did harass Langlade's retreating forces as they returned north along the Illinois River. Langlade's force had learned of the British defeat before they ever reached St. Louis and immediately returned to the Chicago portage, pursued by a force of Americans led by John Montgomery. Langlade's retreat became more perilous when the party passed "through a Band of Indians in the Rebel interest" near Chicago. Fortunately for Langlade's Odawas, two British vessels arrived with "a force sufficient to enable the party to pass with safety." Intimidated and threatened by the local Anishinaabeg at Chicago's portages, Langlade's Odawas returned north to Michilimackinac. The Spanish commander at St. Louis received word about the incident and credited Sigenauk and Nakiowin as leading the southwestern Anishinaabeg in their "courageous resistance made against Langlade" and his forces at Chicago.[88] In the wake of British defeat, Sigenauk and the Chicago Indians emphasized their opposition to incursions from the British along their home waterways, even as they seem to have avoided any outright bloodshed between themselves and their eastern Anishinaabe kin.

Their hostile stance toward the British no longer in question, Sigenauk and his southwestern faction prepared to take the fight into enemy territory. By proxy, this meant a bold move into the homeland of their fellow Anishinaabeg who remained loyal to the British. In the fall of 1781, Nakiowin journeyed south to St. Louis to begin lobbying the Spanish to help them attack the St. Joseph, near the southeastern shore of Lake Michigan.[89] Sigenauk himself arrived in St. Louis the day after Christmas to press the Anishinaabe plan of attack. Sigenauk brought news to Francisco Cruzat, the lieutenant governor of Spanish Saint Louis, that a party of French from Cahokia had already attempted to capture Fort St. Joseph that fall. The expedition had failed when the local Potawatomis rallied under the leadership of the British Indian agent Dagneau de Quindre and captured the small party along Lake Michigan's shoreline. Sigenauk reported the British had "a considerable quantity of all sorts of merchandise," which he intimated would be used for "the expeditions which they are planning against us."[90] As Sigenauk told it, the British had begun supplying Fort St. Joseph as a depot to support further attacks against Spanish St. Louis. While there is no evidence supporting this threat in British documents from the period, Sigenauk proved content to play on Spanish fears to advance his own agenda.

In his vulnerable position at St. Louis, Cruzat dared not refuse his most assertive Native allies. "The urging of the Indian Heturno [Sigenauk], both on his own account and on behalf of Naquiguen" as Cruzat explained, forced his hand. As Cruzat wrote to his superior, his primary reason for approving the raid came from his fear of losing the southwestern Anishinaabeg as allies. He reasoned that "for me not to have consented to the petition of El Heturno and Naquiguen would have been to demonstrate to them our weakness and to make evident to them our inadequate forces." Cruzat feared that not complying would provide "sufficient reason for them to change sides," noting that they "are in the habit of following the strongest."[91] Given the role these Anishinaabe warriors played in threatening British forces at Chicago's portage, Cruzat understood it was vital to stay in their good graces. Should they change sides, or even become neutral, Chicago's portage would transform from a protective barricade guarded by the Anishinaabeg into an open route for further British attacks against St. Louis. Cowed by the power of his Native allies, on January 2, 1781, Cruzat ordered sixty militia under the command of Eugenio Pouré to follow Sigenauk, Nakiowin, and their warriors north against St. Joseph.

As the expedition approached the St. Joseph River valley, Anishinaabe politics took the forefront. Sigenauk and Eugenio Pouré dispatched an Illinois River Potawatomi named La Gesse (the Quail) and a local métis trader, Louison Chevalier, to the St. Joseph Potawatomis' towns. The Potawatomis around St. Joseph, though British allies, gave an audience to this pair with whom they shared kin ties. La Gesse and Chevalier requested that the British-allied Natives stand down and allow Sigenauk's men and their Spanish allies to attack Fort St. Joseph without interference. In exchange, the pair offered to share the plunder of the fort with the St. Joseph Potawatomis. Faced with the choice of fighting their fellow Anishinaabeg or betraying the British for a share of the loot, the decision was an easy one. The St. Joseph Potawatomis agreed to La Gesse's terms.[92]

The next morning, the southwestern Anishinaabeg and their Spanish allies crossed the iced-over river and took the fort without firing a shot. Capturing only a handful of Canadian traders, the Spanish nevertheless conducted an elaborate ceremony to mark their victory.[93] Sigenauk and his southwesterners, meanwhile, conducted their own set of symbolic and political gestures. La Gesse made good on the promises to the local Potawatomis. Having captured the fort without interference, he divided the spoils between Sigenauk's warriors and the eastern Anishinaabeg. According to the later Spanish report, the St. Joseph Potawatomi received a "considerable" amount of plunder for their troubles.[94] The redistribution of goods typically remained a right and obligation of Anishinaabe *ogimaag*, or civil chiefs. By doling out the captured supplies among the St. Joseph Potawatomi as well as their own warriors, Sigenauk, Nakiowin, and La Gesse displayed their magnanimity and authority in the very heart of their Anishinaabe rivals' domain.

Within a day, the allied forces fled back the way they had come, satisfied with their achievement. The Spanish commander believed the raid had sufficiently neutralized the St. Joseph Potawatomis. Given Sigenauk's achievements, he may have been correct. The southwestern Anishinaabe show of force along the St. Joseph River asserted their autonomy, and with very little bloodshed at that.[95] The headmen had also succeeded in demonstrating their own power in the Potawatomis' eastern homeland, sending a message about their unambiguous authority over the southwestern routes, including Chicago and the Illinois River.

The Political Economy of a Portage

With the raid on the St. Joseph, the southwestern Anishinaabeg had distinguished themselves from their eastern kin and asserted their autonomy along the river valleys and portages of Chicago and the upper Illinois River. In the years following the American Revolution, the southwestern Anishinaabeg would maintain their distance from the disruptive warfare to the east while they continued to prosper from their geographic position along Chicago's portages. While most Native peoples north of the Ohio River formed a series of militant alliances to defend against the United States, the Illinois

Country Anishinaabeg remained aloof. Into the 1790s, the southwestern Anishinaabeg maintained arms-length diplomatic relations with the Americans while most other Native peoples engaged in open warfare against the encroaching republic.[96]

The Chicago Anishinaabeg continued to cultivate their alliance with the Spanish at St. Louis as well. Into the 1790s, Spanish agents traveled back and forth to the Chicago portages to maintain these diplomatic ties.[97] These relations served both parties, allowing Spanish officials to monitor their rival empires' activities along their northeastern frontier and providing the Anishinaabeg with another source of patronage, gifts, and political leverage against the British and Americans. Long accustomed to positioning themselves in the diplomatic junctions between empires, the Chicago Anishinaabeg kept themselves strategically distant, but nevertheless engaged, with a variety of regional powers.

Their local interactions with Francophone merchant families also continued, as Chicago's Anishinaabeg welcomed several individual traders into their midst after the American Revolution. In 1782, for instance, a merchant from Cahokia in the Illinois Country named Jean Baptiste Gaffé sent several boatloads of supplies upriver to Chicago.[98] That same year, Jean Baptiste Point de Sable, a French-speaking free Black man involved in the fur trade, settled at Chicago's portage. Having married an Anishinaabe woman named Kitihawa, baptized as Catherine, he secured kinship ties and a welcome reception with the local Potawatomis. Other French trading families, led by Jean Baptiste Maillet, continued to live and trade near Lake Peoria. Maillet, who had accompanied Sigenauk, Nakiowin, and La Gesse on the 1781 raid against St. Joseph, maintained ties with the Chicago Anishinaabeg after the war, even representing them and serving as their interpreter in treaty negotiations with the Americans. These commercial and kin ties remained strategically selective, as Chicago's Anishinaabeg initially only forged relationships with a few traders who could provide goods or serve as diplomatic go-betweens on their behalf.[99] While open to incorporating a few beneficial individuals into their midst through marital ties and commerce, Chicago's Anishinaabeg also guarded their hard-won hegemony over the strategic portage and its local landscape.

A trader's travelogue from 1790 underscored how Chicago's Anishinaabeg had cemented their dominant presence along the route during the latter part of the eighteenth century. This represented a stark change from fifty years earlier, as waves of Anishinaabe migrants had made the river valleys their home and asserted their authority along Chicago's portage route. When Hugh Heward's fur trade brigade of two canoes traveled from Detroit to

Chicago and then down the Illinois River, he noted the omnipresence of Odawas, Potawatomis, and other Anishinaabeg along his voyage.

The Anishinaabe hold over the route through Chicago proved intimidating enough to dissuade some of Heward's voyageurs from ever coming that way again. After portaging over Chicago's pathway and passing the Anishinaabe town at Mount Joliet, two of his *engagés* told Heward "there was so much Danger" that they would not return along Chicago's route. The voyageurs wariness seems deserved—only the previous fall, a group of Potawatomis had pillaged "some of the Traders from Makina going down the Illinois river."[100] Along the route, Heward's party members had to ingratiate themselves to groups of Anishinaabeg they passed. At townsites like the one at the forks of the Kankakee and Des Plaines Rivers, Heward gave gifts of tobacco and other goods to ensure safe passage.[101] Such requisite gifts at bottlenecks and river passages marked a continuation of the tolls that Anishinaabeg groups had extracted along other key travel routes elsewhere in the Great Lakes for over a century.

Even as Heward and his crew trod lightly over the Anishinaabe portages, they relied on the Indians' expertise, labor, and goods to ensure safe passage through Chicago. Throughout their voyage, they paid local Anishinaabeg to guide them over portages and through river channels.[102] In carrying their boats and supplies over Chicago's pathway, Heward employed five Anishinaabe porters to help in the labor, paying them in gunpowder.[103] All these interactions served to benefit the Chicago-area Anishinaabeg as much, or more, than Heward and other traders. Area headmen reaped tolls and gifts from traders like Heward, while individual Indians hired out as guides and porters along the portage and adjacent wetlands of Chicago's myriad waterways to acquire clothing, ammunition, and other important goods from the traders. The southwestern Anishinaabe power at Chicago remained simply the newest iteration in an age-old Indigenous geographic logic that sought to control, and profit from, movement along such portage routes.

While Heward's account confirmed the unambiguous control the area's Anishinaabeg held over Chicago's portages by 1790, Heward's travel narrative likewise highlighted the portages' continued role as a space of movement. The portages provided a key overland connection for Heward and his party of engagés as they worked their canoes from Detroit to the Illinois Country downriver from Chicago.[104] Heward's party left Detroit in early spring to take advantage of higher water, paddling up the Huron River and portaging into the westward flowing Grand River to bypass the lengthy voyage around Michigan's lower peninsula. Party members navigated the route by noting landmarks such as Indian towns, blazed trees, abandoned cabins, wintering camps, and notable heights of land. They exchanged gunpowder and cloth for food and pitch to repair their canoes throughout their journey.[105] From Lake Michigan's eastern shoreline, Heward's voyageurs intermittently sailed and paddled their canoes along the southern basin to reach the mouth of the Chicago River. Coming ashore on May 9, Heward's party stayed at Jean Baptiste Point de Sable's trading house for two days while the men purchased food and acquired new boats from the trader.[106] Such methods used by Heward in 1790 to navigate, secure supplies, and ensure the success of his voyage through Chicago remained the same as those employed by European travelers over the previous hundred years or more as they traversed the interior waterways of the continent, at the behest of Indigenous guides, stewards, and hosts.

Throughout their voyage, Heward and his crew also faced many of the same navigational challenges earlier Europeans had endured when passing through Chicago's waterways. On their journey through the inland waterways of Michigan's lower peninsula, they often became lost and struggled to find portage routes. Many times during the trip, they paused to repair canoes, reapplying gum and pitch to keep their vessels seaworthy. Once they entered Lake Michigan, they remained beholden to wind and waves as they coasted the Great Lake's southern rim, which Heward noted as "dangerous in Stormy weather." The expedition experienced the perils of the expansive lake firsthand on May 4, when a sudden offshore squall "obliged us to make the Shore as fast as possible," before the canoes could fill with water and sink. At other times, the wind prevented party members from even launching their boats, delaying their travel by days.[107] When they finally did reach Chicago, they faced the toil of portaging their boats and supplies between the waterways, a job that required the labor of all the voyageurs as well as five additional Indigenous porters. The work proved so arduous that one engagé became "very saucey & abuseful" during the portage.[108] Despite such hardships, Heward and his crew harnessed the geography of Chicago's narrow continental divide as their best option for moving between the waterways of the Great Lakes and Mississippi River valley. In doing so, they followed the lead of countless Native peoples and other travelers before them who had connected the waters of the continent by portaging their boats over Chicago's pathway.

The Consolidation of Anishinaabe Chicago

Odawa, Ojibwe, and Potawatomi resettlements during the midcentury had established the presence of a southwestern Anishinaabe contingent along Chicago's portage route. The efforts of these southwestern Anishinaabeg after Pontiac's War and during the American Revolution further strengthened their stability along the Chicago, Kankakee, Milwaukee, and Illinois Rivers, confirming a new era of Anishinaabe power southwest of Lake Michigan. Throughout this period of expansion, the southwestern Anishinaabeg harnessed Chicago's strategic geography and prosperous ecology to thrive in this peripheral—but vital—space of movement and connection. By 1790, when Hugh Heward passed over Chicago's portage, the importance of the area as a waterborne, transcontinental crossroads remained clear. But so, too, did the Anishinaabe people's role as Chicago's dominant powerbrokers, controlling movement, providing labor, and facilitating travel between the interior watersheds of North America.

The latter half of the eighteenth century had seen the rise of the southwestern Anishinaabeg as they garnered power from their new position along Chicago's portages, undergirded by their own geographic and ecological knowledge of the area. Throughout the chaotic middle decades of the eighteenth century, Chicago's Anishinaabeg remained hard at work carving out a new frontier of opportunity west of Lake Michigan. The changing power dynamics of French defeat, British occupation, and the American Revolution ushered in a new era of Native politics along the margins of the conflict that unseated older tribal authorities, disrupted expected alliances, and yielded to new geographies of power. A series of local leaders from the Chicago-area towns had managed to assert the autonomy of southwestern Anishinaabeg while also maintaining the stability of their position along the portages. The southwestern Anishinaabeg harnessed these changing circumstances to assert their independence and maintain their distance from European rule, managing to reconstitute a space of Native power at the southwestern fringes of Anishinaabewaki. Their efforts had steered the Chicago-area Anishinaabeg in an upward trajectory around the portage, even as the world around them continued to change.

By 1790, the Anishinaabeg of Chicago found themselves at a crossroads in both space and time. On one hand, they had secured and consolidated their control over the geographic junction of the portages, maintaining a striking continuity in the space's significance as a thoroughfare of trade and movement. As they facilitated travel along the route, they reaped the benefits of such commercial flows as masters over the strategic portages. On the other hand, they had successfully navigated a series of incredible geopolitical upheavals from the seclusion of Chicago's position on the fringes of both Indigenous and imperial worlds. More changes were coming, and the Anishinaabeg along the portages stood prepared to meet them on the sure footing of Chicago's marshy ground.

CHAPTER FOUR

Thoroughfare

American Order and the Threat of a Fluid Frontier

In the last years of the eighteenth century, Chicago's Potawatomis and other Anishinaabeg continued to coordinate movement through the portages. While they reaped the rewards of bustling commercial flows between the Great Lakes and Mississippi River valley, a rising power to the east studied the maps of the region. Military officers and political leaders of the new United States began to contemplate the threats and opportunities provided by the geographic features along their western fringe. Having won a theoretical claim to the lands south of the Great Lakes and east of the Mississippi during the Treaty of Paris that ended the American Revolution, they became even more interested in their western possessions after Congress organized the region into the Northwest Territory in 1787.

Initially, such scrutiny from American officials was merely hypothetical. Most of the territory north of the Ohio River and west of Pittsburgh remained firmly in the hands of Native peoples that held little affinity toward the settler republic to the east. In the final years of the American Revolution, the borderlands between the United States' backcountry settlers and the Native peoples of the Ohio Country had become especially violent.[1] In 1790, the prospect of the United States ever bringing order to the far-flung geography of places like Chicago seemed a pipe dream. Yet, leaders of the young republic such as Henry Knox, George Washington's secretary of war, and Arthur St. Clair, the governor of the Northwest Territory, began planning for future U.S. encroachments into the region. Much of their planning revolved around the geographic potential—or danger—of the region's extensive waterborne routes.

In the years following 1790, as U.S. officials became more interested in the western geography of the Great Lakes and Illinois Country, Chicago loomed large in their plans for governing the region. American officials became convinced that controlling key geographic features, like Chicago's portage, was the solution for frontier stability. Their initial approach to Chicago and other points in the western Great Lakes was not the same as the settler colonial programs of land acquisition and expansion promoted elsewhere in the early republic.[2] In studying the ways these American officials thought about, and at times, obsessed over, the fluid space of Chicago's portage from the 1790s forward, we can see how early U.S. aspirations tracked along a colonizing model similar to the French and British before them. Whereas their European predecessors may have failed in their attempts to control the western Great Lakes, their approach seemed sound to U.S. officials. American statesmen believed they could curb the problems of western security and state weakness along their territory's watery edge by controlling strategic sites of movement, such as Chicago's portage. With this geographic logic, the American government made a concerted effort to bring the space and movement of Chicago's waterways under U.S. authority by securing the adjacent land and building a military post there. These efforts, in turn, would bring the United States into a headlong clash with Chicago's local population—both the Anishinaabeg and resident traders—who still claimed sovereignty over the area and used the portage to facilitate their unrestricted travel and trade between the Great Lakes and the Mississippi valley.

This multifaceted clash remained, at its root, a conflict over geographic visions for Chicago's land and water. U.S. officials first sought to occupy Chicago's local landscape not in efforts to settle the area but as a bid to curtail human movement through Chicago's portages. While local Native peoples, Europeans, and mixed-race métis valued Chicago's fluidity as a conduit for uninhibited travel between the region's waterways, the United States aimed to control the landscape and monitor travel through the portage. From 1790 onward, Chicago's crossroads became a flashpoint of escalating conflict as two disparate visions for the local space, and wider frontier, collided along the continental divide.

The Surveillance State

In the spring of 1789, Henry Knox, the U.S. secretary of war, sent orders for a secret reconnaissance mission into the west. Commissioning Lieutenant John Armstrong to travel to Kaskaskia under the guise of a fur trader, Knox hoped the officer might find a way to explore the lands claimed by Spain on the western bank of the Mississippi.[3] Venturing only as far as St. Louis, however, Armstrong and his superiors deemed the mission too hazardous without further planning and support. Instead, Armstrong spent time in Spanish St. Louis covertly gathering geographic information about the surrounding lands and waterways. Some of his most important intelligence had nothing to do with lands west of the Mississippi. While in St. Louis, he managed to interview a knowledgeable trader who had just come from Lake Michigan by way of the Illinois River. Intrigued by this trader's description of the route that unexpectedly connected the western Great Lakes with Spanish St. Louis, Armstrong recorded the entire meeting and relayed his report on to General Josiah Harmar. Including three pages of detailed description and an annotated map, Armstrong provided the U.S. government with its first thorough intelligence on the portage route through Chicago, arousing interest among Henry Knox and the military officials charged with governing the Northwest Territory.

Based on Armstrong's interview with the unnamed fur trader, U.S. officials became convinced that the portage route through Chicago posed a security threat along their western border. Armstrong's informant had traveled south from the portage in September 1789. Despite the late season, recent rains had flooded the Chicago and Des Plaines Rivers, making "the carrying place so easy at high water," that boats and canoes could pass unhindered "thro a Marsh" from one waterway to the next. The informant stressed the ease of travel along this route, sharing the "remarkable" fact that a boat could "leave the Falls of Niagara, provided the Waters of Chicagou are high, and pass without any Land carriage to the Bay [Gulf] of Mexico, which is not less than 3000 Miles thro the Land of America."[4] Such access through the heart of the continent may have seemed like a boon to the itinerant trader, but to U.S. officials who had no way of controlling the flows of people and goods through Chicago's portage, such waterborne mobility appeared sinister. To make matters worse, the portage route remained in the hands of hostile Indians. Armstrong's informant assured the American that a collection of local Indians— the Anishinaabeg—remained masters over the portage route, indicating a collection of "Indian Villages of Different Nations" on the enclosed, annotated map of the Illinois River.[5]

Other U.S. officials had begun gathering similar information on this open route through Chicago around the same time Armstrong submitted his report. Arthur St. Clair had traveled as far west as Kaskaskia that same spring of 1790 to gain a better understanding of his jurisdiction as the military governor of the vast Northwest Territory. St. Clair's second-in-command, Winthrop Sargent, had accompanied the governor west and, like Armstrong, spent time interviewing local traders to gain an understanding of the region's geography.

Sargent's attempts signaled a double effort to learn the lay of the land and gain information about the Native peoples that controlled the key river routes. With these objectives in mind, his interview with Jean Baptiste Maillet, a French fur trader from Peoria, proved especially revelatory.[6] Maillet described the route north to "the Oak Point carrying place"—the French name for the main Chicago portage—from which canoes could reach Lake Michigan within two leagues of easy paddling through "a narrow passage but good Depth of Water." Maillet also informed Sargent about the various Anishinaabe towns controlling movement through this key bottleneck, detailing the number of warriors and the name of the headman, or ogimaa, for each town. He relayed that "Upon the Chicagou close by Lake Michigan is a town of 30 Poutawatimes," while other "settlements of Poutawatimes" resided at the forks of the Kankakee and Des Plaines, farther up the Des Plaines River, on the Little Calumet and Grand Calumet Rivers, and at the Little Fort settlement headed by Sigenauk north of Chicago. All these settlements totaled to more than 360 warriors by Maillet's estimates. He also made it clear these Anishinaabe headmen maintained a tight control over the river passage and portages, the chief of the largest settlement requiring a toll from "all people going up the Illinois, to give him liquors." While Maillet cast the Chicago-area Indians as "good Poutawatimes" and Odawas, these southwestern Anishinaabeg had already demonstrated their unwillingness to yield their authority in the area. Effectively, the United States maintained no control over the portage and remained powerless to monitor the flows of people and commerce through Chicago's route.[7]

While Sargent gathered his detailed report of the upper Illinois River from Maillet, St. Clair assessed the overall situation at the watery edge of U.S. territory. Meeting with the local French, Spanish, and Indigenous leadership along the Illinois and Mississippi Rivers, he wrote back to Washington to sum up his findings on the region's geography and social landscape. His report illustrated the threats such a mobile and porous frontier posed for American empire in the Old Northwest. The Illinois Country itself remained tied to Spanish holdings across the Mississippi through both commerce and kinship networks that spanned the international border, while French and métis people made up the majority population on the American shoreline.[8] St. Clair described relations from one riverbank to the other as "still very intimate," which allowed for the flourishing of illicit, but seemingly harmless, trade and intercourse. These Spanish and French family networks bothered St. Clair much less than the British traders he found coming down the Illinois River. He reported that the trade of furs from the Illinois Country seemed "almost entirely in the hands of the British."[9] As he soon discovered, fur trade brigades out of Montreal continued to carry large shipments of furs to Canada by way of Chicago's portages. Chicago served as the overland link for such shipments, and St. Clair described how "in the spring of the year the waters of Michigan and the Chicago rise" to a level that allowed for the easy passage of vessels without even a portage. The bark canoes, "which carry a very considerable burden," could transport hundreds of pounds of valuable American furs back to British territory, operated by crews of only four or five men.[10] St. Clair saw this as a direct drain on what could be a flourishing American commerce. Worse yet, with no official presence along the route north, the United States proved incapable of even reaping some revenue from taxes on the merchandise. As one American, Robert Buntin, complained to Sargent several years later, "the English and Spanish Merchants have a great advantage over those of the United States. Their goods come by way of Canada, and free of that duty paid on all Merchandise imported into the United States." Buntin suggested that unless the United States established posts and tax collectors along the Illinois and Wabash Rivers, such a robust flow of illicit trade along the western water routes would continue to rob the American government of taxable exports and take "the bread out of the mouths of our own Citizens."[11]

Even in this early stage of U.S. occupation in the Northwest, St. Clair contemplated how a fort along the Illinois River corridor might throttle the movement of goods and people if necessary. His suggestion back to Washington offered "old Peoria town" as one solution, given the French village there and Jean Baptiste Maillet's apparent loyalties to the United States. Though the Chicago portage remained too far beyond U.S. influence, the French settlements at Peoria sat along a narrow section of the river. St. Clair had never visited the place, but from what he learned from Maillet and other traders, the channel flowed narrow enough that "every boat is obliged to pass within small musket shot."[12] St. Clair could only foresee controlling movement through this waterborne space by force of arms. Other Illinois inhabitants, likewise, suggested the U.S. outfit several gunboats to patrol the Illinois River to "put a check" on the "immense quantity of British merchandize" along its banks and the "baneful influence" that traders from Canada had on the local Native peoples.[13] St. Clair deferred from any immediate action however, given the precarious nature of U.S. authority in the area. While St. Clair diagnosed the overall danger for U.S. power in the west as the openness of river routes and the influx of foreign traders, he and his government remained too weak to rein in the movements of people and goods through the region's fluid geography.

Taken together, however, the geographic information and detailed reporting coming from the west demonstrated a concerted effort on the part of U.S. officials to gain a working knowledge about the riverine nature of their western frontier. Armstrong, Sargent, and St. Clair all gathered intelligence about Chicago's route during their western reconnaissance efforts of 1790, and all came to similar conclusions. Collectively, they reported on the ease of navigation through Chicago's portage and speculated on the threats such mobility posed for American security along the border. While these officials may have overstated the simplicity of the travel through Chicago, their information remained an accurate picture of Chicago's reality in 1790. Given the waterborne mobility of the portage, the area provided a porous route of travel for a variety of people. From an official U.S. standpoint, this thoroughfare of movement, aided by the very geography of the portage, had to be controlled, by force, if the United States could ever hope to bring security to its western frontier.

Claiming Chicago for the United States

St. Clair's attention over the next year would shift to an ill-fated campaign against the confederated Indians along the Wabash, which ended in a crushing defeat for the Americans and a humiliating retreat for St. Clair and his decimated army.[14] For the next several years, American policy in the west reeled in the wake of St. Clair's defeat, while at the local level, Chicago remained an entrepôt for illicit trade, mostly carried out by non-Americans. In the wake of St. Clair's defeat, George Washington appointed General Anthony Wayne to rebuild the U.S. Army in the west, expecting that the general would force stability on the Northwest Territory through a military victory over the Indigenous confederation that had beaten St. Clair in 1791.[15] But like St. Clair, Wayne immediately identified the problems preventing American control over the region as more complex. In Wayne's view, the United States' fragile hold over the region went beyond hostile Native nations, remaining endemically linked to the porous geography of the Great Lakes and the Mississippi River. Like St. Clair, Wayne feared a host of foreign entanglements along the United States' northwestern frontier, and he recognized that the region's waterborne routes allowed for ongoing diplomatic and commercial intrigues by Spanish and British subjects passing through U.S. territory.[16]

Wayne's obsession over the United States' fluid borders came into sharper focus when American troops brought in Joseph Collins in February 1793. Collins had scouted the Illinois River route through Chicago and returned by way of British Canada. Collins claimed he had been recruited by Wayne's second-in-command, James Wilkinson, to spy out the route connecting the Illinois Country with the Great Lakes and ascertain "the situation of the Country, and water courses, & also discover the part in which they wou'd be most vulnerable and easiest of access."[17] According to Collins, during the summer of 1792 he passed north along the Illinois River, traveling undercover as a British trader with an Indigenous guide. Along his way, he met many Canadian and Spanish traders smuggling goods through American territory.

These traders used the Illinois River route because it remained navigable throughout the summer and fall for "boats of large burthen" and offered "an easy current," even when traveling upstream. During spring floods, passage proved even easier, all the way "into Lake Michigan with boats of three tons, or 6000 pounds except a small portage" at "Chikago."

Wilkinson had also instructed Collins to "examine the most proper plans for garrisons or posts on the Illinois river," and while he mentioned likely sites for fortifications at Peoria, the mouth of the Vermilion River, and elsewhere, he assured Wayne that "one of the most capital positions is at the mouth of the river Chikago." He described how the portage facilitated the connection between the Illinois River route and Lake Michigan and assured that Chicago "affords an excellent harbor." He even suggested that a small levee built near the portage could regulate waterflow and provide a canal for waterborne travel year-round. Wayne took careful note of Collins's report. His description of the many "Indians towns & villages" that commanded the route, the numbers of foreign traders using it, and the short portage that allowed for travel between the Great Lakes and the Mississippi further confirmed for Wayne how vulnerable U.S. interests remained in the west. At the same time, Collins's emphasis on Chicago's portage as the local nexus of waterborne movement along the Illinois River route led Wayne to view it as a principal threat, and possible solution, to American woes in the west.[18]

Despite Wayne's convictions about the strategic significance of Chicago and the Illinois River, he remained powerless to exert U.S. authority onto the western route while the Ohio confederacy of Indians held the U.S. Army at bay and rebuffed insincere peace talks with the Americans. Wayne, given his suspicions of Spanish and British traders and agents operating within U.S. territory, viewed even the ongoing Native resistance north of the Ohio River as an outgrowth of foreign interference. As late as July 1794, Wayne continued to collect intelligence from his spies and interviews with Native prisoners that confirmed both British and Spanish involvement in the region's Native affairs. Even as Wayne prepared to lead his army in a decisive campaign against the Native confederacy in the Ohio Country, he shared his worries with Henry Knox that "it is therefore more than probable that the day is not far distant when we shall meet this *Hydra*" of Spanish, British, and Indigenous resistance in battle, "without being able to discriminate between the white & the red."[19]

Wayne's premonitions of Spanish and British collusion proved immaterial when the American army finally closed with the Native confederacy's forces at the Battle of Fallen Timbers in August 1794. Except for a small contingent of Canadian traders who fought with the Indigenous allies, the Native warriors received no overt support from British patrons and soon retreated in the face of Wayne's reorganized American legions.[20] With a U.S. victory at hand, Wayne could now hope to secure a peace treaty and bring control to the Northwest frontier, in his mind eliminating the geographic possibilities of foreign encroachment and Indigenous resistance in the future.

In the aftermath of the Fallen Timbers campaign, Wayne became ever more convinced that western security relied, in part, on a military post at Chicago. Directly after the battle, the general interviewed several Detroit traders captured while fighting on the side of the Native confederacy. One of them, a man named Antoine Lassell, informed Wayne that the hostile Native warriors planned to regroup to the west, possibly on "the Spanish side of the Mississippi, where part of their nation now live." Worse yet, the Shawnee war leader, Blue Jacket, had told Lassell that he "intended to move immediately to Chicago, on the Illinois," which could serve as an ideal location, adjacent to Indigenous allies but also providing easy access to Spanish and British territories, and presumably, support.[21] While there is no evidence that Blue Jacket ever ventured to Chicago, the threatening space of the portage remained clear. Coupled with the geographic intelligence gathered by Armstrong, Sargent, and Collins in the years of uncertainty leading up to Fallen Timbers, Chicago's open route seemed to Wayne to pose a significant threat to U.S. stability in the region.[22]

American officials by no means considered victory certain after the Battle of Fallen Timbers and Wayne remained convinced of a wide array of dangers along this entangled frontier.[23] Without additional fortifications along the key geography of the region, Wayne feared both "British agents" and other "Characters," operating across the expansive Northwest Territory to thwart peace and undermine U.S. conquest.[24] He continued to mistrust possible Spanish interference as well, reporting to Knox the rumor that five armed Spanish galleys had ascended the Mississippi River.[25] In Wayne's view of things, the river channels connecting the Great Lakes with Spanish St. Louis worked against American border security, writing to Henry Knox that such unguarded routes facilitated a "perfect understanding & constant communication between the Spanish Commandant of Post St. Louis' on the Mississippi & the British at Detroit."[26] As late as March 1795, Wayne likewise worried the British would use their mastery of the Great Lakes waterways to reinforce the Native confederacy with supplies and Indigenous allies from the upper Lakes.[27] To Wayne, the only way to ever secure the region as a U.S. territory was to curtail the watery inroads of Spanish and British intrusion.

Desperate letters from the Illinois Country further prodded Wayne to secure a post at Chicago's strategic portage. Since the early 1790s, the Chicago

Anishinaabeg had campaigned down the Illinois River and across the Mississippi against the Osages.[28] Increasingly, these war parties had begun to prey upon outlying farms in the Illinois Country as well, stealing horses, taking captives, and occasionally killing those who resisted their depredations. By 1795, an escalation of violence prompted French and American inhabitants of the Illinois towns to petition Wayne's army for aid. Writing from Kaskaskia, John Edgar and François Canis described their towns' plight, "in a manner, besieged, there being according to the Report above 100 Potowatimies camped on the Spanish side of the Mississippi." These Anishinaabeg had come down from the Illinois River and Chicago, with Sigenauk as their leader. On behalf of the white citizens of the Illinois Country, Edgar and Canis appealed to Wayne "to cause Small garrisons to be established among us as a check upon our Enemies," requesting especially some fortifications north along the "Illinois River which will have the additional good Effect of preventing the Indians from coming down upon us as they now do, in canoes." The image of large war parties of hostile Potawatomis descending upon the Illinois settlements in amphibious raids played on Wayne's continued paranoia over the region's waterways. The Illinois River corridor had become not only a thoroughfare of foreign incursions but also an open road for waterborne assaults by the Anishinaabeg of the Chicago area.

With this in mind, Wayne prepared to include the Chicago portage in his upcoming treaty negotiations. Meeting with the various Native nations of the Northwest Territory at the army headquarters in Greenville, Ohio, in the summer of 1795, Wayne took measures to secure the water routes of the west. His superiors back east had granted him wide leeway in laying out the terms and demands of the treaty, given his "knowledge of the Country."[29] With his research and interviews from 1792 onward, Wayne approached the treaty determined to secure the Chicago portage as a site of U.S. control going forward.[30]

His efforts met with stiff resistance among the assembled Native peoples, even though few representatives from the Anishinaabeg around Chicago attended the treaty.[31] Wayne made his case for a Native cession of land at Chicago based on claims that the French had once maintained a post there. By using this line of argument, Wayne revealed U.S. intentions to rule the Great Lakes in much the same way as their imperial predecessors had tried to, through the control of limited, strategic geography. Wayne's views of that strategic geography in many ways mimicked earlier European interpretations of the land and waters of the region and led him to place a heavy value on the small patch of marshy ground along Chicago's portage route.

But Wayne's line of argument also opened him up to criticism from the assembled Native diplomats. While French maps indeed had portrayed Chicago's portage as a place of imperial power, the reality of French control at Chicago had never materialized. Little Turtle (Mihšihkinaahkwa), the leader of the Miamis, who still claimed the Chicago area as part of their traditional homeland, protested to Wayne's claims of French sovereignty at the portage, stating flatly that "we have never heard of it."[32] If the Americans were going to lay claims to geography in the western Great Lakes as a successor empire to the French and British, Little Turtle intended to at least keep the record straight. A French fort may have appeared on European maps of Chicago, but as the Indigenous leaders assembled in 1795 knew, Chicago had always remained a space controlled by Native peoples. Little Turtle's challenge to Wayne demonstrated that despite U.S. military victories, the region's Native peoples would continue to contest American claims and geographic visions for Chicago's portage going forward.

After much debate back and forth with Little Turtle, Wayne secured a concession for "one piece of land six miles square at the mouth of the Chikago River," for the future purpose of building a fort on the site. Wayne also ensured safe travel for Americans through the portage route, even on lands not explicitly ceded to the United States by the treaty. Included in the official terms, Wayne insisted the "Indian tribes will allow to the people of the United States a free passage by land and by water," along portages throughout the Northwest, and including the route "from the mouth of Chickaga down the Illenois River to the Missisippi."[33] At the same time, the treaty also included an explicit ban on itinerant traders traveling through U.S. territory without official licenses, aiming to curb foreign commerce in the region. This attention to detail regarding Chicago's route highlighted the Americans' growing understanding of both the importance of the portage and the scale of movement through it.[34]

Wayne's report back to the capital the following year further underscored his geographic logic for acquiring the small tract of land near Chicago's portage. American officials and politicians remained concerned about foreign agents and traders moving through American territory and potentially undermining the security of the western territories. Congress wanted to learn the feasibility of defending the border along the Great Lakes and the Mississippi River and enforcing duties on commerce in the region. Responding to this inquiry on the state of the northwestern frontier, Wayne outlined his vision for how the United States might best control the geography and commerce of the Great Lakes, given the cessions in the recent treaty. Overall,

Wayne advocated for the general policy of mimicking French and British models for ruling the region. He emphasized the wisdom of maintaining the same posts that his imperial predecessors had built at places like Detroit and Michilimackinac. Jay's Treaty, negotiated between the United States and Great Britain, made occupying such forts possible for the first time, as the British finally agreed to honor the terms of the earlier Treaty of Paris and turn these garrisons over to the Americans.[35] With now uncontested claims to the southern shores of the Great Lakes, Wayne laid out his final plans to secure the waterborne frontier for the United States.

Wayne emphasized Lake Michigan as a particularly important feature along the border, since "all the Mississippi trade must of necessity pass from Canada" through it. Here, Wayne diverged, slightly, from earlier French and British geographic strategy. He worried that the long-maintained post at Michilimackinac, by itself, proved ineffective for stopping the flow of foreign commerce and people through American territory, equating the waterborne nature of the Great Lakes with a door "rather too wide to close." Instead, Wayne advocated for building a new American post at "a portage called chicago," which led into the Illinois River as "a necessary auxiliary" for controlling movement. By advocating for a post at Chicago, Wayne advanced a new geographic solution to control the untaxed and unmonitored movement through the lakes at a crucial overland bottleneck. Having wrested a piece of land along the Chicago River as part of the cessions at Greenville the year before, Wayne remained optimistic that the United States could now establish its control over the region more effectively than the British and French empires before it. In his view, a post at Chicago's portage would bring some land-based order to an otherwise waterborne and unruly frontier.[36]

Anticipating an American Occupation

On the ground, things would prove more complicated. As Wayne advanced his plan for fortifying Chicago's portage and controlling waterborne movement through the area, the Anishinaabe leadership of the Chicago towns protested. Traveling to Wayne's headquarters on July 16, 1796, a delegation of Potawatomi headmen representing at least four separate Anishinaabe towns around the portage demanded "to know what chiefs gave you our land at Checago."[37] According to Wayne, the Chicago delegation "pestered & interupted" him at his headquarters, demanding "further and particular compensation" for the six-mile tract at Chicago.[38] Wabenaneto, speaking on behalf of the Potawatomi town near the southern portage, pointed out that neither he nor the headmen from north of the Chicago River had attended the Treaty of Greenville. According to the Chicago Indians, their "great chief" was too elderly to have come forward either during the treaty or afterward to represent them properly. In the end, Wayne agreed to pay them directly for the land cession and doled out presents in exchange for some captives the Chicago chiefs had brought along. After six days of haranguing the U.S. commander, the western Anishinaabe leaders expressed their satisfaction at the new peace between the United States and the Northwest Indians, but also reiterated their own distance from the conflict and their lack of consent with the cession of Chicago specifically.[39] While they reaped rewards of presents and other payments from their visit, their journey to Wayne's headquarters also emphasized their continued claims to the lands and waters around Chicago's portage.

The seriousness of the Chicago headmen's visit to Wayne and their ongoing protestations to the U.S. tract at Chicago set the tone for relations going forward. The Anishinaabeg at Chicago, now firmly under Potawatomi leadership, had made efforts since at least the 1750s to maintain a certain level of sovereignty through separation at the portages. Distancing themselves from European posts in other quarters of the Great Lakes since the French regime, they had established a powerful hold over Chicago's strategic geography, reaping all the benefits of a flowing entrepôt of trade while maintaining their detachment from colonial authorities at the local level.[40] A land cession along the Chicago River, even a small one, portended a future U.S. garrison in their midst, and they understood this. Such a militant presence along the portage route threatened to undercut the long-established patterns of waterborne travel and trade that had elevated Anishinaabe power and prosperity at Chicago in the first place.

The Chicago Anishinaabeg had much to lose with an American presence along the portage route. Following the Revolutionary War and Sigenauk's raid against St. Joseph, relations between Chicago's Anishinaabeg and the British had improved significantly. By the 1790s, British traders often passed through Chicago, bringing canoe-loads of goods from Canada through the portage. The Spanish in St. Louis kept well apprised of British activities along the Illinois River, and their best intelligence of the day estimated that in a single fur trading season, British subjects from Canada brought a total of twenty-two canoe-loads of trade goods through Chicago, bound for the Anishinaabe, Kickapoo, and Mascouten towns stretching from the portage south along the Illinois River as far as Peoria. Additionally, traders brought one canoe-load of goods to Sigenauk's town—"Petit Post"—north of Chicago, two more to the Anishinaabe town on the Little Calumet River, and another two to towns on the Grand

Calumet River, all within the vicinity of Chicago.[41] At the same time, the Anishinaabeg around Chicago's portage kept the diplomatic channels open south along the waterways, visiting Spanish St. Louis often, entertaining Spanish agents in their Chicago towns, and receiving yearly gifts from the Spanish government.[42] An official U.S. presence at Chicago's portage threatened to throttle the flows of British commerce from Canada and cut the area's Anishinaabeg off diplomatically from the Spanish in St. Louis.

The Chicago headmen also faced new internal pressures within their towns, a situation that only boded more trouble with the United States going forward. After the disaster at Fallen Timbers and the Greenville Treaty, leading ogimaag among the eastern Anishinaabe towns of St. Joseph and the Huron River wanted to avoid future conflict with the Americans, advising their more confrontational young men to move west. As Thupenenawac, a headman from the St. Joseph valley, explained to U.S. officials, "Too many of our nation aspire to be chiefs—and when they find they cannot be such in their own Country—they fly to the woods."[43] When Thupenenawac said "the woods," he meant the peripheries of Anishinaabewaki—places like Chicago, Milwaukee, and the Illinois River—where western Potawatomi towns offered new opportunities for would-be Anishinaabe migrants and minimal contact with incoming white settlers. Just as earlier generations of Anishinaabeg had relocated to the western towns to escape the imperial influences of the French and British in the mid-eighteenth century, now new migrations of younger warriors, dissatisfied with the defeat at Fallen Timbers and the Greenville Treaty, settled along Chicago's waterways in an ongoing pattern of migration as resistance.[44]

Main Poc was one young leader who heeded the advice of his more tempered elders from the St. Joseph River. Having grown up in the Potawatomi towns east of Lake Michigan, Main Poc had been one of a small number of Anishinaabeg who fought at Fallen Timbers, despite being born with a deformed left hand. After Greenville, Main Poc relocated to the western Anishinaabe towns along the lower Kankakee River, just south of Chicago, where he attracted a following as a powerful *wabeno* (man of the dawn sky), or holy man. His misshapen hand only served to confirm his status among Anishinaabe society, which viewed such physical deformities as a sign of special spiritual powers. Main Poc's town at the Kankakee attracted prominent warriors from the east, including Nescotnemeg and Wabinewa.[45] In turn, these anti-American newcomers forged new bonds with other southwestern Anishinaabe towns in the area, including Sigenauk's village at Petit Post and Campignan's town on the Calumet River. Together, this influx of Potawatomi migrants after the concessions at Greenville facilitated a powerful alliance among the collected Anishinaabe towns in the area, and brought a new, staunchly anti-American population of warriors to Chicago's crossroads.[46]

Despite such forebodings, life for the Anishinaabeg and traders living along the portage continued apace. Jean Baptiste Point de Sable, a French-speaking free Black man, had lived at the mouth of the Chicago River for years, ensconcing himself within the local Anishinaabe society by marrying a Potawatomi woman named Kitihawa, baptized as Catherine. Point de Sable's business interests represented the local nexus of a far-flung transnational trade that stretched from the lower Mississippi to Upper Canada. Passing brigades of fur traders relied on Point de Sable to broker relations with Chicago's Potawatomi headmen, who continued to facilitate movement through the portage.[47] Other métis and French habitants had also congregated around Chicago, drawn by the traffic of the portage but prospering only through kin connections to the area's Potawatomis. Antoine Ouillemette first came to Chicago as a voyageur for the fur trade, but stayed and married Archange Marie Chevalier, a woman of mixed French and Potawatomi descent who maintained close kin ties to the Potawatomi headmen on the Calumet River. After 1790, Antoine Ouillemette made a living at Chicago by selling his labor as a porter, carrying goods and canoes across the portage for passing fur trade brigades.[48] Other French and métis families, such as the Pettells and Le Mais, likewise made their livelihood in the vicinity of Chicago's portage, but maintaining good relations with the local Potawatomis remained a necessity to thrive along this epicenter of Indigenous power.[49] In turn, the Chicago-area Potawatomis benefited from this small contingent of French and métis that provided services as interpreters, blacksmiths, and mediators in ongoing negotiations with traders as they passed through the portage.

Besides the long-distance transportation that passed through Chicago's portage, the local wetlands continued to furnish a large supply of pelts for the area's Indigenous and French population's trade ventures. While Point de Sable's account books have not survived, some of his most prominent business partners kept close records that revealed the types of transactions and scale of commerce taking place along Chicago's waterways. Kakima, a Potawatomi woman from the St. Joseph River, had married an American trader named William Burnett in 1782. Together, the two had worked to extend trade relationships from the St. Joseph River west along the Kankakee marshlands into the Illinois River valley in the years after Sigenauk's raid. The preserved letters and account books of Burnett and Kakima's operation illuminate a still-thriving regional trade, connected through the portages and waterways of lower Lake Michigan.[50]

The Chicago region continued to yield rich harvests of furs for the area's resident traders. In the years after the Greenville Treaty, the Burnetts tallied winter hauls of 1,200 deerskins, 3,500 raccoon pelts, and over a hundred pelts each of muskrat, mink, "cats & foxes," along with smaller quantities of otters, bears, and beavers, all purchased from Indigenous hunters from the St. Joseph, Kankakee, Chicago, Des Plaines, and Illinois Rivers. These they sold at Michilimackinac for around 20,000 livres each summer.[51] While Kakima managed transactions and the extensive farming operation from the St. Joseph River, William Burnett traveled the waterways, wintering along the Kankakee River to purchase furs from the Potawatomis and plying his canoes north to Michilimackinac every summer. He also dispatched engagés, or fur trade employees, to other sites of trade, including Chicago's portage.[52] As early as 1786, the couple had coordinated deliveries of trade goods via boat from Detroit to the mouth of the Chicago River, including "rum, powder, and ball," which they exchanged with Chicago's Potawatomis for their yearly harvest of furs.[53]

Burnett and Kakima employed an array of French, métis, and Anglophone engagés to carry out trade on their behalf at auxiliary locations such as Chicago. By 1786, Jean LaLime had become one of their most prominent representatives, conducting business on their behalf from a trading house at the Chicago portage.[54] While extending the Burnetts' commercial interests along the portage, LaLime married Nokenoqua, a local Potawatomi woman, which facilitated his own prospects in both kinship and trade connections at Chicago.[55] The Burnetts also entered into business with Jean Baptiste and Catherine Point de Sable, who by then had expanded their own operations at Chicago with a trading post, blacksmith shop, smokehouse, and extensive farm raising cattle, hogs, and chickens.[56] Through both more distant trade ventures, such as the Burnetts', or in the localized efforts of the Point de Sable and LaLime households, Chicago's portage had become a commercial center by the latter years of the eighteenth century. In this way, the Native, métis, and European residents of the Chicago area maintained a lifestyle that continued to revolve around the mobility of the portage and the region's Indigenous trade.

Even in its invariable cycle of Indigenous seasonal migrations and fur trade voyages, individuals entwined in the commerce of Chicago began to prepare for an eventual U.S. presence. After news of the United States' land acquisition in the Greenville Treaty reached the region's Potawatomis and traders, Kakima and Burnett made a concerted effort to expand their trading interests at Chicago's portage in anticipation of a new market for their agricultural goods with the American garrison there. With the prospects of an American garrison looming, the Burnetts mobilized to buy Point de Sable's Chicago trading post and farm outright, hoping to extend their trading interest along the Chicago River through two proxies, Jean LaLime, the French habitant, and John Kinzie, an English trader from Detroit with mixed British and American loyalties.[57] In the spring of 1800, Point de Sable sold his holdings at Chicago, with the trading post and other property eventually passing to Kinzie, presumably acting on behalf of the Burnetts.[58]

With a base of trade operations now established at Chicago, Burnett also alerted his Montreal creditors to the possibility of a new American garrison. He predicted this development would prove good for business, as he expected to sell "a good deal of liquors and some other articles" to the future troops since he already had employees and a trading house established there.[59] Indeed, even as it became apparent to these traders that the United States was intent on establishing a presence at Chicago, they remained hopeful to profit from such changing geopolitics. Local magnates like Kakima, Burnett, Kinzie, and LaLime anticipated a new inflow of regional actors with an American garrison at Chicago, but not a fundamental shift in the status quo. After all, other European and Indigenous regimes had come and gone at the portages for over a century without disrupting the overall geographic logic of movement and interaction at the site.

An American Vision of Order

While the western traders and Chicago-area Potawatomis anticipated an American debut along the portage, the politics of the U.S. republic progressed to ensure that reality. In 1800, Thomas Jefferson won the presidential election for the Democratic Republicans, who remained committed to a western design. Moreover, U.S. citizens in the Illinois Country had continued to petition the government to erect forts along the Illinois River route, as Potawatomis from the Chicago area had continued to raid white settlements downstream even in the years after the Greenville Treaty.[60] Main Poc, the spiritual leader of the Kankakee towns, headed many of these raids. Black Bird, Sigenauk's son, had assumed leadership of the town north of Chicago after his father's death and likewise adopted a more hostile stance toward white settlements downstream.[61] Jefferson's aspirations for U.S. expansion in the west combined with the pragmatic imperatives of westerners and earlier frontier officials to make fortifying the Illinois River corridor a continued priority going forward.[62]

In early 1803, Henry Dearborn, Thomas Jefferson's secretary of war, sent orders to Detroit to "dispatch an officer and six men to explore the region surrounding the mouth of the Chicago River."[63] At the same time, the Indian agent and adopted member of the Miami William Wells (Eepiihkaanita) went west to reconnoiter a route between Fort Wayne and Chicago in the summer of 1803.[64] Henry Dearborn ordered Wells to judge the "disposition of area Indians to this intrusion through their territory" and to consult with Burnett about the attitude and activities of the few "white inhabitants," who were "at or near Chikago."[65] Dearborn needed to know whether any of the locals—Native or non-Native— meant to oppose the tenuous U.S. claims to the area. The small detachment of soldiers from Detroit under Captain John Whistler, meanwhile, surveyed the stretch along the mouth of the Chicago River originally ceded in the Greenville Treaty with an eye toward constructing a fort. Dearborn's orders to Whistler reflected the true aims of the future post, which would "control river traffic, exact tariffs, and pacify the local Indians."[66] Dearborn's vision for the Chicago fort implied nothing about future settlement, as the United States only held claim to the small tract ceded by the area's Potawatomis. While Jefferson and other republican thinkers could wax philosophic about settler expansion and yeoman farmers back east, the U.S. Army approached Chicago with the pragmatic objectives of regulating movement through the fluid space of the portage.[67] At places like Chicago, controlling strategic geography remained the goal of early U.S. colonization efforts.

To secure their vision of empire into the future, U.S. national leaders like Dearborn sought to convert Chicago from a portal of mobility in the west into an American center of control. Captain Whistler returned to the area in the fall by way of a Detroit schooner, linking up with a full complement of sixty-nine military personnel and two cannons at the portage, to begin construction on Fort Chicago. The following year, with the post complete, the garrison rechristened it Fort Dearborn after the secretary of war who had envisioned it. The U.S. government had established a presence at the portage, and its agents at the fort would attempt to bring order to this far-flung node of empire (see figure 4.1).[68]

Jefferson's administration, specifically Henry Dearborn, the secretary of war, and William Henry Harrison, the superintendent of Indian Affairs and territorial governor of Indiana, intended Chicago to serve as a new center for U.S. control in the far west.[69] This meant that along with the military presence at Fort Dearborn, the federal government needed local facilities to manage Indian affairs and conduct government-subsidized trade with the Anishinaabeg. In 1805, the War Department ordered Whistler to construct a nearby "agency

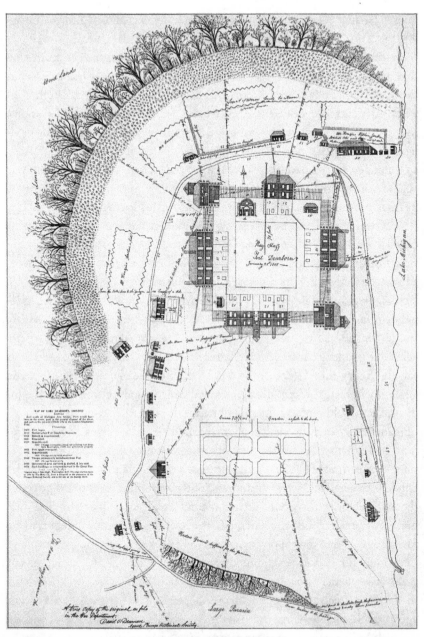

FIGURE 4.1 American heads of state and military officers hoped establishing a garrison at Chicago would help bring an orderliness to the fluid space of the portage. John Whistler, *Map of Fort Dearborn in January 1808 by Captain John Whistler, Commandant of the Fort* (January 25, 1808). Courtesy of the Chicago History Museum.

house." Charles Jouett, the newly appointed Indian agent for Chicago, moved into the crudely built log house with his wife and young daughter shortly after.[70] From there, Jouett coordinated Native relations for an expansive jurisdiction that stretched from the St. Joseph River in the east to the Mississippi River in the west. Within this territory, Jouett remained responsible for distributing annuity payments to appease the Anishinaabeg for their previous land cessions, maintaining good diplomatic relations with local headmen and gathering intelligence about Native dispositions toward the United States. Since the portage continued to act as a necessary link between the region's waterways, Jouett was able to host a variety of Indigenous visitors as they passed through Chicago's nexus of movement and maintain a fairly up-to-date idea of Native activities.[71]

Whistler also deployed his troops to build what the U.S. government called a "trade factory" at Chicago in 1805. President Jefferson's vision for the future of the fur trade in the Great Lakes entailed a series of government-subsidized trading houses dispersed across the frontier. Under this plan, U.S. factors, or government agents, would provide the same function that independent fur traders had throughout the region under the French and British. But under the trading factory program, the U.S. government could partially subsidize goods to ensure a lower, fairer price for Indians looking to sell their furs. In theory, this system would serve multiple federal objectives. Since the government factors themselves controlled which goods to trade with the Indians, they could regulate the sale of harmful wares, such as liquor, while simultaneously peddling "civilizing" merchandise, such as plows, livestock, and farm implements. Politicians like Jefferson also hoped that putting the fur trade in government hands would ease tensions that often arose between independent traders and Native customers. The government-subsidized prices of the trade factories also promised to undercut and eventually outcompete anyone else, funneling the profits of the fur trade away from independent traders, foreign agents, and monopoly companies, and back to the federal government. On a more duplicitous level, trading factories like the one at Chicago meant that Native clients who purchased goods on credit would now owe their debts to the U.S. government, rather than independent merchants. Government agents could then leverage these debts to negotiate future land sales and extinguish Indigenous claims to territory.[72] On various levels, the government motivations to build a trading factory at Chicago revealed an overall design to impose a greater level of state control over the portage and its peoples. As U.S. officials worked to bring order and legibility to the fluid space of Chicago, residents—both Native and non-Native—would begin to express their unease.

Spatial Realities on a Fluid Frontier

Even as the built environment at the mouth of the Chicago River demonstrated a growing U.S presence along the portage route, actual control of the area remained contested. John Kinzie had moved to Chicago's portage shortly after the American garrison arrived, taking up residence with his family in the old Point de Sable trading post north of the river. Together with a new business partner—his half brother Thomas Forsyth—Kinzie worked to expand his own trade connections in the area. Kinzie and Forsyth's citizenship, and loyalties, remained in question, and they used their amorphous status to broker new commercial ventures with British subjects from Canada. With Kinzie as his local associate, Richard Patterson, a British trader from Detroit, began shipping large quantities of liquor to Chicago, selling the spirits to the local Potawatomis through proxies like Kinzie, right under the walls of the American garrison.[73] While Kinzie and Forsyth encouraged this inflow of foreign contraband, Burnett became embroiled in disputes with Patterson over control of the trade in lower Lake Michigan. To Burnett, it seemed the U.S. official presence near the portage did nothing to dissuade such foreign competitors from continuing to infiltrate Chicago's waterways. While Whistler had successfully established a garrison along the crucial portage route, his main objective to curb the waterborne movement remained stymied by the myriad wetlands and waterways of the area.

Local traders and their Anishinaabe wives had initially shown a certain enthusiasm for the prospects of an American garrison at Chicago as they hoped to cash in on monopoly trading rights in the area. The hope had been that if the United States could not be dissuaded from building a fort at Chicago, at least the garrison would impose new trade restrictions on foreign traders and cut down on local competition for merchants like LaLime, Kinzie, and William Burnett. However, even before Captain Whistler's command had completed building the palisade, Burnett had grown distrustful of U.S. state capacity to control foreign trade in the region. Writing to William Henry Harrison, the territorial governor of Indiana, in the fall of 1803, Burnett criticized the inability of U.S. officials to control foreign commerce and competition along the region's waterways. Such negligence, according to Burnett, had left "a large field laid open to the enemies of our country," allowing an influx of British traders, like Patterson, into the region. Burnett railed against the U.S. approach to the Indian trade, which imposed new strictures on Americans while simultaneously doing nothing to limit the activities of foreign traders. Mounting into hyperbole, Burnett speculated that if the Barbary pirates—whom the U.S. Navy was waging war against at the time—suddenly arrived along the shores of the Great Lakes, he had "not the least doubt" that they could easily gain entry to "winter amongst the Indians" and trade within U.S. territory. He became so incensed over the situation that he resigned his commission as the region's justice of the peace.[74]

Rather than enforcing strict control over U.S. territory and waterways, Whistler's garrison remained preoccupied with mere survival. Whistler reported on a host of difficulties that he and his troops faced in their new environmental reality at Chicago. The prairies and oak openings were unlike anything the American troops had faced before. Whistler's troops had brought oxen with them to aid in the work of construction, but the prairie grasses proved too tough for them to chew. He likewise found little corn for sale from either the local traders or nearby Anishinaabeg.[75] While they continued in constructing the fort, Whistler complained that his access to good lumber remained scarce because of the local landscape. Surrounded by sparse savannas and open grasslands, the only timber growing in the vicinity of the fort were the exceptionally hard burr and black oaks, which Whistler's whipsaw could not cut without a steady supply of files for resharpening.[76] Indeed, as late as November 1804, Captain Whistler complained to his superior that good wood proved so scarce locally that he had yet to build himself a dining table.[77]

Chicago's surrounding wetlands also posed a challenge for Whistler and his men. Writing during the summer of 1804 Captain Whistler reported a "touch of the Ague & Fever," among the troops, which had incapacitated "as many as ten or twelve" of his men.[78] The garrison suffered intermittently from bouts of malaria, contracted from large swarms of infected mosquitoes that bred in the wetlands around the portage every summer. The following year, Captain Whistler almost lost his own wife to a serious attack of malaria accompanied by chills, aching, and fever.[79] In the coming years, his letters to his superior back in Detroit included frequent complaints of unpleasant diseases at the post, which remained tied to the sluggish waters of Chicago's wetlands and its large population of malarial mosquitoes.[80]

Another pressing challenge for Fort Dearborn's men was the post's geographic remoteness. Chicago's overland distance from other U.S. posts made it especially difficult to support a military fortification there. Poor overland trails connected Chicago's outpost with Fort Wayne and water routes remained the most practical means of mobility, both regionally and locally.[81] The U.S. government, however, maintained no naval forces on the upper Great Lakes. Military supplies had to be sent along in commercial schooners laden with merchandise bound for trading posts at Chicago and elsewhere along the southwestern coast of Lake Michigan. Even still, the distance from Detroit to Chicago by way of the Straits of Mackinac proved a lengthy voyage, and shipping was relegated to the temperate months of the year. Supplies reached Whistler's men only sporadically. Colonel Jacob Kingsbury, Whistler's superior back in Detroit, commended Whistler for successfully building the fort despite having "no clothing for your men, or even the necessary tools to work with."[82] Even mail, newspapers, and official correspondence had to come by way of cooperative fur traders or Indigenous couriers.[83] William Burnett at St. Joseph, for example, confirmed the receipt of a number of letters and newspapers "for the doctor at Chicagou" in January 1804 and agreed to send them along to Fort Dearborn when he could.[84] Ironically, the military officers and government agents at Chicago remained reliant on the fur traders' and Potawatomis' waterborne mobility as their main connection to the outside world. The geographic realities of Chicago further undermined any grand designs for a truly American vision at the portage, even with its garrison, Indian agency, and governmental trading factory in place.

The remoteness of Chicago from other U.S. army posts led Whistler to deploy his troops in creative ways to make up for their chronic lack of supplies. He mobilized the garrison throughout the year in a variety of seasonal tasks, mimicking the mixed economy of the local Anishinaabeg, métis residents, and white traders around the portage. Whistler supplemented infrequent supplies from Fort Wayne or Michilimackinac with local fare, employing both his family and soldiers in the daily maintenance of vegetable gardens as well as extensive cornfields. He also busied the garrison in tending to his personal herds of cattle and hogs, and each spring, mustered out the soldiers to tap trees and render maple sugar in a nearby grove.[85] Fortunately for the American troops stationed at Fort Dearborn, Chicago's local resources remained plentiful and the post's surgeon, John Cooper, recalled that "grouse and game birds were abundant, as were fish." According to Cooper, during much of the year, the fort's officers employed both "gun and rod" to supplement their rations with protein from the nearby prairies and lake.[86] Soldiers under Whistler found themselves engaged in many of the same pursuits as the neighboring Indians and dependent on much of the same local ecology that inhabitants of the Chicago area had relied upon for decades before U.S. encroachment. In the daily subsistence of Fort Dearborn, the realities of Chicago's environment and geographic distance remained far more important than political objectives touted by officials back east.[87]

But the distance from headquarters also came with a lack of oversight that Captain Whistler proved all too willing to exploit. Whistler often employed troops in the care of his private property and business ventures, rather than garrison duties. He also used his position as the post commander to appoint friends and family members to lucrative positions at the fort. Whistler had brought more than ten family members with him to Fort Dearborn after 1804, and as his interests at Chicago grew, so too did his family involvement. Whistler assigned his own son as the fort's sutler, a designated merchant who sold the soldiers any goods that the army did not supply. This conflated the Whistler family's private and military interests and compromised Captain Whistler's impartiality as the post's commander. It also put the Whistlers at odds with local traders like John Kinzie, who viewed the fort's sutler as their main competition in selling liquor and other luxuries to the common soldiers.[88] Whistler further complicated matters when he appointed his close friend John Cooper, the fort's surgeon, as a co-sutler, to share the profits of the position. To the incredulity of the garrison and the resentment of Kinzie, Whistler issued orders forbidding any soldiers from visiting Kinzie's trading post, compelling them to purchase goods solely from the sutler's store. Using his authority as commanding officer to undercut his son's commercial rivals, Whistler drew the outrage of his second-in-command, Lieutenant Seth Thompson, who saw Whistler's actions as a gross "usurpation of power."[89]

Whistler's actions convinced other government agents at Chicago that the commander had grown irrevocably corrupt so far removed from his superiors. In the spring of 1810, Matthew Irwin, the factor at Chicago's government trading house for the Indians, wrote to Colonel Kingsbury, Whistler's superior at Detroit, to report the abuses at Fort Dearborn. Lieutenant Thompson followed up Irwin's letter with his own list of complaints. The charges Irwin and Thompson leveled against Whistler and his coconspirator, Dr. John Cooper, included gambling with the enlisted men, horse racing, having their personal attendants "severely flogged" without explanation, "selling or giving whiskey to the soldiers at different times," falsifying the fort's muster rolls, and, in general, of "conduct unbecoming of an officer & gentleman." According to Lieutenant Thompson's report, Captain Whistler once took a boat out onto Lake Michigan to partake in a drinking spree away from the garrison. When he returned from his whiskey-soaked cruise the next morning, he remained so "disposed by liquor as to render himself ridiculous to the Officers and soldiers."[90] Clearly, discipline had broken down at this western outpost. Instead of bringing order to a chaotic landscape, Whistler's command at Chicago, cut off from the rest of the U.S. Army, had descended into anarchy and infighting.[91]

Whistler's superiors could not ignore such damning charges, and Colonel Kingsbury ordered his recall from Chicago in 1810. Kingsbury also took measures to scatter the junior officers to various posts to alleviate the feuding that had taken hold of the Chicago garrison. Whistler's superiors looked to restaff the garrison with a more competent commander going forward. That May, Kingsbury sent the young Captain Nathan Heald west to Fort Dearborn, hoping the West Point graduate could bring some eastern order to a western reality.[92]

Chicago's remote locality would continue to undermine U.S. efforts for order at the portage, however. Almost immediately after arriving, Heald expressed his dissatisfaction with being assigned to a post "so remote from the civilized part of the world" and requested a furlough for at least the winter months to travel back east. As he complained to Kingsbury, "I am sorry to inform you I am not pleased with my situation and cannot think of spending the winter here."[93] He would not be the first, or last, Chicago resident to express such a sentiment. Captain Heald received his furlough to go east for the winter, and Philip Ostrander took temporary command of Fort Dearborn in September 1810. Ostrander arrived to find the garrison far from orderly, but according to Heald, he had inherited the disorganized state of things from Captain Whistler.

One pressing matter that Ostrander had to deal with almost immediately involved the prospect of white settlement around the portage. Writing to Kingsbury in Detroit for orders on how to address this issue, he reported that "there are at present some discharged soldiers about this place, who wish to settle near the garrison; on consulting Captain Heald upon the subject, we were of opinion that an indulgence of this nature would be improper." For one, Chicago's prairie environment made timber scarce for the garrison as it was. Ostrander reasoned that white settlement at Chicago would only further tax this valuable resource. But more importantly, Ostrander and Heald argued their job at Chicago, as army officers, remained to control the portage route rather than facilitate settlement. As Ostrander put it, "There can be no civil law be put in force without much trouble and expence, and I never yet knew an old discharged soldier become a good citizen."[94] Ostrander's uncertainty in how to deal with these discharged soldiers who wished to remain in the area reflected the ambiguous nature of U.S. military policy toward settlement in the remote Northwest Territory. In Ostrander's and Heald's view, their assignment at Chicago did not include the implementation of a settled society. They remained an occupying force along a strategic route of movement on a distant frontier, rather than the vanguard of a settled colony.

Against Ostrander's wishes, at least two of these discharged soldiers remained to settle near the portage, as evinced in a petition written by the retired James Leigh to Colonel Kingsbury in the spring of 1811. Leigh requested permission to "make a small improvement" near the western edge of the tract of land belonging to the U.S. government, "near to the Portage distant from the garrison about 3 miles." Before his removal from Chicago, Whistler had promised Leigh the opportunity to "reside on the reserve" after retirement with his wife and father-in-law (another discharged soldier), and Leigh asked Kingsbury to uphold the promise. Leigh had already "built some small house thereon" and cleared about ten acres of land for his "large stock of cattle."[95] Leigh was not the first outsider to seek to settle along the portage in an effort to make a profit, nor was he the first non-Native inhabitant to appreciate the potential of Chicago's local landscape, but his approach to taking up residence in the area raised new concerns for both the U.S. officers and area Potawatomis at Chicago. Unlike other itinerant traders operating along the portage, Leigh came as a settler, meaning to claim the land through his own labors and "improvements." Whereas Point de Sable, Ouillemette, LaLime, and John Kinzie had established themselves at Chicago through kinship and commercial ties among the Anishinaabeg, Leigh brokered no relations with the area Potawatomis. Instead, he sought, and acquired, permission to settle at the portage from distant U.S. officials, against the better judgment of Fort Dearborn's army officers. Leigh's occupation of the land adjacent to the portage signaled a departure from the initial intentions of the American garrison at Chicago and marked a radical break from earlier patterns of settlement and livelihood along the portage.

A Local Insurgence

The encroachment of Americans onto their lands, both locally and regionally, drew the ire of the southwestern Anishinaabeg. By the summer of 1811, Jean LaLime wrote to Fort Wayne about a growing number of "depredations and murders," carried out by Chicago-area Anishinaabeg against U.S. settlements downriver. Main Poc's brother-in-law from the Kankakee had led a group south to steal several horses. In the process, the group had killed a man and captured a woman. Later, this same war party killed an entire family. According to LaLime, their violence lay rooted in renewed concerns among the region's Anishinaabeg that "the Americans were settling on their lands." Main Poc and many of the other Anishinaabeg had become convinced that to prevent this, "they must drive them off, kill their cattle, and steal their horses."[96] Growing desperate in the face of ongoing settler encroachments downriver and continued efforts by U.S. officials in Indiana to secure even more Indian lands, Main Poc's followers on the Kankakee increasingly turned to violent means.

Reporting to the secretary of war just two days after LaLime's sobering letter, however, Captain Heald at Fort Dearborn revealed his own naivety toward the growing resentment of the Anishinaabeg. Unaware, Heald relayed that "the Indians in the neighborhood of this post . . . appear to be perfectly friendly & well disposed towards the United States."[97] The garrison at Chicago, meant to curb hostile war parties in the region, seemingly proved a weak deterrent to Main Poc's raids as the fort's commander remained oblivious to growing tensions in the area.

The next winter, as Anishinaabe warriors from the Chicago area continued to raid south along the Illinois and Wabash Rivers, LaLime shared more reports that "war is the common talk of the Potawatomis," and he worried that "they are forming plans to attack this garrison."[98] The flow of Indigenous traffic through the portage only heightened such possibilities, as rumors circulated that hostile Ho-Chunk and Odawa warriors from Wisconsin moved through the area, easily navigating the space despite the U.S. garrison nearby.[99]

Heald and the Dearborn garrison awoke from their stupor only once violence erupted at the local level. The flashpoint of the violence, unsurprisingly, became Leigh's farm, adjacent to the portage itself and the only site of white settlement in the area. On April 6, 1812, Jacob Leigh, the son of James Leigh, and a soldier named John Kelso, were at work in the fields with two other men when eleven Native warriors approached them "armed with guns, tomahawks, & knives." While Leigh and Kelso fled downriver toward the fort, the attackers shot, knifed, and scalped the other two white laborers. Leigh and Kelso raised the alarm with the garrison, and Heald quickly gathered the few trading families from the north shore of the river into the fort for protection, simultaneously firing a cannon to alert the soldiers on leave outside the post of the danger. Seven soldiers who had rowed upriver earlier in the day to fish fled back toward the fort but stopped at the Leigh farm along the way to find the remains of the two men "shockingly butchered."[100] While the perpetrators had fled the scene, Heald realized the precarious nature of the Americans' security beyond the fort walls and ordered all further interactions with the local Anishinaabeg halted until further notice. The escalating tensions around the portage had finally spilled into open bloodshed at Chicago between the area's Native peoples and the U.S. garrison there. That the Indigenous assailants chose Leigh's farm at the head of the portage as their target for the attack was no coincidence. This was no random act of violence. Rather, the Indigenous assault on Leigh's place demonstrated their growing awareness, and frustration, over American ambitions to transform Chicago's portage and surrounding landscape into a place controlled by the U.S. military and white settlers rather than an open route of movement under the continued authority of the region's Native peoples.

The area's Native people were not the only ones who continued to resent the growing U.S. presence along the portage. Tensions remained high among local traders as well as the Anishinaabeg. Though Whistler's monopoly over the garrison's trade had been broken with his recall in 1810, old resentment stirred, and new rivalries formed between John Kinzie and various government agents at Chicago. John Kinzie became particularly embittered toward Jean LaLime, his former associate. The two had worked closely as fellow fur trade employees of William Burnett when they first came to Chicago's portage around the turn of the century. LaLime and Kinzie had cooperated to purchase the original Point de Sable trading post along the river. But in the years since the two men had come to Chicago, they had taken divergent paths. Kinzie continued to operate as a trader, with commercial relations among the local Anishinaabeg as well as British Canada. By 1812, LaLime was working as the United States' interpreter at Chicago. Given LaLime's official capacities at the fort, Indian agency, and government trading factory, he had grown close with Matthew Irwin and several other U.S. officers who opposed Kinzie's activities around Chicago.

Kinzie, like the other European and métis traders and local Anishinaabeg, remained committed to a certain vision of Chicago as a fluid space of movement, centered around the portage and unimpeded by official U.S. sanctions. In the wake of Heald's no-contact order, Kinzie continued to trade with the area Potawatomis who visited his trading post north of the river. He and his partner, Thomas Forsyth, had also continued, through familial connections to British Canada, to import foreign goods to the portage despite protests from Matthew Irwin, the government factor. Kinzie also challenged U.S. officials like Irwin for meddling in local affairs. Irwin continued to meddle, writing many letters of complaint to his father, the governor of Pennsylvania, and his political connections in Washington. Irwin and his allies accused Kinzie and Forsyth of being "pretended Americans" and "dangerous characters" who maintained connections to British Canada by way of the water routes north. On several occasions, Irwin reported Kinzie smuggling in goods from Canada, and when Captain Heald failed to punish him, Irwin, through letters to the secretary of war's office, questioned Heald's ability to command.[101] Irwin and his allied American officials condemned Kinzie, Forsyth, and other local traders for the fluid nature of their trade and kinship networks, which harnessed the portage to maintain linkages to British Canada and the local Indigenous communities.

The feud between Kinzie and the government agents came to a head in June 1812, only two months after the attack at Leigh's farm. Jean LaLime had been openly "defending the pure motives and just dealings" of Matthew Irwin and the U.S. trading factory at Chicago, enraging Kinzie.[102] Kinzie sharpened a butcher knife and crossed the river to the fort to challenge LaLime to a duel. LaLime may have drawn a pistol, according to some reports, but Kinzie rushed him and struck him down with a blow to the chest. The two "fell down together" and Kinzie stabbed LaLime "to the heart," killing him.[103] In the aftermath of the fight, Kinzie fled to Milwaukee, fearing reprisals from the Americans at the post. Irwin also fled Chicago, fearing that his life might be next. Kinzie returned to Chicago after several weeks of hiding out, where Captain Heald exonerated him from all charges.[104] The young Heald, uncertain of how to navigate the disorienting local politics of Chicago, deferred to the local influence of John Kinzie, whose ties with both Anishinaabeg and Europeans at the portage made him a valuable asset to the fort's commander, especially in the absence of the Americans' now-dead interpreter. Despite Heald's exoneration, Kinzie's attack on LaLime demonstrated that the struggle at Chicago broke down not along lines of racial distinction but rather between those who actively promoted a state-designed approach to controlling the local landscape and those who remained determined to maintain an older model of interpersonal and geographic fluidity at the portage.

The outbreak of war between Great Britain and the United States in the same month as LaLime's murder precipitated an even bloodier clash between Potawatomi interests and the growing American presence at the portage. Main Poc had led a group of overtly hostile Potawatomis from the Kankakee to join up with other Indigenous warriors inspired by Tecumseh and Tenskwatawa's Nativist movement, heading east to ally themselves with British forces at Malden in Canada. Robert Dickson, the British Indian agent for the western Great Lakes, meanwhile began gathering support from Ho-Chunks and Dakotas to the north and west of Chicago, distributing gifts of blankets, liquor, guns, and ammunition to garner favor with Indigenous leaders.[105] Even more troubling, John Kinzie claimed that after going into hiding following LaLime's murder, Potawatomi friends of his had warned him on July 10 "that the Indians were hostile inclined and only waiting the Declaration of War to commence open hostilities."[106] Whether Heald and the U.S. garrison could trust such information from a man who had only a month earlier killed their interpreter remained up for conjecture, but nevertheless the rumors of open hostility proved concerning for the American troops. All this activity, so near to Chicago and involving local Indians, made the U.S. presence at Fort Dearborn all the more tenuous by the summer of 1812.

Within the first month of hostility, the British and Indians confirmed the security fears that U.S. officials had harbored since the 1790s about the region's waterways. On the early morning of July 17, a party of British, Indians, and independent voyageurs crossed the Straits of Mackinac in canoes to capture the American Fort Mackinac in northern Michigan.[107] British reports confirmed that without "the co-operation of the Fur Traders, who were fortunately there at the time, with a considerable number of canoe-men in their service" the bold maneuver would have been impossible. More troubling still to Americans, the "blow was effected by persons engaged in the trade to the Mississippi and other districts beyond Michilimackinac"; the very traders who had been passing through the Chicago portage for years.[108] U.S. officials learned that the orders for Robert Dickson to gather Indigenous allies for the assault had passed directly through the Chicago portage with a party of Ojibwes earlier that year.[109] The mobile people of the lakes and the fluid geography of the region now openly threatened the security of the entire U.S. frontier.

For General William Hull at Detroit, this underscored the vulnerability of Chicago as well. Without a waterborne supply line through the Straits of Mackinac, the United States could not hope to support a garrison so far afield. On July 29, General Hull wrote to John Armstrong, the secretary of war, that he would "immediately send an express to Fort Dearborn with orders to evacuate that post and retreat to this place or Fort Wayne."[110] Captain Heald received the orders to evacuate on August 3, and several days later William Wells, the Indian agent from Fort Wayne, arrived with an escort of U.S.-allied Miamis to assist Heald in negotiating a withdrawal from the hazardous borderland.

A Potawatomi, Winamac, had delivered Hull's orders to Heald, and news of the pending evacuation spread quickly through the Anishinaabe towns of Chicago and the Illinois River. Within days of the order's delivery, crowds of Potawatomis, Ojibwes, Odawas, Sauks, Ho-Chunks, and Kickapoos arrived to watch the abandonment of a fort that had tried to impose a new order along their ancient portage. These onlookers gathered on the prairies surrounding the fort, traveling from as far as Milwaukee and Tippecanoe in hopes of acquiring the large surplus of goods that the U.S. garrison would leave behind. Such "unusually large" numbers of Native peoples worried Heald and his advisers, and both Kinzie and Wells cautioned Heald against the dangers of an overland march, given the disposition of the Anishinaabe leadership.[111] The safety of the retreat became even more questionable when Black Partridge, one of the local Potawatomi headmen, turned in his peace medal to Captain Heald on August 13 as a warning. Handing over this symbolic token of alliance with the American government did not bode well.[112] While Wells and Kinzie worked on the garrison's behalf to secure a guarantee of safe passage from the six hundred to seven hundred Native people gathering outside the fort, Heald busied his men with destroying excess gunpowder and liquor before the final withdrawal.

The American garrison evacuated the fort on the morning of August 15 in a spirit of apprehension. John Kinzie, placing his family in a boat along the river for safety, chose to march with the retreating force hoping his presence and personal connections among the Potawatomis might allay any violent encounters.[113] Meanwhile, William Wells, who had grown up among the Miamis of Kekionga, embraced an Indigenous custom and painted his face black in preparation for what he worried would be a doomed retreat.[114] The column of sixty-eight soldiers and thirty Miami scouts, accompanied by supply wagons, women, and children, made it almost two miles along the lakeshore before a large contingent of Potawatomis appeared on the sand dunes above, led by Black Bird and Nescotnemeg. Realizing the hostile intentions of the assembled Anishinaabeg, Heald ordered his men to advance up the dunes when the Indigenous combatants opened fire. According to Kinzie, the Potawatomis made "good use of their rifles at long distances," and "poured so deadly a fire into them," that nearly half the Americans fell in the first few minutes of fighting.[115] Following up their withering volleys, the force of several hundred Native warriors then rushed down the dunes, overwhelming Heald's small command. Sergeant Walter Jordan had accompanied William Wells and the Miami scouts from Fort Wayne. He remembered the day Heald's garrison left Fort Dearborn as "the most Limentable Day I ever saw." When the fighting began, Jordan witnessed Wells fall and watched as a warrior named White Raccoon killed Wells and mutilated his body.[116] Heald and his second-in-command, Lieutenant Linai Helm both received multiple gunshot wounds but reformed their remaining men around a small hillock. From there, they watched as the Potawatomis attacked the unprotected baggage train, killing or capturing the women and children and looting the wagons. The Indigenous warriors had routed the Americans, and Black Bird advanced to demand the remaining troops surrender. Heald agreed, the soldiers laid down their weapons, and the Potawatomis led them back toward Chicago.[117] The American captives returned to the Indigenous encampments in time to watch Black Bird and the other leaders set fire to Fort Dearborn, the agency house, and the U.S. trade factory. With a sweeping victory at hand, the Anishinaabeg made short work of wiping away any physical trace of U.S. control along the banks of the Chicago.

Anishinaabewaki Restored

In the days following the violence, it is striking to note the sudden return to normalcy around the portage. The Potawatomis divided the survivors, such as Linai Helm and Walter Jordan, as captives among the participating warriors and took them back to Anishinaabe towns along the Illinois, Des Plaines, and Kankakee Rivers the morning after the engagement.[118] John Kinzie had survived the battle unscathed, and in the days following, he reported that Black Bird showed him and his family "kindness"—neither taking them as captives nor threatening them further.[119] The Native combatants left Kinzie's trading post standing on the north bank of the river, along with several other cabins belonging to the métis and French habitants.

Incredibly, in the first few days following the battle, John Kinzie actually resumed business at his post, with transactions appearing in his account book from a number of regional French and métis traders.[120] His business partner, Thomas Forsyth, arrived at the Chicago portage from Peoria on August 16, coming to assess the damage and assist Kinzie in further negotiation with the area Potawatomis.[121] In the wake of the battle, the area's traders carried on in safety. Individuals like Kinzie and Forsyth, and the French and métis families living along the Chicago River, had not threatened the Potawatomis' hegemony over the portage route, nor curbed their waterborne movement through it. They were not the target of the attack. For the Anishinaabeg of the area, the eruption of bloodshed at Chicago had not been a clash between whites and Indians but rather a concentrated act of violence against the U.S. government and its imposition of authority along the portage's local geography.

Downriver, American settlers and territorial authorities in the Illinois Country worried the portage, now unquestionably under Potawatomi control, would serve as the portal for British and Indian attacks against Illinois, Indiana, and even the Missouri Territory. Governor Ninian Edwards of Illinois ordered militia units to patrol the Illinois River in gunboats, and as far away as St. Louis, the Americans began to throw up defenses. But Chicago's Potawatomis had little interest in carrying on a larger conflict on behalf of the British or eastern Indians. With Fort Dearborn destroyed and its garrison scattered, many of the area Potawatomis returned to their usual pursuits of farming, hunting, and fishing, rather than mobilizing for war.[122] The actions of the Chicago Potawatomis in the months after the battle reveal the limited nature of the conflict. Their belligerence remained rooted in the very localized struggle to control the portage and its surrounding geography, rather than the geopolitics surrounding the War of 1812 or Nativist movements to the east. Content in their hegemony over the portage after the fall of Fort Dearborn, the Potawatomis of Chicago rode out the rest of the war in relative peace.

The United States' first attempt to conquer Chicago's portage had failed in a spate of bloodshed and burning. Believing they could bring order to a waterborne region through the control of limited, strategic geography, American leaders had seen to the building of Fort Dearborn. The garrison's presence had led U.S. officials into a headlong clash with Chicago's Anishinaabeg and resident traders in the years leading up to 1812. But the United States would prove relentless, seeking again to bring the portage under its control in the years following the War of 1812. These future efforts would differ significantly from this initial attempt to fortify the portage along the logic of their European predecessors. Americans would return to the area hoping not only to control the movement of the regions' peoples but to overcome the land and water of Chicago itself. These renewed attempts to bring Chicago under U.S. control would signal more struggles and violence going forward. When the United States returned to Chicago after the war ended, it intended a total conquest of both the people and the nature of the portage.

CHAPTER FIVE

Floodgates

Rendering Indigenous Space into an American Place

In July 1815, the fur trader John Kinzie wrote to Lewis Cass to petition the U.S. government to return to Chicago. Cass served as the governor of the vast Michigan Territory, which at the time covered the Michigan peninsulas as well as the Wisconsin shoreline and what would become northern Illinois. The War of 1812 had finally drawn to a close despite continued trepidation of British and Native alliances and raids along the United States' western frontier. Now Kinzie requested the governor to revisit the prospects of an American presence along Chicago's portage. According to Kinzie, the method to "establish peaceable and friendly intercourse" with all the Indigenous nations spanning from Michilimackinac to the Illinois territory was to concentrate American efforts around Chicago once again. Kinzie, well versed in the actual Indigenous economy of the region, explained that although Native peoples like the Anishinaabeg and Ho-Chunks seemed dispersed throughout the Lake Michigan basin, they were all "compelled to emigrate at certain seasons to the waters of Chicago, Illenois, and Fox Rivers" in order to secure "a proper supply of game."[1] Kinzie was right that the Chicago portage remained a necessary node of Indigenous seasonal migration patterns. Chicago's waterscape served as an intersection for Anishinaabeg mobility, facilitating hunting expeditions on the Illinois prairie; winter fur trapping across the wetlands of Chicago, Calumet, and the Kankakee, and travels to do business with trading firms farther afield, down the Illinois River and in St. Louis.

It seems unlikely that Lewis Cass had much interest in the motives behind Indigenous subsistence patterns, but Kinzie hoped his suggestion would still strike a chord with Cass, seeing that such a thoroughfare could bolster U.S. control throughout the region. Kinzie asserted that "Chicago is all important to the Illenois Country as it is the key of communication and has command of the trade over a vast Territory," and he suggested that a fort be rebuilt at the place of old Fort Dearborn. A military post at the site was not only necessary for the control of the Native population, however. Kinzie called for a garrison at Chicago "to protect trade" and hoped that "keeping such an armed force" at Chicago "will awe medling intruders into silence."[2] These meddling intruders were the British fur traders and royal agents who had supposedly incited

Native violence in the recent war and had worried U.S. policy makers along their western border for decades. Kinzie crafted his letter to address the borderlands concerns that officials like Cass had obsessed over in the past, using the logic of a political economy that revolved around the fur trade and prioritized securing U.S. borders in the Great Lakes region.

But the vision that Kinzie laid out in 1815 was already an outdated one, at least in the minds of U.S. politicians like Lewis Cass. Officials had tried to bring their vision of order to the west by fortifying the portage at Chicago. Their efforts had failed spectacularly in the disintegration of an orderly garrison, the outbreak of war, and the culminating violence of the U.S. defeat at the Battle of Fort Dearborn in 1812. After the war, officials like Cass in Michigan, William Clark in St. Louis, and Ninian Edwards, the territorial governor of Illinois, would begin to formulate a new strategy for order in the west—one that relied less on a frontier of fortifications and military expenditures. After 1815, western officials would find new energy, and new financial backing, from state and federal policies of "internal improvements" that aimed to overhaul, or in many cases create new, networks of mobility and infrastructure. At the national level, this impulse aimed to unite the country into one coherent network of transportation, agriculture, and commerce, eliminating sectional divides and strengthening the overall political economy of the union. But in the Great Lakes region, the fervor for infrastructure became wrapped up in the same official impulses to bring control to a disorienting borderland in which the economy still revolved around Indigenous transportation systems and Native power. In the years after the War of 1812, American officials would look to public works and internal improvements as a new solution to thwart this old balance of power that favored intercultural exchange and Indigenous authority at sites like Chicago.

A New Type of Conquest

That old geographically informed balance of power had blunted U.S. authority at Chicago before the war and led to a disastrous American defeat at Fort Dearborn during the conflict. And in the interim, it meant that any preliminary U.S. designs to reestablish a foothold at Chicago after the war would require Anishinaabe consent. After the burning of Fort Dearborn and the scattering of their American prisoners in August 1812, the majority of the area's Anishinaabeg rode out the War of 1812 in peace. Downriver, the Illinois River Potawatomis at Peoria began renewing diplomatic ties with the Americans that winter to negotiate some sort of reconciliation in the west.

Indeed, when U.S. officials returned to Chicago following the war, they did so only under the auspices of rekindled alliance with the area Anishinaabeg. Already by the winter of 1814, before the war had officially concluded, U.S. Army officers received reports that "at the head of Lake Michigan all is quiet" and "the Potawatimies are all tired of War and say they will fight the Americans no more."[3] Indeed, the Illinois River Potawatomis had begun actively pursuing peace with the Americans. Black Partridge, a Potawatomi leader speaking on behalf of the towns around Peoria, assured William Clark in St. Louis that "your young man can pass safe up and down our rivers."[4] While the Anishinaabeg had asserted their claim to their rivers with their targeted violence against Fort Dearborn in 1812, by 1814, they had grown willing to entertain a new alliance with the Americans, one on equal footing and with a clear recognition of Anishinaabe sovereignty over their waterways.[5]

The U.S. government never won a military victory against the Chicago Anishinaabeg. When the Americans did return to Chicago, it was not as conquerors but rather through the negotiated consent of the region's Native people. But the U.S. return to Chicago would bring a new, more insidious, approach to the American occupation of the Great Lakes region. The period following the bloodshed of the War of 1812 saw no outright conflict between the incoming Americans and Chicago's Anishinaabeg. Rather, a period of government investment into internal improvements defined the renewed efforts of U.S. officials to conquer Chicago's perplexing space. Unexpectedly, this process would prove far more effective than the prior attempts at colonization carried out by generations of French, British, and American predecessors.

In the end, Chicago did not transform into a Euro-American place because of a militant occupation nor a settler invasion but by an overhauling of its geography through a series of government-backed "improvements" to the land and waters around the portage. Between 1816 and 1836, American officials pursued a series of public works projects at Chicago, meant to tame the marshland southwest of Lake Michigan and transform the upper Illinois Country into a productive landscape that would serve the U.S. republic while directly undercutting the power of Chicago's Anishinaabeg and local Euro-American traders.

Though American politicians never succeeded in gaining full sanction for a coherent federal policy of funding public works projects, the piecemeal approach of local, state, and national support for infrastructure that unfolded at Chicago proved enough to secure the midcontinent for the settler republic.[6] In 1816, a military detachment returned to Chicago and rebuilt Fort Dearborn, the agency house, and a government trading factory. That same year, a delegation of American officials led by William Clark negotiated for a twenty-mile-wide corridor of land between Lake Michigan and the Illinois River for the future construction of a proposed canal. Statehood and land surveys along the upper Illinois River followed in 1818. By 1822, Congress granted the ceded lands between Lake Michigan and the Illinois River for a federally endorsed canal. The following year, the Illinois state legislature created a Canal Commission to oversee this government works project. In 1828 and again in 1831, Congress earmarked funding for harbor improvements to make Chicago a viable shipping port for vessels larger than canoes for the first time. The state of Illinois funded the construction of the first bridges across Chicago's muddy river in 1832. The Chicago-Vincennes Road, assayed that same year by state surveyors, would connect the frontier outpost to American settlements in southern Indiana over 250 miles away. In 1833, government contractors and army engineers had begun construction of both the harbor and a set of piers. By 1834, the speculation around these improvements and the promised canal caused a land rush that saw Chicago's settler population grow from around 100 people to 3,000 in the span of one summer. By 1836, only twenty years after U.S. officials returned to Chicago, government investment rendered the ancient portage route irrelevant by breaking ground on the Illinois and Michigan Canal. This infrastructure project promised to unite the waters of Lake Michigan with the Illinois River, and ultimately the Mississippi, in a rebuilt system of artificial waterways designed to serve a new settler economy in the region.[7] All these so-called improvements served U.S. interests while stripping the southwestern Anishinaabeg of their power, which since the 1740s had remained grounded in their control of movement through Chicago's amphibious geography.

This transformation of Chicago's local environs significantly reshaped the balance of power between the area's Anishinaabeg and Euro-American traders, tipping the scales in favor of incoming government officials and, eventually, eastern land speculators, businessmen, and migrants. Before, the dynamics of power at Chicago had proven a careful game of negotiating movement through the portage and navigating the natural geography of Chicago's marshes, waterways, and low-lying prairies. But after the return of a U.S. presence to the portage in 1816, American officials sought to change the rules of that game. They did so not by directly confronting the power of Chicago's Anishinaabeg but by switching out the board on which the game was played. By overhauling the local geography and, in turn, the regional patterns of movement, American officials sought to bring a more legible, *winnable* topography to Chicago—one that officials could understand, and control, better than the older Indigenous portage.[8]

By viewing Chicago's transformation in this way—as a two-pronged conquest of both the local people and the local space—we can better understand U.S. efforts, and eventual successes, in subjugating the continental interior. Many historians have written on the government-backed era of internal improvements and building projects that created infrastructure, bolstered capital, and propelled change in both transportation and commerce in the interior. These scholars focus on the "transportation revolutions" and "market revolutions" that reshaped the political economy of the continental interior into an interconnected nation, benefiting expanding agricultural settlers and facilitating the growth of interior cities. These state-backed efforts to improve infrastructure, facilitate agricultural settlement, and increase market productivity transformed Chicago as they transformed many places across the Great Lakes and the Mississippi valley.[9] Other historians have highlighted the devastating process of Indigenous dispossession and the land cession treaties that eventually led to Indian Removal in the 1830s. Looking at how white migrants to the interior demanded Native lands and divested Indigenous peoples from their sovereignty and land ownership, these narratives demonstrate how racialized policies benefited settler colonists at the expense of Indigenous peoples like Chicago's Anishinaabeg.[10] Very few historians have taken these two processes—internal improvements and Indigenous dispossession—as parts of the same whole: a fundamental alteration of the interior geography, backed by state investment and authority, that served to undercut Indigenous power on the continent and eventually transformed spaces like Chicago's portage from a site of Native hegemony to a place of American authority and settler prosperity.[11] At Chicago, and elsewhere in the Old Northwest, historians must consider the implications of Native dispossession and internal improvements as mutually reinforcing strategies that shifted the balance of power in favor of the United States.

The American Return to Chicago

Initially, the Anishinaabeg and local traders both had some reason for relative optimism about the new U.S. post at Chicago. Captain Hezekiah Bradley returned to the site of the destroyed Fort Dearborn with a complement of soldiers and began rebuilding the garrison in July 1816.[12] John Kinzie, longtime trader along the portage, had requested the return of a garrison there in part to drive up business. Soldiers with government pay meant more customers for Kinzie's trading post across the river and more ready cash on hand.[13] Chicago's Anishinaabeg, as well, saw some potential advantages to their new arrangement with the Americans. In their minds, they had clearly established their upper hand when they routed the garrison and burned the fort years earlier.[14] But with a new post at Chicago, they could expect subsidized goods from the U.S. trading factory and the ear of the local Indian agent, Charles Jouett.[15] Within a year of the official U.S. return to Chicago, the local Anishinaabeg had leveraged Jouett to relocate a portion of the annuity payments for the Potawatomis to Chicago, further benefiting from a local government presence near the portage.[16] In all, the Americans reestablished Fort Dearborn and their presence along the portage with little conflict in 1816, the local traders and Anishinaabeg of Chicago seizing the development to gain further advantages at the site. Even as U.S. officials returned to Chicago under the good graces of resident traders and area Anishinaabeg, local administrators warned that the original source of tension between the Illinois Country Anishinaabeg and the American government persisted. U.S. official impulses to claim and control the landscape in a legible fashion threatened to ignite further conflict on the upper reaches of the Illinois River.

U.S. officials' interest in surveying and securing lands along the headwaters of the Illinois River had intensified as ideas for a canal connecting the waters of Lake Michigan and the Illinois River circulated in the nation's capital in the years following the War of 1812. Albert Gallatin had initially proposed the plan for such a public works project in 1808 as secretary of the treasury. After peace came to the frontier, promoters of the Illinois Territory, and internal improvements more generally, took up the cause to connect the waterways of Lake Michigan and the Illinois River through Chicago's local geography. As early as 1814, the Baltimore news magazine, *Nile's Weekly Register*, crowed about the potential for such a canal through Chicago, which would in effect unite New York's western shipping with that of New Orleans "by inland navigation." In the American public's view, such a route seemed like "a stupendous idea," that promised to dwarf any other artificial waterway in the world in importance.[17] For the U.S. government then, surveying the lands between Lake Michigan and the Illinois River had become vital, given new enthusiasm at the state and federal levels to pursue a canal project through Chicago's ancient portage.[18]

Such efforts to map and mark off a canal route between the watersheds threatened to spark yet another conflict between the United States and area Anishinaabeg. Indeed, while the Anishinaabeg had expressed their desire for peace with the United States after the conclusion of the War of 1812, they remained violently opposed to any survey efforts along the upper Illinois River. Captain Whistler, now commanding the post at Fort Wayne, relayed news in

May 1815 that the "Indians to the northwest ... will not suffer the lands in the Territories of Michigan Indiana & Illinois to be surveyed or settled." According to Whistler, these Anishinaabeg contemplated renewing the war if surveyors threatened the region.[19] Such fears prompted William Clark to suggest the Indian Department equip mounted infantry as a guard for the surveyors, since "the Indians on the Illinois have been for several months discontented and restless ... much opposed to the surrender of that country."[20] In the fall of 1815, George Graham of the War Department wrote to Brigadier General Thomas Smith to request a military guard for the party of surveyors at work in northern Illinois, due to notable "symptoms of dissatisfaction" that "had been manifested by the Potawatamies, in consequence of the surveys that were ordered."[21] That winter, the trader Thomas Forsyth spent time with eight lodges of Kickapoos and Potawatomis north of Peoria and likewise reported that "they appeared to be much alarmed about surveying their lands."[22] These Native peoples understood such surveys as the precursor to future land cessions and a threat by the U.S. government to alter the geographic balance of power.

In 1816, Indian agents relayed ongoing "reports of contemplated hostility by different parties of Indians." Native warriors directed their animosity specifically toward the efforts of the land surveyors. One agent, Benjamin Parke, shared the contested process of surveying the lands between the Illinois and Mississippi Rivers, reporting that "many of the surveyors marks, in that tract, have been defaced and the likeness of Indian warriors a tomhawk in one hand and a spear in the other, substituted—an intimation that cannot be mistaken." Parke blamed the vandalism on Potawatomis and Kickapoos and added somberly that "the surveying and settlement of the same lands" may provoke the region's Native peoples into open war.[23] Indigenous groups like the Anishinaabeg, still the dominant force along the upper reaches of the Illinois River and Chicago's portage, understood surveying as a clear tool of the settler republic to map, claim, and control terrain, and expressed strong concerns against such efforts in their homelands.[24]

Throughout 1815 and 1816, officials in the Indian Department laid plans for a treaty with Chicago's Anishinaabeg at Portage des Sioux, just north of St. Louis, to prevent the threat of open war. The treaty negotiations of 1816 would prove particularly delicate, since the Meskwakis and Sauks had also claimed some of these same lands north of the Illinois River and yielded them to the U.S. government by an earlier 1804 treaty at St. Louis.[25] The treaty commissioners worried that without a separate cession from the Illinois Country Anishinaabeg, the surveyors appointed to map the region for military bounty lands and the canal corridor "will meet with some serious opposition."[26]

In August 1816, Ninian Edwards, governor of Illinois, William Clark, territorial governor of Missouri, and Auguste Chouteau, a prominent St. Louis fur trader, acted as hosts to "the chiefs and warriors of the united tribes of Ottawas, Chippewas, and Potawatomies, residing on the Illinois and Melwakee Rivers, and their waters, and on the southwestern parts of lake Michigan." Journeying from towns southwest of Lake Michigan, the Anishinaabeg gathered to meet the U.S. treaty commissioners at the designated site. But these southwestern Anishinaabeg, suffering from low quantities of food, munitions, and other supplies since the war, found themselves in a difficult position to bargain with the Americans.[27] The government treaty commissioners used this to their advantage. Ninian Edwards, recognizing their current poverty and knowing the history of prosperity among the Chicago Anishinaabeg, used the proposed canal to appeal to their sense of commerce. He argued that a modest land cession, of only twenty miles wide along the proposed canal route between Lake Michigan and the Illinois River, would increase the flows of trade through their lands. The Chicago-area Anishinaabeg would be enriched once again, this time by a canal route rather than a portage running through their territory.[28]

When this line of argument alone failed to yield results, a promise of a $1,000 annuity payment for the next twelve years, delivered directly to the Anishinaabeg at Chicago and the Illinois River, convinced the poorly supplied Anishinaabeg to accept the U.S. proposal. On August 24, 1816, the southwestern Anishinaabeg signed away their hold over Chicago's portage "to a point ten miles north of the west end of the Portage, between Chicago creek, which empties into Lake Michigan, and the river Depleines, a fork of the Illinois; thence, in a direct line, to a point on Lake Michigan, ten miles northward of the mouth of Chicago creek; thence, along the lake, to a point ten miles southward of the mouth of the said Chicago creek." In all, the treaty did not signify an exceptionally large land transfer compared to other treaties of the day. The Chicago-area Anishinaabeg had managed to retain hunting and fishing rights in the relinquished lands, and the U.S. commissioners guaranteed Indigenous rights to continue living in and moving through the space so long as it remained the property of the government. Nevertheless, the treaty marked a turning point in the balance of power at Chicago. With the United States gaining a specific claim to the ground between the watersheds, the Anishinaabeg had lost direct control over their strategic portage. In the years to come, with the portage secured as federal property, members of the U.S. government, Illinois officials, and local boosters would aim to reshape the terrain of the portage to suit American visions for Chicago and the wider region.[29]

Preparing a Settler State

While the treaty commissioners hurried to placate the Chicago-area Anishi-naabeg and confirm a land cession, U.S. military personnel made plans to scout the upper reaches of the Illinois River with an eye toward the canal project and new fortifications. During the War of 1812, the U.S. military had assembled a select group of topographical engineers to survey terrain and de-sign fortifications. In the aftermath of the conflict, the federal government directed this elite corps into an expanded role, mapping the Northwest Terri-tory and Louisiana Purchase, plotting strategic positions for fortifications, and drawing up proposals for public works projects along the frontier.[30] In 1816, Major Stephen H. Long, one of the army's six topographical engineers, led a trip up the Illinois River to assess the site of Fort Clark, on Lake Peoria, and conduct the first scientific report on Chicago's portage and continental divide. His findings provided encouragement for the wider government proj-ect of transforming the Old Northwest into U.S. space, and specifically high-lighted the value Chicago's geography offered for the American republic.

As for the wider region, Long encouraged a general federal investment in public works across the region, as "no part of the United States can afford greater facilities, or exhibit more powerful inducements for the construction of public canals and national roads."[31] Long described how waterways bi-sected the lands south of the Great Lakes in ways that promised feasible wa-terborne transportation, reckoning that "these may be considered the cords which will unite the northern interests of the country, and will eventually be-come the most important links in the grand chain that surrounds the whole." Of all these rivers, Long highlighted four as the most important: the Ohio, the Mississippi, the Illinois, and the Chicago, "being the channels through which trade and all kinds of intercourse will be kept up."[32] Long was prescrib-ing a grand design of connections in the west, echoing back to suggestions made as early as French explorations under Jolliet and La Salle, which relied on the waterways of the region to hold together an imperial, and in the U.S. case, a federal, system. As with his colonial predecessors, Long saw Chicago as essential to such connections and control.

In his report, Long confirmed the proximity of the watersheds across Chi-cago's low-lying continental divide and provided a blueprint for the feasibil-ity of a canal connecting Lake Michigan with the Mississippi drainage. As Long described it, a "small lake of the prairie" lay between the Des Plains and Chicago Rivers in most seasons—called Mud Lake by locals. Parties of fur trading and Indigenous canoes maintained a channel of water between the two rivers through this muddy expanse during the wetter seasons in order to avoid carrying their vessels across the divide. Long reported how "boats pass and repass with facility between the two rivers." Even in the driest parts of the year, Long estimated "the Portage is seldom more than 3 miles," and the overall elevation across the continental divide never exceeded twelve feet above water level.[33]

This geography, as Long noted, had proven abundantly conducive to canoe travel through Chicago. However, there remained some problems, from his point of view as a military engineer commissioned to analyze the utility of the site for U.S. security and settlement. The mouth of the Chicago River, where it entered Lake Michigan, remained choked with sand. The submerged sandbar provided no more than a couple feet of depth. This was a nonissue for Native-built canoes with their shallow draft but signaled a clear problem for larger American vessels that could run aground in such shallow waters. Long proposed sinking piers on both sides of the river mouth and removing the sand between them, confident that through this alteration, "the river would be rendered a safe and commodious harbor for shipping."[34] Long advised altering the local waterscape in a way that would benefit larger-scale, Euro-American shipping, rather than Indigenous techniques.

Even more ambitious, Long identified "a canal uniting the waters of the Illinois with those of Lake Michigan" as "the first in importance of any in this quarter of the country." Long was confident that the "water course which is already opened between the river Des Planes and Chicago river needs but little more excavation to render it sufficiently capacious for all the purposes of a canal." Long detailed the sluices, embankments, dams, and locks necessary to sufficiently alter the portage between the watersheds into a canal.[35] His proposal laid out plans to renovate the space that had served for centuries as a passage between the waters, transforming it from an Indigenous portage to an American-built canal of engineered control.

This first engineer through Chicago's portage also offered a recommendation on where Chicago's Anishinaabeg factored into such plans. Long's report, subsequently printed for the public in Washington D.C.'s *National Register*, articulated the mutually reinforcing program of Native dispossession and public works that would define federal policy in the Northwest going forward. He expressed confidence that "should it be the desire of the general government to adopt a system of internal improvement . . . no part of the country presents a more favorable opportunity" than the lands south of the Great Lakes. Long clearly understood that this would serve to benefit both the government and white settlers, at the expense of the area's Anishinaabeg. The lands, as Long noted, "are or will be the property of the public. Upon a part of them the Indian claims are already extinguished, and the remainder must sooner or later fall into the hands of the United States."[36] As early as 1817, Long was expressing the two-fold approach that would work to transform Chicago, with Native dispossession and a rewriting of the topography through internal improvements going hand in hand to secure U.S. control of the continental crossroads.

Soon, politicians in the nascent state of Illinois became new and aggressive stakeholders in the negotiation of Anishinaabe lands intended for the canal project. Illinois gained statehood in 1818, and legislators immediately launched efforts to claim Chicago and outline a plan for altering its space. In the bill ratifying Illinois statehood, territorial delegate Nathaniel Pope advocated to push the upper boundary of Illinois north to take in Chicago and the surrounding western shoreline of Lake Michigan, writing that the importance of this measure was "too obvious to every man who looks to the prospective weight and influence of the state of Illinois." Pope's goal was to "gain, for the proposed State, a coast on Lake Michigan" and thus secure "the perpetuity of the Union." The *Intelligencer*, a Washington-based newspaper, reported Pope's plan, writing, "The facility of opening a canal between Lake Michigan and the Illinois River . . . is acknowledged by everyone who has visited the place. Giving to the proposed State the port of Chicago (embraced in the proposed limits), will draw its attention to the opening of the communication between the Illinois river and that place, and the improvement of that harbor." Pope and other Illinois leaders had begun formulating extensive plans for Chicago's local environment as well as how these on-the-ground schemes would connect to their statewide, regional, and even national ambitions for uniting the country in a new geographic network that aimed to replace older, Indigenous forms of mobility through the portage (see figure 5.1).[37]

Clashing Visions for the Future of the Old Northwest

Anishinaabe leaders would have their chance to challenge these official U.S. ambitions for the Northwest when Lewis Cass assembled the united bands of Potawatomis, Odawas, and Ojibwes at Chicago itself in August 1821 for a new land cession treaty. Cass came to the negotiations as the Michigan territorial governor, with the intention to wrest the last lands south of the Grand River, in Michigan, from the Potawatomis and fellow Anishinaabeg. Alexander Wolcott, the Indian agent at Chicago, had convinced Cass that selecting Chicago as the treaty site, distant from the St. Joseph River valley (Cass's prime target in the land cession), might make the gathered Anishinaabeg more willing to

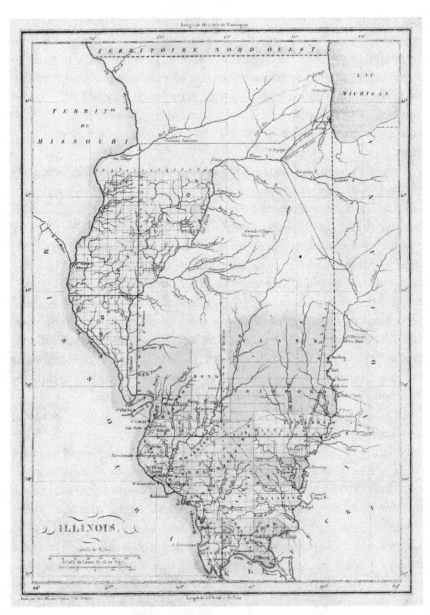

FIGURE 5.1 This French map, depicting the new Illinois state boundary, also portrays the Chicago portage as a prominent detail within the 1816 land cession and proposed canal route. Note that at the time of statehood, most settlement in Illinois concentrated in the south. Alès, *Illinois* (Paris, 1819), Hermon Dunlap Smith Collection. Courtesy of the Newberry Library.

part with their eastern lands.[38] Wolcott proposed that the government commissioners might further smooth the upcoming talks by bribing the Anishinaabe spokesmen prior to the meeting, assuring Cass, "It will be requisite to bribe their chief men by very considerable presents and promises; and that should be done, in part at least, before the period of treating arrives."[39] Knowing the Anishinaabeg commitment to protecting their territory, the American officials openly conspired to corrupt the treaty dealings before they even began.

Despite Wolcott's machinations to relocate the treaty to Chicago and buy off the Anishinaabe leadership, the collective Anishinaabeg offered stiff resistance to Cass's insinuations and demands come treaty time. Though Cass articulated the official U.S. stance that the Anishinaabeg possessed "an extensive country" that remained underutilized and uncultivated and had "very little" game left for their use, the Anishinaabe leadership protested. While Mitchell—described as "an old Pottowattomie Chief" in the treaty proceedings—assented that "our lands are scarce of game, and afford us a scanty subsistence," he asserted that the Anishinaabeg were "not of the race of men who complain." Though game had become scarcer "upon our lands, they will still support us, and poor as they are, we feel loth to part with them."[40] According to the Potawatomi leader, the Anishinaabeg had always found a way to thrive in their homeland and would continue to do so into the future.

Other leaders echoed these sentiments, most notably Metea, a headman from the Fort Wayne area. Responding to Cass's tired critique of underdeveloped lands and the desire of the United States to purchase and improve them, Metea expressed frustration. "Our country was then very large," he told Cass, referring to the era when Americans first came to the Great Lakes. But now, "it has dwindled away to a small spot, and you also want that." Metea reminded Cass they had already given some lands to the U.S. government for settlement, but "you are never satisfied. . . . You are gradually taking away the country which is our only inheritance. Treaty after treaty is called, and piece after piece is cut off from it." Metea, in clear terms, articulated what he saw as an unending desire of the United States to transform his homeland into settled space.

Metea understood the crux of this process too. This was not a military conquest, per se—Metea assured Cass that there was no animosity between the Anishinaabeg and the Americans; "we bring no bad passions with us." Instead, he understood that the foundation of their disagreement lay in how they used the lands and waters differently. Cass and the treaty commissioners sought to possess the land in order to transform it for their settler republic.

Metea lamented that while he and his fellow Anishinaabeg utilized their lands in various ways, for "hunting and fishing" as well as movement and cultivation, the Americans were interested in overhauling the land for a political economy of settler agriculture. He protested to Cass that the transformation of the land occurred so rapidly, the "ploughshare is driven through our tents before we have time to carry out our goods, and seek another habitation."[41] Both locally, at sites from Chicago to the St. Joseph River and Metea's home near the Maumee portage, and regionally, Metea forewarned how American intentions to transform the geography of the lower Great Lakes directly threatened the Anishinaabe way of life.

Even though Metea, Mitchell, and other assembled Anishinaabeg protested Cass's demands and argued their point for nearly two weeks, the U.S. commissioners eventually secured the signatures needed to make the treaty a success for the government. The negotiations concluded on August 26, but the terms remained so uncertain that Lewis Cass deemed it best to depart aboard a vessel bound for Detroit before the final conclusion of ceremonies. The lower-level commissioners waited until he left to distribute whiskey to the Anishinaabeg, for fear of reprisals, and then only did so "8 or 10 miles into the prairie" and away from the American garrison at Fort Dearborn.[42] Clearly, by the end of the haranguing, both sides remained aware of the uneasy conclusion to negotiations and incompatible visions for the region's landscape going forward. The treaty had resolved little. While none of the current concessions affected Chicago's local vicinity, Cass's line of argument signaled U.S. ambitions for the future transformation of the region. Likewise, his success in securing land cessions east of Lake Michigan would bode ill for Chicago's Anishinaabeg in their efforts to defend their remaining hold over its western shore.

Regardless of the self-serving rhetoric of U.S. officials like Cass in discrediting the region's productivity, both the fur trade and the older forms of movement through Chicago continued to operate viably for thousands of the regions' Indigenous people and Euro-American fur traders. Shortly after the American garrison had rebuilt Fort Dearborn in 1816, John Kinzie returned to his trading house on the north shore of the river to reopen business. The Detroit-based partnership of Conant & Mack likewise sent John Crafts to Chicago to establish a fur trade depot at the head of the portage that year.[43] By September 1816, Jacob Varnum arrived as the government factor for the reestablished U.S. Trading Factory at Chicago, commissioned to collect furs from Indigenous customers in exchange for goods at a subsidized rate.[44] And by 1819, the American Fur Company, the ascendant U.S.-based trading monopoly of John Jacob Astor, deemed the commerce in furs sufficient to relocate its agent Jean Baptiste Beaubien to Chicago to capitalize on the nexus of trade.[45] Another French trader, Joseph Bailly, established a post along the nearby Calumet River in 1824. In the years after 1816, the cycles of the fur trade proved as consistently productive as they had before the war, drawing old and new traders to Chicago's nexus of wetlands and canoe routes.[46]

New Pressures on an Ancient Space

Despite the still-thriving movement of canoes and furs through Chicago's portage, by 1825 the stakes for internal improvement in the western Great Lakes had risen even higher. In that year, the Erie Canal opened, connecting the waters of the Hudson River with Lake Erie. Under the encouragement of Governor DeWitt Clinton, the New York state legislature had backed and bankrolled the ambitious project to the tune of $7 million.[47] The canal circumvented the need for the labor-intensive portage around Niagara Falls, providing the first waterborne route between the Atlantic Seaboard and the Great Lakes that did not rely on portages. Now, if a canal could be dug at Chicago, American boosters realized the United States could prosper from an internal transportation network of uninterrupted waterways connecting the Mississippi River valley with the Great Lakes and Eastern Seaboard in one comprehensive system of waterborne movement. Adding to the clamor for further investment in Chicago's waterways, after 1825 new waves of immigrants followed the Erie Canal westward to Buffalo and then on to Michigan, Indiana, and Illinois, looking for land and demanding access to eastern markets through still more improvements to transportation.[48]

At first, few of these immigrants lingered in the vicinity of Chicago's wetlands, moving farther downriver into the more promising bottomlands of central and southern Illinois.[49] The permanent population of white settlers around the portage remained strikingly low for a site of such national attention. As late as 1826, the state of Illinois only recorded thirty-five eligible voters at Chicago—meaning white male citizens—while only fourteen held taxable property in the area, most of it associated with fur trade storehouses.[50] Yet at the state level, promoters of Illinois's future within a connected "American system" of infrastructure and transportation continued to advocate for changes that would at once promote settlement at Chicago and overhaul it into a hub of settled commerce.

State-level officials continued to stress the need for a canal to replace the older portage connecting Chicago's waterways. Illinois congressman Daniel P. Cook took up matters at the national level, advocating for federal investment in the canal project while marshaling appeals from Illinois citizens, state legislators, and the Illinois governor to supplement his own petitioning.[51] By the mid-1820s, growing support for internal improvements at the national level and the continued demand for infrastructure from boosters downriver, in the Illinois state government, drove efforts to transform Chicago's landscape at the local level.[52] State legislators began funding road projects, bridges, and ferries at Chicago in anticipation of still further investment, growth, and construction to come.

Such developments did not go unnoticed by the Chicago-area Anishinaabeg, who remained uneasy about American designs toward Chicago. Anishinaabeg living near Chicago; along the Michigan shoreline; and in the Des Plaines, Kankakee, Illinois, and Fox River valleys did not forcefully resist many of these activities, though they still greatly outnumbered American settlement around the portage.[53] Kinship bonds with local traders and French residents at Chicago helped maintain friendly relations at the local level. As late as the 1820s, Chicago remained a multicultural and multiracial space as a trading outpost in which Indigenous headmen, métis middlemen, leading merchants, and government officials cooperated as power brokers in an intimate community that still revolved very much around the fur trade and waterborne traffic through the portage.[54] Such cooperation between Anishinaabe leaders and local U.S. agents dissuaded most of the area's warriors from blatant attacks on American settlers in the region. Their limited use of violence throughout the late 1820s into the 1830s, however, was telling.

These local Native people directed their aggression, instead, against two elements of incoming white settlement: agriculture and infrastructure, rather than white migrants themselves. They wrecked settlers' crops and raided vegetable gardens. Potawatomis along the Illinois River set fire to haystacks to disrupt autumn harvests. They slaughtered hogs, stole horses, and mutilated cattle—all representations of the shifting agricultural landscape of the Illinois prairie that so vexed Chicago's Anishinaabeg.[55] Illinois governor Ninian Edwards equated these "daring robberies" and other depredations with "actual war" between the Illinois River settlements and Chicago.[56] When the barracks at Fort Dearborn caught fire in June 1827, allegedly from a lightning strike, a visiting group of intermarried Ho-Chunks and Potawatomis led by an ogimaa named Big Foot drew suspicion from white Chicagoans. Big Foot's party watched the conflagration and refused to aid the Americans in extinguishing the flames.[57] In 1832, approximately 150 Anishinaabeg warriors from Chicago raided south toward Starved Rock and the southern terminus of the proposed canal route, killing livestock and threatening to cut communication lines between Illinois and Lake Michigan. Meanwhile to the west of Chicago, another group of disgruntled Anishinaabeg burned a bridge at a river crossing.[58] These acts of low-level violence maintained a certain spatial logic. The Anishinaabeg around Chicago, while avoiding open war with the Americans, directed their attacks on the manifestations of the changing landscape around them. Bridges and building projects altered space, while livestock and crops threatened to unseat Indigenous engagement with the wetland and prairie environments. Native individuals that participated in such raiding, vandalism, arson, and theft signaled how well they understood these targets as direct threats to an Anishinaabe way of life in the region.

By 1827, the Anishinaabeg around Chicago's portage had every reason for uneasiness. In March 1827, the federal government granted a portion of the canal lands acquired from the 1816 treaty to the Illinois state government. The idea was for the land, running five miles inland on either side of the proposed canal route to be sold in lots, with the proceeds from such sales going toward the actual canal construction. As Edmund Roberts, one of the state's canal commissioners, reported to Ninian Edwards, the country had begun to settle rapidly. An influx of wagons arrived from Indiana that spring, and by summer, would-be immigrants had begun disembarking from ships off of Chicago's sandbar in anticipation of the coming land sales. Roberts expressed his hopes that the state lots would sell quickly, and at high prices, based on land speculation over the proposed canal route.[59] It was no coincidence that a particularly destructive spate of Native depredations came in the months after these first government land sales along the canal route, as Chicago's Anishinaabeg faced the prospects of escalating white settlement along the route they had relinquished in 1816. Even still, the influx of agrarian settlers had only come after significant government backing in the form of land sales and the promise of further internal improvements in the Chicago area.[60]

These newly arrived agrarian settlers, while enthusiastic about the prospects of a canal at Chicago, approached the local environment with more skepticism. Chicago's portage, and much of the landscape around it, remained a wetland and low-lying "wet prairie." This held an advantage to both Indigenous people who knew how to exploit the various natural resources of wetlands environment and the fur trade brigades who relied on seasonal flooding to ease their passage between the Chicago and Des Plaines Rivers every spring.[61] But to incoming Americans, Chicago's "swamps" signaled a clear impediment to progress. White migrants to the west viewed wetlands environments as unhealthy ecosystems that bred malarial mosquitoes and inhibited agricultural development.[62]

As incoming waves of Americans arrived from the east, the changing understandings of the land and geography of Chicago became evident. Juliette Augusta Magill Kinzie, John Kinzie's daughter-in-law, relayed the story of "a gentleman who visited Chicago" during the land speculation boom of 1833. Passing over the wetlands near the portage path on horseback, he complained about being "up to my stirrups in water the whole distance." The fluid condition of the area, which in years past would have been a boon to canoe transportation and mobility, was lost on the eastern land speculator. He failed to see the value in the amphibious portage, declaring that he "would not have given sixpence an acre for the whole of it."[63] Another visitor, John Tipton of neighboring Indiana, noted how a "Grate part of the Prairie [was] low and wet" around Chicago.[64] Jeremiah Smith, a federal surveyor commissioned to map and assess the nearby Kankakee marshes, deemed the entire area as "rather uninviting to the capitalist and speculator."[65] For Chicago's space to hold value for such eastern land developers, the wetland environment would have to be destroyed and remade into a place worth purchasing and settling.

Smith's recommendations were for immediate drainage south of Lake Michigan, in both Illinois and Indiana, to eliminate such swamps. The state of Indiana took the first steps in these efforts in 1832 with the passage of a "ditch law" that formed citizens' committees at the local level to assess drainage necessities for wetlands areas. Several years later, Illinois went even further, empowering local governments and property owners to form field committees, in order to direct drainage efforts such as ditches, dikes, and levies.[66] Efforts to reshape the waterlogged landscape around Chicago would continue in the decades to come, ultimately overhauling nearly 367,485 acres of wetlands in the immediate area—the most in any part of the state.[67]

With the ongoing drainage of the wetlands came new threats to both the viability of the fur trade and the Anishinaabe way of life. Reductions in aquatic vegetation, most notably wild rice, removed an important food source for Chicago's Anishinaabeg, who harvested the abundant grain every autumn from the meandering river channels of the Chicago and Des Plaines Rivers. The loss of wild rice, and wetlands in general, also diverted waterfowl migrations away from Chicago's local vicinity, reducing hunting quarry as well. Concurrently, with the loss of wetlands, even the furbearing aquatic mammal populations that had endured the height of the fur trade now faced local eradication from habitat loss. Writing in 1831, James Hall, the editor of the *Illinois Monthly Magazine*'s first volume, reported that "the beaver and otter, were once numerous, but are now seldom seen."[68] Such population decreases remained more a matter of habitat loss than hunting, as the fur trade around Chicago had continued unabated for over a hundred years until wetlands drainages commenced and aquatic mammal populations began to decline.[69] Later scientific evidence further suggests that Chicago's local fish numbers began to drop rapidly around this same time due to the loss of spawning and breeding grounds within the wetlands, removing yet another subsistence resource for local inhabitants, both Native and non-native.[70] Despite this observable ecological loss, newly arrived immigrants to Chicago and Illinois state officials continued to promote a program of wetland drainages as they worked to expand salable farmland along Chicago's future canal route and townsite.[71]

An Invasion of Infrastructure

But promises of a coming canal and arable land alone could not solidify an American hold over Chicago's intractable local geography. Because of the hydrology of Chicago's sluggish river and the prevailing winds and currents along the lake, there remained no natural harbor. As Major Stephen Long had noted in his initial assessments of 1817, the river diverted south along the lakeshore and even its outlet remained only a few feet in depth, due to the submerged sandbar there. Gurdon Hubbard, the fur trader so well versed in Chicago's riverine geography, later described the river's mouth as "a shifty one," constantly changing with the seasons and dependent on lake level, winds, and gales.[72] The combination of muddy inland waterways, varying marshlands, and shifting sands along the shoreline made Chicago's waters and landscape unpredictable, and thus, unnavigable for Euro-American shipping and nautical knowhow. The most confounding element, the sandbar at Chicago's mouth, ensured that water levels at the entrance of the river remained only a few feet deep, even though the rest of the river flowed at depths of twenty feet or more farther upstream.

This had never prevented Indigenous watercraft from entering and exiting the river, since the minimal draft of canoes and bateaux allowed them to navigate even the shallowest of water. Larger vessels that came to Chicago had always anchored about a half mile from shore, relying on canoes and other light boats to ferry their cargo and passengers to shore. While this makeshift solution had worked for the few fur trade schooners and packet ships that had visited Chicago from the mid-eighteenth century onward, it remained inefficient for anything other than the occasional disembarkation. If American officials and commercial boosters were to enjoy their reimagined transportation hub at Chicago, the river's mouth would have to be fundamentally reconfigured to allow ships to enter.

The first attempts to render the river navigable came from the only local representation of state capacity: the army garrison stationed at Fort Dearborn. Though ill-equipped and lacking the technical expertise of engineers, Captain John Fowle, the post's commander, attempted to utilize the seasonal floods to his advantage, mustering the garrison out in the spring of 1828 to dig a ditch, by hand, across the isthmus north of the river's sand-blocked mouth. After an hour of digging with the river in flood stage, Fowle's men succeeded in opening a new channel from the Chicago River into Lake Michigan, which swelled to an opening fifteen feet wide and eight feet deep in the days after the excavation. But the crude solution filled in with sand shortly after as river levels stabilized and lake currents once again shifted the shoreline.[73] Despite the best local efforts to reshape Chicago's littoral space, the area's geography remained insurmountable.

Restructuring Chicago's space would require an unprecedented level of government investment and state capacity. In 1829, Illinois state officials began jockeying to secure federal support for harbor improvements at Chicago.[74] Finally, in 1831, federal funding and expertise made way for the first step in this overhaul: the construction of a lighthouse to guide navigation through the notoriously treacherous waters off Chicago's river mouth. Congress allotted $5,000 and commissioned Samuel Jackson to build a fifty-foot stone structure along the southern shoreline near Fort Dearborn. Construction of the lighthouse finished just in time for the typical gales of autumn to begin pounding the lakeshore. And Jackson had built the lighthouse on sandy soil. As one Chicago resident described, by late October, large cracks had begun showing in this first of Chicago's public works. Jackson directed repairs and the removal of some of the more precarious stones, but within a week of the first fissures appearing, "down tumbled the whole work with a terrible crash." While some town residents assured themselves Jackson's poor handiwork had led to the collapse, Jackson himself argued that shifting sands beneath the structure had led to the disaster, declaring "that a light-house can not be made to stand here." Given the infamous transience of Chicago's sandy shoreline, Jackson's explanation may have held merit. Whether by incompetence, the precarious local landscape, or some combination of the two, Chicago's first public works project had failed with a resounding thud.

Government investment in internal improvements would not relent, despite such setbacks. Within six months, federal funds financed the construction of a second lighthouse near the same spot. This time, the forty-foot structure stayed upright. Following the lighthouse project, the federal government earmarked an additional $25,000 for harbor improvements in 1833.

That summer, Major George Bender, commanding at Fort Dearborn, supervised the beginning of construction—this time with the added skills of army engineers sent from the east.[75] Charles Hoffman, a visitor to Chicago at the time, recounted his observations of the ongoing project in the *New York American*, describing the work as "going on briskly" with efforts to cut a permanent channel through the sandbar at the river's mouth and construct wharves along either side of the passageway, in order to hold back the shifting sands of the lakeshore. Hoffman shared a booster's optimism with his East Coast audience, predicting that "when finished the river will have a straight and direct channel into the lake, so that any vessel can come into the river and unload anywhere in town." Hoffman echoed the consensus of earlier surveyors and engineers that this project would save the time and expense of unloading and transporting cargoes from ships onto lighter watercraft to bring ashore while also mitigating "the danger of anchoring in the lake." Once the U.S. Department of Engineers completed the project, Hoffman concluded, Chicago would "be a safe harbor" where "vessels will be able to run up the south branch a mile or two, and up the north branch some farther."[76] For the first time, it seemed, Chicago's waterways would become navigable to more than just Native-built canoes and other fur trade watercraft, opening up access to sloops, brigs, and other European-style vessels.

Hoffman's projections held true, in one sense. In 1834, after construction had begun, two brigs and two schooners managed to clear the bar and enter the river. Once the engineers had completed the piers and established a permanent navigational channel later that summer, the number of large ships entering Chicago's river exploded to well over one hundred vessels by 1835 and continued to increase from there. With the prospects of a viable lake port at Chicago, local leaders utilized still more federal dollars to purchase a dredging boat in order to maintain the channel and keep the shifting sands at bay (see figure 5.2).[77] Despite such efforts, sandbars continued to form at the mouth of the river on a seasonal basis and after large storms. The harbor and piers therefore required ongoing dredging, repair, and extension. As seasonal gales tore along the lakeshore, the U.S. engineers tasked with maintaining the harbor found it necessary to construct still longer piers as well as break walls, dikes, and levies to secure the constantly moving coastline.[78] By the end of the decade, the U.S. Congress had poured over $192,600 into continued harbor upkeep and improvements at Chicago—an incredible effort to create an artificial harbor at a site that, based on cartographic logic of the day, should have proven a naturally advantageous place for waterborne movement.[79] Even with so much investment in harbor infrastructure, Caroline Kirkland,

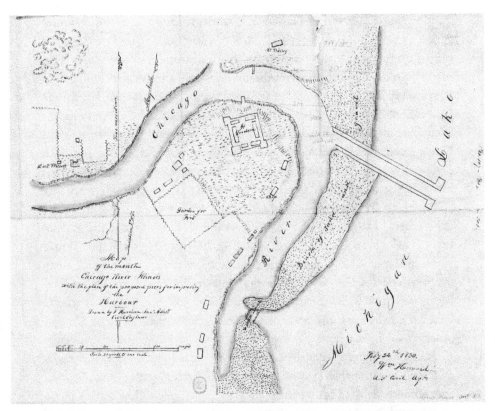

FIGURE 5.2 This manuscript map demonstrates the proposed alterations to Chicago's waterways in the early years of U.S. hegemony and public works efforts. William Howard, *Map of the Mouth Chicago River Illinois with the Plan of the Proposed Piers for Improving the Harbour* (1830), Graff Collection. Courtesy of the Newberry Library.

who visited Chicago more than a decade later, still described it as the "worst harbor and smallest river any great commercial city ever lived on."[80] But Chicago's boosters, both locally and farther afield, remained determined to remake these sluggish waterways and the sand-choked lakeshore into a site of settled commerce and control. Repeatedly, the solution for this major overhauling of the natural geography came only through government investment at the federal level, and on a scale that earlier imperial powers had been unprepared to leverage in such a far-flung location.

Reshaping Chicago's Human Geography

Chicago's regional- and state-level promoters, however, viewed investment in infrastructure as only one component to their overall vision for Chicago's transformation. These same individuals who advocated for a conversion of Chicago's space used threats of Indigenous violence during the late 1820s and early 1830s to push for the expulsion of the southwestern Anishinaabeg from Illinois. Ninian Edwards, as governor of Illinois, saw the livestock killings and crop thefts perpetrated by Chicago's Anishinaabeg in 1827 as an excuse for the government "to drive them from their present residence."[81] Chicago-area Anishinaabeg actually aided American forces in two near wars with regional Native groups, the Ho-Chunk, Sauk, and Fox, in 1827 and again in 1832 with Black Hawk's War. Yet vocal proponents of internal improvements, such as Edwards, saw in these brief conflicts a pretext to push for the total removal of Native peoples from the valuable geography around Chicago's route. Their efforts drew further support from the new arrivals to Chicago in the wake of Black Hawk's War, as speculation over Chicago's valuable geography spread in the east and waves of migrants began to arrive to purchase acreage along the proposed canal route.[82] For the first time, machinations at the state level began to sync up with local demands to both develop the topography around Chicago and remove the area's Anishinaabeg from their homeland in the process.

Calls from Illinois leaders like Ninian Edwards joined a growing movement on the national level as politicians, businessmen, settlers, missionaries, and moralists debated Native Removal as a federal policy. The push for Indigenous expulsion at the national level brought together a peculiar coalition of proponents ranging from those who wished to extinguish Native land title for sheer profit to those who embraced a misguided sense of paternalism, deeming removal as a benevolent step to free Native peoples from the corrupting influence of white society. On all sides, the debate remained fraught in a racist ideology of hierarchies and "civilization." But in the lower Great Lakes, most advocates for Native Removal remained motivated by a desire to seize Indigenous lands for further development above all else.[83]

To characterize the calls for Native Removal as a general demand for arable lands, driven by would-be settlers alone, misses the point. The need for strategically significant pieces of geography for the purposes of internal improvements, rather than a general land hunger of settler colonists, drove Indian policy and Native dispossession. Advocates of Anishinaabeg expulsion at Chicago and elsewhere throughout the Midwest remained interested in acquiring specific tracts of land for federally funded infrastructure that would strengthen the Union's hold over the region.[84] The neighboring state of Indiana, for instance, petitioned Congress six times between 1829 and 1831 to extinguish Native titles to lands that state officials argued "impede a system of internal improvements essential to the prosperity of our citizens."[85] In Illinois, similar demands from state officials like Edwards called on the federal government to negotiate with Anishinaabeg in Illinois for further land cessions, which could in turn be sold to raise additional funding for the Illinois and Michigan Canal project.

This process of government-backed dispossession in the service of internal improvements held particular appeal to regional officials, who looked for a federal solution to what they perceived as twin challenges in bringing order to both the space and peoples of their western territories. Lewis Cass, speaking as the territorial governor of Michigan and the superintendent of Indian Affairs, told one group of Potawatomis in 1826 that the white settlers "cannot have roads and taverns and ferries, nor can we communicate together," because the Anishinaabeg stood in the way of further internal improvements and public works projects. Cass made clear that this need for a communications network went beyond information transfer and included public roads and other transportation efforts that could facilitate commerce and encourage further settlement. According to Cass's racially exclusive vision for the landscape, the very presence of Anishinaabeg on strategic sites of development served as a hindrance to American society.[86]

Regional officials like Edwards and Cass saw Native dispossession at Chicago and elsewhere as a prerequisite step toward their overall visions to remake the region into a coherent network of canals, harbors, roads, and other government-built infrastructure. Cass and his collaborators in Illinois and Indiana made the explicit argument that internal improvements could serve as the strategic method for ensuring more land cessions and, eventually, total expulsion of Native peoples across the Mississippi River. These regional officials understood this to work as a reinforcing feedback loop of land development and land cessions. Cass and his cosigners explained the logic in a letter to the secretary of war in 1826: "what is more important to us," describing an adjacent roadbuilding project in Indiana, is that such public works "will sever their [the Potawatomis'] possessions and lead them at no distant day to place their dependance upon agricultural pursuits or abandon" the lands. This, they argued, made such infrastructure projects' eventual importance twofold in both a "pecuniary or political view."[87] In other words, these government officials understood what they were doing. As Cass put it in an open letter promoting Indigenous expulsion in 1827, "We hold no fellowship with those to whom the sound of the Indian's rifle is more attractive than that of the woodman's ax." Developing the landscapes of the American interior was tantamount to conquest for promoters of Indian Removal like Cass, who envisioned white settlers "spreading with exterminating force over the forests and prairies of the west."[88] As they undermined Native lifeways by overhauling Indigenous geography, these government officials believed their public works projects would make life untenable for the region's Anishinaabeg. This, in turn, would lead to more land cessions that could further settlement and eventually push Native people to accept removal.[89] Once the Indian Removal Act set a precedent for the policy at the federal level in 1830, official proponents for removing the Anishinaabeg from Chicago arranged for a final land cession treaty to wipe out Indigenous claims to all lands southwest of Lake Michigan.

The final excuse U.S. officials needed to justify their aims at removal in the Chicago region came when Black Hawk led his Sauk, Meskwaki, and Kickapoo followers back across the Mississippi River to reoccupy their traditional lands in northern Illinois. Black Hawk had remained a staunch opponent to U.S. territorial acquisitions for decades, arguing that "land cannot be sold" if Native peoples continued to "occupy and cultivate it."[90] When Black Hawk brought his people back from the western shore of the Mississippi in 1832 to escape conflicts with the Dakotas and Menominees, however, American settlers in Illinois raised the alarm. After Black Hawk's retinue clashed with Illinois militia at the Battle of Stillman's Run that May, U.S. officials summoned federal and state troops to intercept the Native refugees. In the coming months, thousands of armed forces descended on Chicago as their base of operations, though no fighting ever threatened the Chicago area itself. Chicago's Anishinaabeg, in a last-ditch effort to maintain their own status as brokers in the region, actually aided American forces as scouts, diplomats, and spies throughout the crisis. By early August, government forces had pinned Black Hawk's people down along the banks of the Mississippi River, while armed steamboats like the *Warrior* patrolled the river to cut off their retreat. In an attack that targeted not only warriors but the many women and children of Black Hawk's following, U.S. troops brought a violent end to the last armed Indigenous resistance in the Illinois Country at the Bad Ax massacre. In the aftermath, it mattered little to U.S. officials that Anishinaabeg from Chicago had helped in the American cause. With Black Hawk imprisoned and his comrades defeated, government agents saw their opportunity to vacate all remaining Native lands east of the Mississippi River, including those still held by Anishinaabeg near Chicago.[91]

Their objectives clear, officials from Illinois and neighboring states across the Great Lakes assembled the Anishinaabeg at Chicago in 1833 to secure the area for the United States and eliminate an Indigenous hold over the strategic site. Local Indian agents Thomas Owen and William Weatherford acted as commissioners for the treaty, alongside Michigan's territorial governor and superintendent of Indian Affairs, George Porter.[92] The commissioners hosted perhaps the largest gathering of Native people ever to assemble on Chicago's prairies, with estimates from the time ranging as high as 8,000 Anishinaabeg from Illinois, Wisconsin, Indiana, and Michigan.[93] One observer later described the assemblage: "far and wide the grassy Prairie teemed with figures" ranging from mounted warriors, French traders, whiskey peddlers, and métis interpreters to women, children, dogs, and livestock. Central to the negotiations were the "grave conclave of grey chiefs seated on the grass in consultation."[94] The Native leaders had a reason for gravity. The calls for this treaty had made clear the goals of the negotiations—total expulsion of the Anishinaabeg from lands south and west of the Great Lakes, with little leverage to deny these U.S. demands.

The treaty negotiations took about two weeks, as the Indigenous headmen brokered the best deal they could for the sale of their remaining lands in the region. It had become widely accepted among Chicago's Anishinaabeg by this point that their lands in the Great Lakes would be forfeited regardless of how much they protested. As American building projects commenced along the shoreline, canal lots sold for incredible prices, and game grew scarce in the depleted marshlands, some of Chicago's Anishinaabeg came to view expulsion as the best decision in a slate of unappealing options.[95] Prominent interlocuters for the Anishinaabeg, mixed-heritage traders and interpreters such as Billy Caldwell and Alexander Robinson, saw the treaty at Chicago as their last chance to acquire both personal and tribal wealth from government annuities and proved eager to broker a deal with American officials.[96] In the end, the united Anishinaabeg southwest of Lake Michigan ceded over 5 million acres in exchange for almost $1 million in immediate payments and future annuities, plus $130,000 in goods and $56,000 in cash for previous overdue land claims.[97] This was a massive outpouring of federal funds to secure the agreement of the assembled Anishinaabeg to remove west of the Mississippi River within three years.

Local Chicago boosters interpreted the process of Native Removal in terms of further infrastructure development. Reporting on the conclusion of the treaty, Chicago's start-up newspaper the *Chicago Democrat* rejoiced at the acquisition of lands "of the highest importance and value," both for further public works projects that would benefit Chicago's bid as a commercial center and for "agricultural purposes" in its hinterland. There was no mistaking the motivations behind the 1833 Treaty of Chicago, nor its intertwined ambitions for the region after the Anishinaabeg signed away their final cessions. In that same debut issue of the *Chicago Democrat*, the newspaper declared the significance of Chicago as a site transformed. Tracking its metamorphosis from a seat of Indigenous power into a place of state and commercial ambitions, the newspaper's editor lauded "the very spot of ground where we are now writing" that "a few months since was the abode of the savage; and where are now seen a long line of habitations for white men."[98] The local geography of Chicago still mattered, but now it served as the foundation on which to build a white settler state.

Ceremonies of a Changed Landscape

After the Anishinaabe cessions of 1833, change came quickly to Chicago. In the years after the final treaty, further investments in infrastructure, more land sales by the state and federal government to incoming settlers, and an influx of eastern capital would drive Chicago's ongoing development into a place of American commerce and power.[99] In 1833, Chicago's citizens officially incorporated their outpost as a town, with the population spiking from 150 to 350. Within another year, the population around the mouth of the Chicago River had ballooned to 1,800 residents, and by the end of the decade, the town limits alone contained 4,500 inhabitants with even more spilling into the surrounding prairies and river valleys.[100] As this flood of new settlers arrived, they brought with them further demands for continued harbor improvements and, especially, pressure for forward progress on the canal project.[101] The Illinois state government responded by passing legislation in 1834 to finally open up funding for the canal and appoint commissioners to oversee construction of the project.[102]

Excitement around the canal drove up land prices, as investors from the east and newly arrived migrants speculated on the future viability of real estate in Chicago proper and along its crucial portage-turned-canal route. The feasibility of the canal project rose with the speculation in land around Chicago. Canal dreams fed purchases of government lands at higher prices, which in turn provided more funding for the canal project. The *Chicago Democrat* reported the local celebration when the Illinois legislature passed the final Canal Bill in January 1836, confirming actual construction on the project for that summer. Land buyers and sellers alike predicted heady times to come.[103] By the spring of 1836, land prices had risen so high in Chicago that a local newspaper, the *Chicago America*, boasted about one plot of land that had "risen in value at the rate of one hundred per cent per Day on the original cost ever since 1830, embracing a period of five years and a half."[104] The rising real estate sales of nearly 285,000 government-held acres along the canal route by late spring, plus additional loan shares sold in eastern states, provided the state of Illinois with sufficient funding to begin the physical transformation of Chicago into the reconstructed landscape that state and then local boosters had clamored for since at least 1818.[105]

Finally, on July 4, 1836, the Illinois State Canal commissioners gathered at the newly renamed "Canal Port," near the terminus of the portage path along the southern branch of the river, to conduct an elaborate groundbreaking ceremony for the canal. Chicago's dignitaries had purposely chosen the date as emblematic of ascendant American power at the portage, and they conducted "imposing ceremonies" to symbolize the changing political economy of the interior. William Archer, the lead canal commissioner from the Illinois state government, lifted the first shovel-load of mud from the riverbank and then turned the digging over to Gurdon Hubbard, the longtime fur trader of Chicago's wetlands. As one of the few Indian traders who had successfully transitioned to a leading citizen and businessman of Chicago's fledgling town, Hubbard's spadework served to represent Chicago's own transformation. The groundbreaking ceremony, for these gathered officials, represented the physical manifestation of Chicago's transition away from the old portage geography that had bolstered the fur trade and Indigenous power in the region.[106] Actual construction along this section of the route would stall in the months after the festivities due to the difficulty of directing workers and supplies through the marshy terrain, but nevertheless, state and local officials had demonstrated their full intentions to see the project of overhauling Chicago's geography through to its end.[107]

It was no coincidence that such dignitaries broke ground on the Illinois and Michigan Canal the same summer the southwestern Anishinaabeg conducted a final ceremony of departure along the Chicago River. In August 1835, as many as 5,000 Anishinaabeg from Illinois, Indiana, and Wisconsin encamped one last time outside Chicago to accept their annuity payments and collect themselves for their supervised expulsion west of the Mississippi. Assembling on the north bank of the river, these Anishinaabeg prepared to bid farewell to the land and water they had controlled so well, for so long. John Caton, the local justice of the peace, witnessed this ceremony, what he called "the last war dance ever ... on the ground where now stands this great city," and recorded in detail the farewell ritual of the southwestern Anishinaabeg.

Over eight hundred Anishinaabe men joined in what Caton equated to "a sort of funeral ceremony" complete in "all the grandeur and solemnity possible," as they grieved the loss of "their native soil." The Anishinaabe mourners had stripped and painted themselves for the occasion and proceeded along the bank of the river while they wailed, sang, beat drums, and brandished war clubs. "They advanced, not with a regular march, but a continued dance," all along the north side of the river, before crossing over to the south side and performing in front of Fort Dearborn as well. Caton thought he read "fierce anger, terrible hate, dire revenge," and "unaffected rage" in the expressions of these Anishinaabeg as they prepared to depart Chicago's muddy shoreline for the last time. In an evocative ending to their procession, the Anishinaabeg departed the town of Chicago over one of the newly constructed public bridges that state funds had furnished along the river. As officials from Chicago and greater Illinois pursued their ongoing vision for a remade environment with construction of the canal on the regional level and a number of public works on the local level, Chicago's Anishinaabeg took their mournful leave of a rewritten landscape.

Both the Anishinaabeg and the government officials that had unseated them as masters over Chicago's crossroads conducted their ceremonies of 1836 in ways that revealed the true nature of the struggle for—and subjugation of—Chicago and the Old Northwest in the nineteenth century. Differing visions for Chicago's geography had defined the incompatibility of American and Indigenous understandings of Chicago's significance as a site of power and prosperity in the interior. Anishinaabeg had harnessed the wetlands environment, the waterways, and the portage in their decades of dominance at Chicago. In their departure from Chicago, their ceremonial march had outlined the course of that influential riverscape. The Americans, on the other hand, had signaled their contrasting goal for the area's topography. With their groundbreaking celebration for the canal, they effectively promised to destroy the old geography of Chicago's portage and overhaul the area into a new center of U.S. prosperity. Federal and state commitment to public works projects like the canal, manifested in both policy and financing, allowed these officials to reshape the local geography in ways that had remained incomprehensible to earlier imperial powers. This investment and overhaul of Chicago's space defined their efforts to both dispossess Native peoples from Chicago and gain the upper hand in the region. In 1836, both the Anishinaabeg and the state officials understood that the root of power remained a matter of controlling Chicago's muddy ground, either as the network of waterways and portage paths or the overhauled landscape of American visions.

A Blasphemy against Nature

By 1836, it seemed Chicago's boosters at the national, state, and local levels had achieved the ultimate coup. They had conquered a space that for centuries had proven intractable to European imperial ambitions. Chicago's portage had eluded earlier French, British, and American officials who had seen in the portage a way to control the interior. Chicago's Anishinaabeg had maintained a better hold over the geography, utilizing their clear understanding of waterborne transportation and ecological knowledge to maintain influence in the region from the mid-eighteenth century into the 1820s. But as U.S. officials, businessmen, and incoming settlers began to reshape the landscape and waterways of Chicago's portage in the decades after the War of 1812, Anishinaabeg who had long held sway over the means of movement through the area began to lose control over Chicago's strategic geography. As white Americans reenvisioned Chicago's space, and the U.S. government invested unprecedented amounts of money into reconstructing Chicago's geography to match such visions, Chicago's Native peoples found themselves in a disorienting and disempowering new landscape. This landscape, rather than being reliant on Indigenous knowledge and transportation technology, lent itself instead to white settler logics of movement, exchange, and production. By 1836, when most of the region's Anishinaabeg were forced west of the Mississippi River, Chicago had been remade into a site of white settler authority and potential prosperity.

Despite this apparent victory of American government and commerce at Chicago, the road ahead would prove rocky for the U.S. settler republic. Land speculation around Chicago would spark a local real estate bubble—and bust—in 1837 at the same time an economic downturn in the east would prompt a major monetary crisis for Illinois. Many investors in Chicago property went bankrupt in the fallout. Canal construction progressed only in fits and starts, partly because funding sources dried up, and partly because Chicago's watery geography did not. Chicago's environmental reality complicated human ambitions yet again as the marshland through which the canal would pass delayed digging during much of the year and floods reversed the construction's progress.[108] In 1841, largely because of overcommitment to the canal project, the state of Illinois filed for bankruptcy and work on the canal halted indefinitely. It would take until 1848, after years of rebuilding state finances and private confidence, for workers to finally open the artificial waterway connecting the waters of the Illinois River and Lake Michigan. In a final, wry turn, the fever pitch for continued internal improvements and transportation made the Illinois and Michigan Canal irrelevant nearly as soon as construction finished. Competition from railroads undercut the canal's expected profits and quickly replaced the water route as the conveyance of choice for Chicago commerce. Passenger service on the canal ended by 1853 and a rail line, the Chicago, Rock Island, and Pacific Railroad, took its place of prominence along the old portage route.

In the end, whatever success did come about in Chicago as a nineteenth-century American city transpired not because of its local geographic advantages but in spite of them. Business interests and government initiatives continued to pour money into Chicago's infrastructure to completely overhaul the landscape and waterways of the area despite natural impediments and economic deterrents. In the decades after the canal project, further efforts would include more harbor improvements, wetlands drainages, railways, telegraph lines, lakeshore break walls, and ground elevation projects.[109] Such works culminated in an engineered reversal of the Chicago River's flow to direct sewage down the Illinois River rather than into the lake, which took from 1871 to 1900 to succeed.[110] By the turn of the twentieth century, Chicago's American residents, backed by government investment and aided by further technological advances, had rendered Chicago's wetlands, prairies, oak openings, and even its waterways wholly unrecognizable compared to the portage geography of less than a century before. Proud Chicagoans rejoiced in such a city, which to them represented both a triumph over natural geography and an enduring proof of American greatness in the interior. Robert Herrick, writing of Chicago's rise as a city in 1898, proclaimed, "It stands a stupendous piece of blasphemy against nature," delighting in the manufactured success of this recently wrought American place (see figure 5.3).[111]

Such revelers in Chicago's incredible growth as a nineteenth-century city have clouded our understanding of this history as chiefly a conquest over nature. Yes, government backing, commercial investment, and settler population growth all contributed to a radical expansion of Chicago as a cityscape that overtook local terrain and led to environmental subjugation farther afield. But the growth of Chicago's infrastructure and this conquest over nature also came at a human cost.

Chicago as an American city replaced not only a natural environment of wetlands, prairies, and lakeshore but also a human geography that had facilitated a much longer history of mobility, prosperity, and exchange in the Great Lakes interior. Only by returning the intertwined history of Chicago's Native Removal to its more recognizable story of urban growth and infrastructure

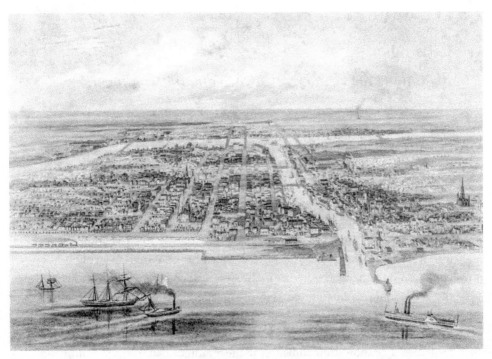

FIGURE 5.3 This bird's-eye view of Chicago from 1853 depicts a thoroughly overhauled topography and waterscape, with bustling shipping, a shoreline railroad, gridded streets, bridges, and the now clearly demarcated land and waterways of the settler city. George Robertson, *Chicago, Ill.* (1853). Courtesy of the Newberry Library.

development can we appreciate the full ramifications of the American conquest of the continental interior during the nineteenth century. This remained both a story of Indigenous dispossession and internal improvements, of an expanding settler political economy and environmental degradation. Chicago, as an important site in the earlier Indigenous world of the interior and later as a place of American prosperity, provides a way of retracing these mutually reinforcing processes of conquest at the local level. Rather than an exceptional case, Chicago's local trajectory demonstrates the wider transformation of both the human and physical geography of the American interior during the era of U.S. expansion across the continent.

Closings

Rethinking the History of Chicago and the Continent

In July 1893, a young historian sat in Chicago, frantically finishing a presentation he would deliver later that day. Frederick Jackson Turner recalled to a friend the "final agonies" of "getting out a belated paper." Many of his historian colleagues had ventured out to the sights and sounds of the city's Columbian Exposition, meant to celebrate the four hundredth anniversary of Columbus's arrival off the coast of the Americas. The entire event marked a national, even international, commemoration of four centuries of European exploration, conquest, settlement, and culture in the Americas and specifically highlighted the ascendancy of the United States. The afternoon of July 12, 1893, while Turner rushed to finish his talk, Buffalo Bill Cody had offered to host an impromptu performance of his Wild West Show adjacent to the fair's summerlong festivities. The show drew audiences in the hundreds that day as it glamorized the sweep of U.S. violence and conquest in the American West.[1]

Across from the exciting spectacle of Cody's show, Turner took the podium as the last of five historical lectures that evening. He presented to an audience of around two hundred members of the American Historical Association, no doubt hoping to impress the crowd with his ambitious claims. Unlike the preceding speaker, who had offered an address on the intricacies of early lead mining in western Illinois, Turner went big. He proposed to offer his audience not just a snapshot of America's past but an explanation for the whole of American culture, society, and politics—a grand narrative for thinking about, and understanding, the trajectory of American history up to that time. Taking as his point of entry the "closing" of available lands in the West according to U.S. Census data, Turner argued that American history and identity could be understood only through its engagement with a frontier zone of available land, which had now finally ended an era of expansion and development.[2] In other words, Americans' engagement with the frontier, as both a place and process, had defined "the first period of American history," and with its closing, U.S. citizens were entering a new, uncertain future. For all Turner's ambitions of grandeur, his bold claims fell flat that night. Another panelist remembered "the audience reacted with the bored indifference normally shown a young instructor from a backwater college."[3] Not one person in the audience even bothered to ask a follow-up question for Turner during the Q&A.

Yet Turner left a larger impact than he could have ever anticipated that evening at the Chicago world's fair. In the years after 1893, as he continued to dig into, promote, and expand his "Frontier Thesis," his ideas would gain traction.[4] His notions influenced not just fellow American historians but politicians, cultural commenters, and everyday people who came to subscribe to a view of the United States as uniquely expansionist, uniquely individualized, and uniquely suited for democratic greatness all because of a distinct frontier past. Turner's thesis fueled the flames of American exceptionalism, offering a seemingly well-grounded historical interpretation of how the United States had risen to power on the continent, and later, the world stage. For more than a century since Turner's paper at Chicago, historians and Americans have grappled with his ideas and legacy as we try to make sense of the past, and our own identity as a nation.[5]

By now, generations of scholars have called out many of the obviously troubling elements to Turner's view of American history. In Turner's original thesis, women went unnoticed and Native peoples remained absent except as foils for America's "procession of civilization" across the continent.[6] Unspoken violence between settlers and Native Americans formed one of the central questions and "political concern" of successive American frontier zones, according to Turner.[7] But Native people as forceful actors did not factor into his description of the frontier as a transformative element in the history of the continent. Later historians have corrected this neglect with vigor as we learn increasingly more about the complexity and dynamism of Native history on a continent that remained dominated by Indigenous peoples at least until the nineteenth century.[8] Likewise, the frontier in American history is no longer viewed in the progressive tones embraced by Turner's subscribers. Now, historians highlight the shocking violence and destructiveness of the United States' conquest throughout North America within zones of contact and conflict.[9] Despite such revisions, Turner's thesis has proven hard to fully dismiss. As late as 1998, John Mack Faragher, an American historian and careful critic of Turner, still described Turner's paper delivered at Chicago in 1893 as "the single most influential piece of writing in the history of American history."[10] Something beyond mere exceptionalism has kept Turner's thesis around.

Taking Turner's setting into consideration, and examining Chicago's long transition into an American place, offers some clues into the persistence of his frontier explanation. He gave his address at a site that had functioned as a space of contact and colonization from the seventeenth through the nineteenth centuries. It had operated as a crossroads of movement for the region's Native peoples as well as incoming French, British, Spanish, and, eventually, white American settlers. Chicago's portage fit alongside other bottlenecks of geography Turner referenced in his speech, such as the Cumberland Gap and the South Pass. These were spaces where, as Turner proposed, one could trace the progression of the history of the wider continent in localized form over time. Turner offered an account of that history which examined the ways these subsequent waves of people engaged with the spaces of the North American continent as its central narrative. Given the palimpsest of regimes that tried to harness Chicago's key geography over centuries of contact and struggle, Chicago presented a worthy site for Turner to launch his bold ideas about the spatial elements of frontier development.

Turner also spoke with the awareness of someone who had grown up in Portage, Wisconsin. Named for the carrying place that connected the waters of the Fox and Wisconsin Rivers, Turner came of age along a waterborne route of mobility not unlike Chicago's own portage. Even as a professor, Turner frequently went on canoe camping trips in the summers between school years, portaging his canoe with friends and family through the upper reaches of the Great Lakes drainage. He was tangibly in tune with the importance of waterborne mobility and the influential role geography played in humans' relationship with the region's environment (see figure E.1).[11]

Turner continued to develop these environmental and spatial elements in his subsequent scholarship. Adopting an old moniker, he referred to the American settlers of the United States' Midwestern frontier as "men of the Western waters," emphasizing the significance of watersheds and waterborne mobility in American state formation in the trans-Appalachian west.[12] Turner underscored the "vast physiographic provinces, each with its own peculiarities" that lay "across the path" of settler migrations, each providing "a special environment for economic and social transformations." Particularly attentive to the Great Lakes region, Turner placed Chicago as the reigning center of the Middle West, arguing that an attention to the "physiographic" space mattered more to understanding the settlement of this region than any political boundaries. More savvy or self-conscious than some modern historians give him credit for, Turner expressed clear awareness that this region of overlapping environments and watersheds was the United States' first "imperial domain," equating the conquest of the Midwest's prairies and lakes with the later "colonial history and policy" of the United States as a global empire.[13]

FIGURE E.1 Frederick Jackson Turner "on the portage," in Turner's family photo album from a 1908 canoe camping trip to the Boundary Waters and Lake Nipigon. Frederick Jackson Turner Papers. Courtesy of the Huntington Library.

Given Turner's personal connections to the environment and geography of the Great Lakes region, his emphasis on the spatial elements of the frontier makes sense. His assertion that the United States' rise was a story of humans' relationship to land, water, and space resonated with the history of Chicago, and hints at his most enduring, and usable, understanding of American history. If anything, his thesis lacked the environmental and geographic specificity that would have bolstered his arguments about space. His thesis, meant to encapsulate all of American expansion, remained too broad and nebulous to highlight the specific geographic significance and environmental particularity that had proven so influential to the history of contact, interaction, and colonization at sites like Chicago.[14] Yet, boiled down and stripped of its racist triumphalism, Turner's thesis argued that the United States' conquest of space had remained the fundamental driver of American culture, identity, politics, and society for its first century or so of existence. Delivering his paper over Chicago's transformed landscape, Turner's argument—at least when it came to space—was on solid ground.

It is Turner's attention to space and the influence of the American environment—even if vague and unspecific—that has sustained elements of his thesis for so long.[15] He had identified many of the key components in the story of U.S. expansion, but his formula for how these forces came together to explain American history needed further sorting. While he argued that settlers advanced as the primary drivers for the conquest of continental space, he neglected to factor in the strength of the American state in actively reshaping that space and facilitating settler expansion.[16] As Turner had phrased it, the "history of the colonization of the Great West" could be summed up with a simple equation—geographic space plus American settlement equaled U.S. development as a nation-state. Had he delved back into the details of Chicago's transition from an Indigenous space to an American place, he would have found all the parts of his equation. But it had played out in reverse to Turner's thinking. Rather than settler expansion into space, which then led to U.S. state formation, the U.S. development of infrastructure, government, and public works had transformed Chicago's space, facilitating "American settlement westward."[17] Turner stood on a podium overlooking a changed landscape. The process of conquest indeed remained intimately linked with the spatial geography of sites like Chicago, but the state, in the form of the U.S. government, had been the driving force in overhauling space and making way for white settlement there and elsewhere across the continent.

Frederick Jackson Turner was not the only one who had ventured to Chicago to make sense of the continent's history during the world's fair. Simon

Pokagon, a member of the Pokagon Potawatomis, also came to the city that summer as the invited guest of Chicago's mayor. Simon Pokagon traced his Anishinaabe heritage back to the St. Joseph River valley. His father, Leopold Pokagon, had signed the 1833 Treaty of Chicago but managed to navigate his people safely through the era of Native removals and maintain their hold over lands in southwest Michigan. In 1893, Chicago dignitaries asked Pokagon to join in the fair's events as a stand-in for historical Anishinaabe figures. The exhibition featured a series of reenactments from Chicago's early history, including the Battle of Fort Dearborn and the signing of the 1833 treaty that saw Indigenous land rights extinguished by the American government. It is hard to imagine the indignity Simon Pokagon experienced in having to "play Indian" through a series of re-created scenes that had proven so central to his people's own dispossession in the Great Lakes region.[18]

But Pokagon used the opportunity afforded by Chicago's crowds to offer a different interpretation of what had happened at Chicago's portage in the preceding centuries. Accepting the mayor's invitation to stand in for Chicago's Indigenous population, Pokagon nevertheless subverted his play-acting with some actual history. In preparation for his visit to the Columbian Exhibition, Pokagon authored a pamphlet to advance an Anishinaabe account of history across the Americas in the four centuries since Columbus. Pokagon distributed these pamphlets over the course of the fair. Titled, alternatively, the "Red Man's Rebuke," or the "Red Man's Greeting," Pokagon's pamphlet encapsulated an account that gets us nearer to a true picture of Chicago's history than Turner's thesis alone can provide.[19]

Pokagon did not have his booklet printed on paper, but rather chose sturdy birchbark to convey his message. As he explained to his readers in the introduction, this was a conscious choice, "out of loyalty to my own people" who had used the "manifold bark" of "this most remarkable tree" for centuries, to mark down their most important, even sacred, messages. Pokagon understood birchbark to hold powerful properties not simply as a medium for communication, though. He also cataloged the ways his people had utilized bark from the birch tree for their houses, utensils, attire, tools, and religious symbols.[20] It was indeed Anishinaabe mastery of birchbark canoe construction that had endowed them with so much waterborne power at Chicago and across the Great Lakes in earlier times. Pokagon designed his birchbark pamphlets to reflect that power (see figure E.2).[21]

Invoking these earlier times, Pokagon found little to celebrate at Chicago in 1893. As he pointed out to readers, "where stands this 'Queen City of the West,' once stood" a hub of Anishinaabe power. Pokagon described how Chicago's

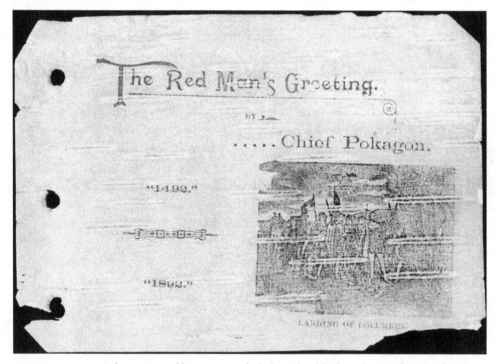

FIGURE E.2 Pokagon printed his message of rebuke on birchbark—a direct reflection of the Anishinaabeg and their enduring power and presence across the Great Lakes region. Simon Pokagon, *The Red Man's Greeting, 1492–1892* (Hartford, Mich.: C. H. Engle, 1893). Courtesy of the Newberry Library.

ground had operated for the Anishinaabeg as "the center of their wide-spread hunting grounds," and a network of mobility that stretched across the Great Lakes, to the Mississippi and even Missouri River valleys, as well as south to "the great salt Gulf." But incoming white settlement had torn asunder this incredible system of Indigenous mobility that had coursed through the local space of Chicago's portage. Explaining this process in terms strikingly similar to Turner, Pokagon described how "the cyclone of civilization rolled westward; the forests of untold centuries were swept away; streams dried up; lakes fell back from their ancient bounds, and all our fathers once loved to gaze upon was destroyed, defaced, or marred." Whereas Turner gilded his narrative in the language of progress, Pokagon told a story of stark environmental and human destruction. Paraphrasing an earlier poet, Pokagon grieved, "The plow is on our hunting grounds. The pale man's ax rings through our woods, the pale man's sail skims o'er floods; our pleasant springs are dry."[22] Incoming white settlement had destroyed Native societies not just through force of conquest but through the degradation of the continent's environments as well.

Pokagon meant the bit of poetry to mourn the wider changes over the Great Lakes land- and waterscape, but his imagery served as a ready epitaph for Chicago's local space as well. Where once had persisted a fluid geography of land and water that the Anishinaabeg had harnessed to craft their own ascendancy in the borderlands, now the urban landscape of Chicago's refurbished space held sway. An artificial harbor welcomed more ships annually than the saltwater ports of New York City, Philadelphia, and San Francisco combined.[23] A labyrinth of wrought-iron railways steamed in resources from a vast hinterland of extraction. Street grids mapped and controlled movement across the dry land while canals diverted waterways into manageable flows. As Pokagon concluded, looking around Chicago's cityscape, "the eagle's eye can find no trace" of the Indigenous terrain that had once defined the area.[24]

While Pokagon's lament left little room for optimism, he did not intend for it to read as yet another declensionist account of waning Native history. This was a declaration of Indigenous persistence despite the alteration of Indigenous space.[25] As the opening lines of the birchbark proclaimed, "On behalf of my people, the American Indians, I hereby declare to you, the pale-faced race that has usurped our lands and homes, that we have no spirit to celebrate with you the great Columbian Fair now being held in this Chicago city." Pokagon brought his birchbark scrolls to Chicago specifically to call out the injustice of past actions and present circumstances. No doubt, the distribution of his message throughout the summer made more than a few fair-goers uneasy in their celebrations of Columbian legacies. Pokagon, even while participating in the festivities of the exposition, had come to Chicago to assert an Indigenous *present* at the portage, as well as a Native past. This Anishinaabe leader would not let Americans celebrate their continent's past without acknowledging the Indigenous experience inherent in that story. Pokagon made unmistakably clear with his dissemination of birchbark manuscripts that any effort to understand American history—at Chicago or elsewhere—had to take Native peoples into account.[26]

Taken together, Turner and Pokagon bring us closer to an accurate understanding of how the historic grounds of Chicago, and the wider North American borderlands, transformed from initial contact into the present day. Colonization and settlement on the continent took place as an ongoing, two-part conquest of both space and people, each facilitating and reinforcing the other. While Turner had downplayed the violence against always-present

Indigenous people in his interpretation of that history, he had accurately articulated the importance of environment and a spatial element to the trajectory of American expansion across the continent. Development and infrastructure, propagated by both settlers at the margins and, more importantly, by government officials at all levels, had worked to transform space across a continent.

Pokagon gets us further, reasserting the centrality of Indigenous peoples to any history of North America's transformation in the centuries leading up to 1893 and beyond. He also expressed an awareness of the environmental and geographic dimensions of this conquest. In his writings, he mourned a vanishing landscape rather than the disappearance of Native people, who remained very much a part of the human geography of the Great Lakes in 1893, as they still do today. His insistence that Indigenous peoples be factored into any understanding of American history brings us closer to a full picture of Chicago before and after U.S. incursion into the region. Pokagon's presence at the world's fair reminded Americans that U.S. expansion entailed not just an advance across space but also a violent invasion and the undermining of an Indigenous world. Together, Turner's thesis and Pokagon's rebuke highlight the terms of U.S. conquest in North America as a twofold process that saw a transformation of geography at the same time the American government worked to divest Native peoples of their lands. Examining Chicago's long, localized history works to rejoin these two major strands in the story of North America's borderlands, and the eventual conquest of the continent by the United States.

Chicago's history, far from exceptional, offers a striking example of a process that played out repeatedly upon the continent. The relationship between the conquest of space and the subjugation of Native people just happens to remain more traceable in a location like Chicago. Here, confounding waterways, a marshy ecotone, and a continental divide between watersheds complicated European visions of order for centuries before the United States was able to render Chicago controllable and, in turn, dispossess the area's Anishinaabeg of their homeland. Even before this eventual outcome, however, the control and manipulation of Chicago's space as a site of settlement, interaction, and movement played a leading role in the history of Indigenous and European relations at the portage and the wider continent. Reintegrating this human story of interaction within the spatial setting of contact offers a fuller depiction of how Native peoples and Europeans coexisted with one another in the places of early America. It also provides a clear explanation for how Euro-Americans, during the era of U.S. expansion, went about subduing both the peoples and spaces of a continent. The history of contact in North America remains a story of specific places of human interaction and the ensuing conquest of those spaces and peoples during the rise of the United States.

Coda

Pokagon and Turner offer two interpretations of the same contested history—alternatively celebrating or lamenting how Chicago and the wider continent shifted from a vast space of Indigenous power to a landscape of settler hegemony. But there is still one key player in this drama yet to offer closing remarks. Chicago's metropolis arose in an environmental zone of lakes, rivers, prairies, oak groves, and mud. The lands and waterways of Chicago persist, despite human engineering that has brought the settler city forward to the present day. Chicago's cityscape may have usurped the native ecosystems of this region decades ago. But whether such an engineered space of human "progress" can long endure remains uncertain. Chicago now sits astride several ecological crises that can all be traced back to its beginnings as a city built upon a low divide between fluctuating wetlands and intertwined rivers.

Chicago has existed as a city for less than two centuries yet faces several environmental challenges that may make its next hundred years a harrowing ordeal. Just downstream from Chicago, in the Illinois River, the once-promising canalways now teem with the biological threat of millions of invasive Asian carp. Electric barriers hold these fish back, but the very waterways that destroyed the portage and made way for U.S. power now threaten to open the door to an aquatic disaster. If these invasive species were to somehow breach the watersheds and enter the Chicago River and wider Great Lakes, the ecological and economic impact would be devastating. These invasive species, the Silver and Bighead carp especially, can eat up to half their body weight in plankton a day, stripping a waterbody of its balanced food web. Where they have secured a niche in the Mississippi drainage, they have come to dominate the riverine ecology. Along some stretches of water, invasive carp make up more than 90 percent of the fish diversity as they crowd out native species. If they were to enter Lake Michigan and spread, scientists fear they would destroy the $7 billion fishing industry and the $16 billion recreational boating industry on the Great Lakes, with a cascade of effects for local and regional economies, as well as tribal fishing harvests farther north. Chicago's riverine bottleneck again serves as the last barrier to invasion, and now, ironically, the best solution may be to sever the connection between the two watersheds and close off the canal channels connecting Chicago to the Illinois River.[27]

Add to this looming fisheries apocalypse the fact that Chicago, the city, is sinking. The rate varies across Chicagoland, but along the lower Lake Michigan basin, the land has sunk somewhere between four and eight inches, on average, over the last hundred years. The effect, in the long run, will lead to an encroaching lakeshore as the waves of Lake Michigan lap farther inland along Chicago's coast. Coupled with this slow sinking of the land and the advancing Great Lake, Chicago is seeing higher-than-average rainfall. Floodwaters continue to rise across Chicago neighborhoods with each passing year. And while city planners tried to account for flooding issues in an earlier era, the 109 miles of tunnels they built to account for storm runoff and overflow sewage may still not be large enough to control Chicago's unconquerable deluges. The low-lying wetlands on which city boosters built their metropolis may not be the foundation of prosperity they once imagined.[28]

Human activity at the continental and global levels has exacerbated these environmental concerns. The troubles over Chicago's sinking skyline, lapping floods, and threatening waterways have only increased in the era of human-induced climate shifts. It turns out that the long-term consequences of destroying carbon-capturing swamps, carving drainage ditches through the lowlands, and breaking the tallgrass prairie of Chicago's ecosystems (to say nothing of the cycles of industrialization, urban abandonment, and rapacious consumerism) may yet prove disastrous. State-backed infrastructure wrested Chicago and the continent from its original owners and wrecked its natural environments, but the victory of "progress" may prove temporary. A reckoning looms for Chicago and the society it has come to represent. It remains too soon to say whether technology or investment can continue to bolster this space as a hub of modernity or whether one day the great settler city will succumb to the mud it presumed to conquer.

Acknowledgments

This book has been a good bit of time in the making. In writing it, I have relied on a considerable number of friends, relatives, mentors, and experts who have proven more prescient, forbearing, and enthusiastic about the work than I could have ever been on my own.

First thanks go to my advisers at Notre Dame, Patrick Griffin and Jon Coleman. Patrick's dogged faith in my abilities allowed me to see myself as a historian in my own right. Jon's wit, close reading, and probing questions made me think about my project, and the field of history, in new and challenging ways. Other mentors during my time in graduate school offered key thoughts and words of encouragement along the way. I thank Rebecca Tinio McKenna, Thomas Kselman, Annie Coleman, Paul Ocobock, Alex Martin, and Brad Gregory as exemplars of how to teach and lift up graduate students. Darren Dochuk was an especially generous teacher to me and gave me my first crack at getting something published. Katie Cangany served as an early adviser before leaving South Bend, but her dedication to my learning never waned. I thank her for her close reading of my work and sustained interest in my project. Our mutual admiration of the Great Lakes has been an ongoing source of conversation and inspiration in the years since she left Notre Dame. Katie Jarvis also deserves thanks for her role on my dissertation committee. But more than that, it was from Katie that I learned how to captivate a lecture hall full of students. I still deploy many of the tips and tricks she taught me in inspiring my students today.

Separate from my formal training in graduate school, Mark Schurr, of Notre Dame's Anthropology Department, and Madeleine McLeester, allowed me to accompany them and briefly participate in an active archaeological site near the forks of the Kankakee and Des Plaines Rivers. The visit was a wonderfully eye-opening experience in fieldwork and helped me to think through the precontact realities of the region. When I was on-site, Joe Wheeler of Midewin National Tallgrass Prairie generously gave me a personal tour and shared his knowledge of the ecology and history of the area.

While advisers and teachers in graduate school played a key role in my development as a scholar, I was also fortunate to be part of a collaborative graduate program at Notre Dame. A cohort of genuinely decent individuals set the tone for my time in South Bend, and I thank Sejoo Kim, Suzanna Krivulskaya, Anna Holdorf, Anna Vincenzi, and Andy Mach for camaraderie in the many classes together. My peers in the Atlantic World Working Group were as good a group to get a beer with as they were to talk eighteenth-century historiography, and our regular workshops were a highlight of my early years at Notre Dame. A special nod to Sam Fisher, who initially organized the group and who has continued to be a close friend and reader of this manuscript in more recent years. It is good to have such a merry soul with which to share writings and witticisms. Jay Miller likewise read key parts of this manuscript in its various forms. Jay is also to be thanked for housing my canoe, the SS *Dorothy Parker*, along the banks of the St. Joseph River for several years during graduate school. Finally, Philip Byers became a close friend and confidant during graduate school despite his journalistic tendencies, and I am happy that our friendship has persisted well past graduation.

Before I ever ventured to graduate school, I had some incredible mentors and friends in my corner at Gettysburg College. Michael Birkner, more than anyone, taught me how to think and write like a historian in his Methods course. Timothy Shannon helped foster my interest in early American history and was an early advocate for me on the road to graduate school. He has continued to be invested in my success to the present day. Josh Poorman and Alex Skufca are two of the best friends a guy could ask for, and each has weighed in on crucial elements of this project at various points. Ian and Sam Isherwood grew into fast friends before I left Gettysburg, and they, too, have become trusted confidants when it comes to my work as a historian.

No good historian could do his or her work without access to archives and, more importantly, access to the great minds of archivists and librarians who know the ins and outs of such repositories. I benefited from the expertise of Brian Leigh Dunnigan and other staff at the William L. Clements Library at the University of Michigan, where I enjoyed a Jacob M. Price Visiting Research Fellowship. An adjacent week of work in the Bentley Library on a Bordin-Gillette Fellowship churned up some archival gems as well. Generous support from a variety of libraries and institutes helped to fund the project, and I am indebted to the Institute for Scholarship in the Liberal Arts at Notre Dame and the Phillips Fund for Native American Research through the American Philosophical Society for funding various research trips over the course of the project. A John H. Holliday Dissertation Fellowship from the Society of Indiana Pioneers helped sponsor final research for this project in 2019. I also wish to thank library staff at the Clements Library, Notre Dame's Hesburgh Library, the Great Lakes-Ohio Valley Ethnohistory Collection at Indiana University, Ohio History Connection Archives and Library, the John Carter Brown Library, the Library and Archives Canada, the Huntington Library, the Chicago History Museum, and the Newberry Library for their help with my queries throughout the research process and their final help with images and permissions.

A berth at the Newberry Library as a graduate Scholar-in-Residence during the academic year of 2017–18 allowed for a perfect venue to begin my work in earnest. Besides camping out at an appropriate locale for a Chicago-centered project, I was able to interact and share work with a rich pool of fellow historians and researchers that indelibly improved the book. I want to thank Keelin Burke and Brad Hunt for facilitating the scholarly programs during my time at the Newberry, and Jim Ackerman for sharing his knowledge of maps within the library's collections. Fellow scholars who shared their time and encouragement over the course of that year included Woody Holton, Susan Johnson, Liesl Olson, Loren Michael Mortimer, Ryan Taycher, Tim Soriano, and Kelly Wisecup. I especially appreciated the ways Julie Pelletier and Susan Sleeper-Smith took me under their wings as I worked to bring Native voices to the fore of the project. Ben Johnson and Michelle Nickerson were generous enough to feed me and let me stay over at their home several times when taking the South Shore Line late at night proved impractical. They even put me up and let me house-sit for a month's worth of research one summer, and it was at their desk overlooking their back garden that I drafted the first chapter of the manuscript.

Chicago's local cohort of history enthusiasts and experts helped bring forth details that would have otherwise escaped my attention. I owe a great debt to John Swenson, who shared his extensive notes about the Chicago portage with me over phone calls and visits. I thank Craig Howard and Richard Gross for their ongoing work on La Salle and the Margry translations at the Detroit Public Library. Timothy Kent generously shared many of his published materials with me, enlivening my work on the French fur trade era. Ann Durkin Keating's excellent works, and a few email exchanges along the way, enriched my local knowledge of Chicago during the American period. No one has done more to emphasize the maritime elements of Chicago's past than Ted Karamanski, and I hope this book helps advance that interpretation along. I also want to thank individuals like Patrick McBriarty, Christopher Lynch, Josh Coles, and Gary Mechanic in their enthusiasm for Chicago's early history. It is such local energy that makes writing a place-based history so rewarding.

This book has continued to improve through the thoughtful input I received from fellow historians at conferences, workshops, and other forums over the past seven years. Many of the people I have spoken to and collaborated with on panels, Q&As, and email exchanges have shaped the project in big and small ways. I want to thank Eliga Gould, Ned Blackhawk, Alec Zuercher Reichardt, Laura Keenan Spero, Max Edelson, Christian Koot, Bryan Rindfleisch, Nathan Holly, Josh Piker, T. H. Breen, Jessica Choppin Roney, Mark Peterson, Zach Bennett, Wes Bishop, Elizabeth Ellis, Dan Richter, Gabe Paquette, Josh Reid, Jacob Lee, Charles Prior, and Sam Truett for their helpful commentary along the way. Bob Morrissey has become a de facto mentor and friend over the years, and I thank him for the many occasions when he has taken the time to offer guidance and professional wisdom. Cole Jones became a good friend during my brief stint living in Lafayette, Indiana, and I appreciate his ongoing scholarly generosity. Pekka Hämäläinen graciously served as an outside member of my dissertation committee and subsequently read early drafts of the entire manuscript. Michael McDonnell also generously read drafts of many of these chapters. Michael Witgen and Rob Harper both read portions of what became chapter 3 and offered crucial insights that helped shape the nuances of my arguments there. Over the process of researching and writing this book, I have been humbled by both the depth of knowledge and the generosity shown by my fellow historians. Many of the attributes of the final product are a result of their expertise, and any flaws that remain are solely my own.

Moving to a new place is never easy, and moving to Lubbock, Texas, amid a global pandemic seemed a potentially isolating prospect. Even still, new friends and colleagues at Texas Tech made that transition not only smooth but rewarding. My thanks to Alan Barenberg and Abby Swingen, Daniella McCahey, Lucia Carminati, Catharine Franklin, Paul Bjerk and Ange Baziwane, Jake Baum and Emily Skidmore, Matt and Lindsay Johnson, Erin-Marie Legacey and Ben Poole, Miguel Levario, Mark Stoll, Aaron and Leah Sanchez, and Gretchen Adams. Dale Kretz and Katie Moore helpfully read and commented on final drafts of the introduction. Mayela Guardiola, Valorie Duvall, and Amanda Chattin have provided the logistical support to get me squared away at Texas Tech, and to facilitate much of the final work on this project. Sean Cunningham has been everything a new faculty member could ask for in a department chair, and I appreciate the warmth he and his wife, Laura, have shown us as friends as well.

As the project took shape into a book, Ben Johnson became a relentless champion of UNC Press and the David J. Weber Series in the New Borderlands History as a potential home for the manuscript. I am indebted to him and Andy Graybill for their encouragement as series editors. I am also grateful to Debbie Gershenowitz for her work as editor, shepherding the project forward. Thanks also to the editorial team at UNC who helped to sharpen prose and correct errors as needed.

Final thanks go to my family, who have been a part of my fascination with this project before it ever became an academic pursuit. Growing up, I inherited an enthusiasm for the natural world from my mother, as well as a love of reading and learning. All those passions shaped this narrative. My father instilled in me a love of history, and some of my earliest memories involve sitting on a stump, in the middle of our woods, while he captivated me with stories of Ohio's frontier past. Dad would have loved to have been a history teacher, and it is my privilege to live out our mutual dream now. Thanks also to my siblings, Mary, Danny, and Rachel Nelson, who have kept me grounded and in good company my entire life. My two sisters are more loved than they know, and Danny has grown from a younger kid brother into a trusted best friend and fellow history buff. The genuine interest in my work from Harish and Ami Doshi in recent years has also been a blessing, as they welcomed me into their lives as a son. When I began this project in graduate school, I had four living grandparents. Since then, all have passed, but I know they would be proud of my progress. I especially wish Grandpa Nelson were still here to read the project in its final form. His sustained interest in my work and unwavering support continue to drive me even now that he is gone.

My immediate family has been the energizing force behind my final push toward completion. Priya Grace arrived just in time to offer input on final edits. She oversaw the submission process from her baby bouncer during her early months of life. Given what I already know about her personality, I anticipate that she will have plenty more substantial suggestions for future writings. But neither the project, nor its author, would be what they are today without the gentle support of Shruti Nelson. Tougher than me yet kinder, Shruti often provided the determination and strength I needed to navigate graduate school, the academic job market, and the final writing. She won't tell you this, but I know she has sacrificed much for the sake of my dreams. I hope it was worthwhile. As for me, she continues to make every effort worth it.

Notes

Abbreviations

Ayer MSS   Edward E. Ayer Manuscript Collection, Newberry Library, Chicago

BHL   Bentley Historical Library, University of Michigan, Ann Arbor

CHM   Chicago History Museum, Chicago

CISHL   *Collections of the Illinois State Historical Library*. 38 vols. Springfield: Illinois State Historical Library, 1903–1978

GLOVE   Great Lakes-Ohio Valley Ethnohistory Collection, Erminie Wheeler-Voegelin Archives, Glenn A. Black Laboratory of Archaeology, Indiana University, Bloomington

GP   Papers of General Thomas Gage, William L. Clements Library, University of Michigan, Ann Arbor

JP   James Sullivan et al., eds., *The Papers of Sir William Johnson*. 14 vols. Albany: University of the State of New York, 1921–1962

JR   Reuben Gold Thwaites, *The Jesuit Relations and Allied Documents: Travels and Explorations of the Jesuit Missionaries in New France, 1610–1791: The Original French, Latin, and Italian Texts, with English Translations and Notes.* 73 vols. Cleveland, Ohio: Burrows Brothers Company, 1896–1901

KP   Jacob Kingsbury Papers, Chicago History Museum, Chicago

MPHC   *The Historical Collections of the Michigan Pioneer and Historical Society.* 40 vols. Lansing: Michigan Historical Commission, 1874–1929

NYCD   E. B. O'Callaghan, Berthold Fernow, and John Romeyn Brodhead, eds., *Documents Relative to the Colonial History of the State of New-York: Procured in Holland, England, and France.* 15 vols. Albany, N.Y.: Weed, Parsons, & Co., 1853–1857

TP   Clarence Edwin Carter, ed., *The Territorial Papers of the United States*, 28 vols. Washington, D.C., 1934–1975

WB   Wilbur Cunningham, ed., *Letter Book of William Burnett* (St. Joseph: R. W. Patterson, for The Fort Miami Heritage Society of Michigan, 1967)

WHC   Reuben Gold Thwaites, *Collections of the State Historical Society of Wisconsin.* 20 vols. Madison: State Historical Society of Wisconsin, 1855–1911

WMQ   *The William and Mary Quarterly*

WP   Anthony Wayne Family Papers, William L. Clements Library, University of Michigan, Ann Arbor

Introduction

1. Adapted from the modern definitions provided in *Merriam-Webster Dictionary*, "portage," www.merriam-webster.com/dictionary/portage.

2. This scene is reconstructed from the memoirs of Gurdon Hubbard. Gurdon Saltonstall Hubbard, *The Autobiography of Gurdon Saltonstall Hubbard, Pa-Pa-Ma-Ta-Be, "the Swift Walker"* (New York: Citadel Press, 1969), 32–33.

3. Hubbard, *Autobiography*, 32–33. For the abundance of wildlife, see 40.

4. Hubbard, *Autobiography*, 41–43. For more on the grueling work of such portage travel, see Ronald S. Lappage, "The Physical Feats of the Voyageur," *Canadian Journal of History of Sport* 15, no. 1 (1984): 30–37; Carolyn Podruchny, *Making the Voyageur World: Travelers and Traders in the North American Fur Trade* (Lincoln: University of Nebraska Press, 2006), 114–133.

5. For Chicago as an early space of cross-cultural encounter and cosmopolitanism, see Jacqueline Peterson, "Goodbye, Madore Beaubien: The Americanization of Early Chicago Society," *Chicago History* (1980), 98–111. For nearby parallels, see Lucy Eldersveld Murphy, *Great Lakes Creoles: A French-Indian Community on the Northern Borderlands, Prairie Du Chien, 1750–1860* (New York: Cambridge University Press, 2014).

6. For Hubbard's marital liaisons, see Hiram Williams Beckwith, *History of Iroquois County, Together with Historic Notes on the Northwest, Gleaned from Early Authors, Old Maps and Manuscripts, Private and Official Correspondence, and Other Authentic, Though, for the Most Part, out-of-the-Way Sources* (Chicago: H. H. Hill, 1880), 4–6. For the extensive literature on these "country marriages," or interethnic sexual relations during the fur trade, see Jennifer S. H. Brown, *Strangers in Blood: Fur Trade Company Families in Indian Country* (Vancouver: University of British Columbia Press, 1980); Sylvia Van Kirk, *Many Tender Ties: Women in Fur-Trade Society, 1670–1870* (Norman: University of Oklahoma Press, 1983); Susan Sleeper-Smith, *Indian Women and French Men: Rethinking Cultural Encounter in the Western Great Lakes* (Amherst: University of Massachusetts Press, 2001). For a broader theoretical and comparative assessment of these relationships, see Ann Laura Stoler, "Tense and Tender Ties: The Politics of Comparison in North American History and (Post) Colonial Studies," *Journal of American History* 88, no. 3 (December 2001): 829–865.

7. Historians studying the history of Euro-Indigenous encounters and eventual U.S. conquest typically point to a shift in power dynamics sometime in the early nineteenth century. For prominent examples that track such a watershed in the Great Lakes and Mississippi River valley during the era of the early republic, see Richard White, *The Middle Ground: Indians, Empires, and Republics in the Great Lakes Region, 1650–1815* (Cambridge: Cambridge University Press, 1991); Eric Hinderaker, *Elusive Empires: Constructing Colonialism in the Ohio Valley, 1673–1800* (Cambridge: Cambridge University Press, 1997); Gregory Evans Dowd, *A Spirited Resistance: The North American Indian Struggle for Unity, 1745–1815* (Baltimore, Md.: Johns Hopkins University Press, 1992); Kathleen DuVal, *The Native Ground: Indians and Colonists in the Heart of the Continent* (Philadelphia: University of Pennsylvania Press, 2006); Michael A. McDonnell, *Masters of Empire: Great Lakes Indians and the Making of America* (New York: Hill and Wang, 2015), esp. 318–327; Jacob F. Lee, *Masters of the Middle Waters: Indian Nations and Colonial Ambitions along the Mississippi* (Cambridge, Mass.: Harvard University Press, 2019). Several recent works place this shift firmly in the 1830s. See Christina Snyder, *Great Crossings: Indians, Settlers, and Slaves in the Age of Jackson* (Oxford: Oxford University Press, 2017); Claudio Saunt, *Unworthy Republic: The Dispossession of Native Americans and the Road to Indian Territory* (New York: Norton, 2020).

8. Speech of John C. Calhoun, February 4, 1817, *Annals of Congress*, House of Representatives, 14th Cong., 2nd sess., 853–854. Space as the "first enemy" of states and nations hear-

kens back to Fernand Braudel's classic work on the Mediterranean basin as a unit of study. Fernand Braudel, *La Méditerranée et le Monde Méditerranéen à l'époque de Philippe II* (Paris: Colin, 1949). For a translation, see *The Mediterranean and the Mediterranean World in the Age of Philip II*, trans. Sian Reynolds (New York: Harper & Row, 1976), quote on 355. Braudel's conception of a Mediterranean world united by waterways has shaped my own approach to North America's Great Lakes region. For the role of space in imperial history, see Lauren Benton, "Spatial Histories of Empire," *Itinerario* 30, no. 3 (2006): 19–34. For land and space as the central issues of U.S. dominion in the Old Northwest specifically, see Peter S. Onuf, *Statehood and Union: A History of the Northwest Ordinance* (Bloomington: Indiana University Press, 1992); Andrew Shankman, ed., *The World of the Revolutionary American Republic: Land, Labor, and the Conflict for a Continent* (New York: Routledge, 2014).

9. For more on "legibility" and the state, see James C. Scott, *Seeing Like a State: How Certain Schemes to Improve the Human Condition Have Failed* (New Haven, Conn.: Yale University Press, 1998).

10. For internal improvements during the early republic, see John Lauritz Larson, *Internal Improvement: National Public Works and the Promise of Popular Government in the Early United States* (Chapel Hill: University of North Carolina Press, 2001); John Lauritz Larson, "'Bind the Republic Together': The National Union and the Struggle for a System of Internal Improvements," *Journal of American History* 74, no. 2 (1987): 363–387; Daniel Walker Howe, *What Hath God Wrought: The Transformation of America, 1815–1848* (New York: Oxford University Press, 2007). For the power of the state in such processes, see Carter Goodrich, *Government Promotion of American Canals and Railroads, 1800–1890* (New York: Columbia University Press, 1960). The literature on Native Removal is extensive, but for northern dispossession efforts, see John P. Bowes, *Land Too Good for Indians: Northern Indian Removal* (Norman: University of Oklahoma Press, 2016). Recently, Claudio Saunt has drawn compelling connections between Indigenous expulsion in the Southeast and the expanding geography of plantation slavery. See Saunt, *Unworthy Republic*. Michael Witgen has likewise explored the political economy of plunder and exploitation that transformed the Old Northwest. See Michael Witgen, *Seeing Red: Indigenous Land, American Expansion, and the Political Economy of Plunder in North America* (Chapel Hill: University of North Carolina Press, 2022). The standard older work on the motivations behind Removal is Reginald Horsman, *The Origins of Indian Removal, 1815–1824* (East Lansing: Michigan State University Press, 1970). See also Snyder, *Great Crossings*, 126–129; Paul Frymer, *Building an American Empire: The Era of Territorial and Political Expansion* (Princeton, N.J.: Princeton University Press, 2017); Karim M. Tiro, "The View from Piqua Agency: The War of 1812, the White River Delawares, and the Origins of Indian Removal," *Journal of the Early Republic* 35, no. 1 (2015): 25–54; John T. Fierst, "Rationalizing Removal: Anti-Indianism in Lewis Cass's North American Review Essays," *Michigan Historical Review* 36, no. 2 (2010): 1–35; Leonard A. Carlson and Mark A. Roberts, "Indian Lands, 'Squatterism,' and Slavery: Economic Interests and the Passage of the Indian Removal Act of 1830," *Explorations in Economic History* 43, no. 3 (2006): 486–504; Patrick Griffin, "Reconsidering the Ideological Origin of Indian Removal: The Case of the Big Bottom 'Massacre,'" in *The Center of a Great Empire: The Ohio Country in the Early Republic*, ed. Andrew R. L. Cayton and Stuart D. Hobbs (Athens: Ohio University Press, 2005), 11–35.

11. For the idea of "imaginative geography," as a driving force of expansion, settlement, and cultural encounter, see Elliott West, *The Contested Plains: Indians, Goldseekers, and the Rush to Colorado* (Lawrence: University Press of Kansas, 1998). For an argument about disparate visions for the landscape as the central contention between Native peoples and incoming Europeans, see James H. Merrell, *Into the American Woods: Negotiators on the Pennsylvania Frontier* (New York: Norton, 2000).

12. Casting the rise of U.S. hegemony as a two-pronged conquest offers an expansion of Max Weber's classic definition of state capacity as a "monopoly on the legitimate use of physical force," demonstrating that physical force can be deployed against both people and environments. Max Weber, "Politics as a Vocation," in *From Max Weber: Essays in Sociology*, ed. H. H. Gerth and C. Wright Mills (1946, repr., New York, 1958), 77–128, quote from 78. For more on "state landscaping" as a key facet of state power from earliest human history onward, see James C. Scott, *Against the Grain: A Deep History of the Earliest States* (New Haven, Conn.: Yale University Press, 2017), esp. 23.

13. For military explanations, see David Curtis Skaggs and Larry L. Nelson, eds., *The Sixty Years' War for the Great Lakes, 1754–1814* (East Lansing: Michigan State University Press, 2001). For diplomacy, Colin G. Calloway, *Crown and Calumet: British-Indian Relations, 1783–1815* (Norman: University of Oklahoma Press, 1987); Timothy D. Willig, *Restoring the Chain of Friendship British Policy and the Indians of the Great Lakes, 1783–1815* (Lincoln: University of Nebraska Press, 2008). For racial violence, Patrick Griffin, *American Leviathan: Empire, Nation, and Revolutionary Frontier* (New York: Hill and Wang, 2007); Peter Silver, *Our Savage Neighbors: How Indian War Transformed Early America* (New York: Norton, 2008). For economics, Richard White, *The Roots of Dependency: Subsistence, Environment, and Social Change among the Choctaws, Pawnees, and Navajos* (Lincoln: University of Nebraska Press, 1983); John Reda, *From Furs to Farms: The Transformation of the Mississippi Valley, 1762–1825* (Ithaca, N.Y.: Cornell University Press, 2016). For population, James Belich, *Replenishing the Earth: The Settler Revolution and the Rise of the Anglo-World, 1783–1939* (Oxford: Oxford University Press, 2009). For settler-colonialism as an explanatory concept, see Patrick Wolfe, "Settler Colonialism and the Elimination of the Native," *Journal of Genocide Research* 8, no. 4 (2006): 387–409; Lorenzo Veracini, *Settler Colonialism: A Theoretical Overview* (New York: Palgrave Macmillan, 2010). In American context, see Frederick E. Hoxie, "Retrieving the Red Continent: Settler Colonialism and the History of American Indians in the US," *Ethnic and Racial Studies* 31, no. 6 (September 2008): 1153–1167; John Mack Faragher, "Commentary: Settler Colonial Studies and the North American Frontier," *Settler Colonial Studies* 4, no. 2 (2013): 1–11. For its application in explaining the U.S. conquest of the Great Lakes, see Bethel Saler, *The Settlers' Empire: Colonialism and State Formation in America's Old Northwest* (Philadelphia: University of Pennsylvania Press, 2015). For comparisons and critiques of the explanatory power of settler colonial processes around the world and in U.S. context, see Jessica Choppin Roney, "Containing Multitudes: Time, Space, the United States, and Vast Early America," *WMQ* 78, no. 2 (April 2021): 261–268; "Forum: Settler Colonialism in Early American History," *WMQ* 76, no. 3 (July 2019): 361–450; Charles Prior, "Beyond Settler Colonialism," *Journal of Early American History* 9, nos. 2–3 (2019): 93–117; Gregory Evans Dowd, "Indigenous Peoples without the Republic," *Journal of American History* 104, no. 1 (2017): 19–41; Belich, *Replenishing the Earth*.

14. For Anishinaabe survivance strategies and challenges in the Great Lakes region, see Ben Secunda, "In the Shadow of the Eagle's Wings: The Effects of Removal on the Unremoved Potawatomi" (PhD diss., University of Notre Dame, 2008); John N. Low, *Imprints: The Pokagon Band of Potawatomi Indians and the City of Chicago* (East Lansing: Michigan State University Press, 2016); Witgen, *Seeing Red*. For "survivance," as an Indigenous concept, see Gerald Vizenor, *Fugitive Poses: Native American Scenes of Absence and Presence* (Lincoln: University of Nebraska Press, 1998), 15. Some scholars have recently proposed new terminology for Native Removal as perpetrated by the United States during the early republic—descriptors like deportation, expulsion, elimination, extermination, and ethnic cleansing all have the potential to more accurately capture the forced nature of Indigenous dispossession during this era. I have chosen to use *removal* and *expulsion* interchangeably, as well as dispossession more broadly. *Removal* remains the most recognizable term for many students of U.S. history, while *expulsion* gets closer to the forcible nature of these efforts by the U.S. government. I have not embraced the use of *deportation* since this has the potential to reify U.S. sovereignty in Native homelands. For these terms, their historical usage, and their limitations, see Saunt, *Unworthy Republic*, xii–xv; Emilie Connolly, "Fiduciary Colonialism: Annuities and Native Dispossession in the Early United States," *American Historical Review* 127, no. 1 (April 2022): 223, fn. 2.

15. For an excellent geographical view of how space and place shaped encounter in this age, see D. W. Meinig, *The Shaping of America: A Geographical Perspective on 500 Years of History*, vol. 1, 4 vols. (New Haven, Conn.: Yale University Press, 1986). See also Terry G. Jordan and Matti E. Kaups, *The American Backwoods Frontier: An Ethnic and Ecological Interpretation* (Baltimore, Md.: Johns Hopkins University Press, 1989). For two foundational works that foreground environmental factors in the history of colonialism, see Alfred W. Crosby, *Ecological Imperialism: The Biological Expansion of Europe, 900–1900*, 2nd ed. (New York: Cambridge University Press, 2004); William Cronon, *Changes in the Land: Indians, Colonists, and the Ecology of New England* (New York: Hill and Wang, 1983).

16. For this later story of Chicago as a hub of commercial and ecological exploitation, see William Cronon, *Nature's Metropolis: Chicago and the Great West* (New York: Norton, 1991).

17. While ubiquitous, portages remain understudied by modern historians as a distinct geographic feature in the landscapes of early America. A few, older exceptions include Archer Butler Hulbert, *Portage Paths, the Keys of the Continent* (Cleveland, Ohio: A. H. Clark, 1903); T. C. Elliott, "The Dalles-Celilo Portage; Its History and Influence," *Quarterly of the Oregon Historical Society* 16, no. 2 (1915): 133–174. The best overview of portage networks in the Great Lakes region remains Claiborne Skinner, "The Sinews of Empire: The Voyageurs and the Carrying Trade of the 'Pays d'en Haut,' 1681–1754" (PhD diss., University of Illinois, Chicago, 1991). For an assessment of the continued importance of historic portage sites in economics and geography, see Hoyt Bleakley and Jeffrey Lin, "Portage and Path Dependence," *Quarterly Journal of Economics* 127, no. 2 (2012): 587–644. While portages were a key feature to Indigenous geographic logic across North America, there are other examples of portages outside Native America. Notable examples occurred in the ancient Mediterranean, the Viking hinterlands, and interior Russia. David K. Pettegrew, *The Isthmus of Corinth: Crossroads of the Mediterranean World* (Ann Arbor: University of Michigan Press, 2016); Rune Edberg and Lennart Widerberg, "In the Wake of the Vikings through Russia:

Going Upstream, Crossing Portages, Avoiding Anachronisms," *International Journal of Nautical Archaeology* 46, no. 2 (2017): 449–453; Robert Joseph Kerner, *The Urge to the Sea: The Course of Russian History: The Role of Rivers, Portages, Ostrogs, Monasteries, and Furs* (Berkeley: University of California Press, 1946).

18. Angela Pulley Hudson, *Creek Paths and Federal Roads: Indians, Settlers, and Slaves and the Making of the American South* (Chapel Hill: University of North Carolina Press, 2010); Laurence M. Hauptman, *Conspiracy of Interests: Iroquois Dispossession and the Rise of New York State* (Syracuse, N.Y.: Syracuse University Press, 1999); John Lauritz Larson, *Laid Waste!: The Culture of Exploitation in Early America* (Philadelphia: University of Pennsylvania Press, 2019). For more on the ways mobility factored into resistance as well as settler programs of conquest, see Matt Cohen, *The Networked Wilderness: Communicating in Early New England* (Minneapolis: University of Minnesota Press, 2010); Katherine Grandjean, *American Passage: The Communications Frontier in Early New England* (Cambridge, Mass.: Harvard University Press, 2015); Alejandra Dubcovsky, *Informed Power: Communication in the Early American South* (Cambridge, Mass.: Harvard University Press, 2016); Christine M. Delucia, *Memory Lands: King Philip's War and the Place of Violence in the Northeast* (New Haven, Conn.: Yale University Press, 2018), esp. 49.

19. This view of Chicago follows the work of others who have looked for sites of Indigenous power across the borderlands of early America in what we might call "Native Grounds"—spaces dominated by Indigenous peoples for much of the colonial period. For prominent examples, see DuVal, *Native Ground*; McDonnell, *Masters of Empire*; Lee, *Masters of the Middle Waters*; Elizabeth A. Fenn, *Encounters at the Heart of the World: A History of the Mandan People* (New York: Hill and Wang, 2014).

20. John Riley, *The Once and Future Great Lakes Country: An Ecological History* (Montreal: McGill-Queen's University Press, 2013), 5; Grahame Larson and Randall Schaetzl, "Origin and Evolution of the Great Lakes," *Journal of Great Lakes Research* 27, no. 4 (2001): 518–546.

21. Hubbard, *Autobiography*, 41.

22. James H. Merrell, "Coming to Terms with Early America," *WMQ* 69, no. 3 (2012): 535–540; Merrell, "Second Thoughts on Colonial Historians and American Indians," *WMQ* 69, no. 3 (2012): 451–512. See also Susan Sleeper-Smith's use of the term in her recent *Indigenous Prosperity and American Conquest: Indian Women of the Ohio River Valley, 1690–1792* (Chapel Hill: University of North Carolina Press, 2018), and "A Note on Terminology" in Gregory Ablavsky's *Federal Ground: Governing Property and Violence in the First U.S. Territories* (Oxford: Oxford University Press, 2021).

23. To get a sense of the various Native peoples who operated around Chicago and continue to claim the region as part of their homeland, see the reports of the U.S. Claims Commission, such as Joseph Jablow, *Illinois, Kickapoo, and Potawatomi Indians*, American Indian Ethnohistory. North Central and Northeastern Indians (New York: Garland Publishing, 1974); Erminie Wheeler-Voegelin and Emily J. Blasingham, *Indians of Western Illinois and Southern Wisconsin: Anthropological Report on the Indian Occupancy of Royce Areas 77 and 78* (New York: Garland Publishing, 1974), as well as "A Guide to the Peoples of Chicago's Portage" in this work.

24. As M. J. Morgan has argued when discussing similar patterns of mobility in the Illinois Country, "human passage" through specific environments was often as historically

significant as human settlement. M. J. Morgan, *Land of Big Rivers: French and Indian Illinois, 1699–1778* (Carbondale: Southern Illinois University Press, 2010), 11–12. See also Vizenor, *Fugitive Poses.*

25. Terry Straus, *Native Chicago* (Chicago: Albatross, 2002); James B. LaGrand, *Indian Metropolis: Native Americans in Chicago, 1945–75* (Urbana: University of Illinois Press, 2002); R. David Edmunds, ed., *Enduring Nations: Native Americans in the Midwest* (Urbana: University of Illinois Press, 2008); Rosalyn R. LaPier, *City Indian: Native American Activism in Chicago, 1893–1934* (Lincoln: University of Nebraska Press, 2015); John N. Low, *Imprints: The Pokagon Band of Potawatomi Indians and the City of Chicago* (East Lansing: Michigan State University Press, 2016).

26. Only recently have historians begun exploring the important waterborne contact zones that facilitated colonial incursions and fostered Indigenous resistance. So far, most of these works have continued to focus on the coastal waterways of the continent, but increasingly, scholars are pushing their analysis upstream into the North American interior. See John R. Gillis, *Islands of the Mind: How the Human Imagination Created the Atlantic World* (New York: Palgrave Macmillan, 2004); Andrew Lipman, *The Saltwater Frontier: Indians and the Contest for the American Coast* (New Haven, Conn.: Yale University Press, 2015); Joshua L. Reid, *The Sea Is My Country: The Maritime World of the Makahs, an Indigenous Borderlands People* (New Haven, Conn.: Yale University Press, 2015); Matthew Bahar, *Storm of the Sea: Indians and Empires in the Atlantic's Age of Sail* (Oxford: Oxford University Press, 2018). For attention to interior waterways, see Zachary M. Bennett, "'A Means of Removing Them Further from Us': The Struggle for Waterpower on New England's Eastern Frontier," *New England Quarterly* 90, no. 4 (2017): 540–560; Lee, *Middle Waters.* For parallel ways in which waterscapes shaped power dynamics between Africans and Europeans, see Kevin Dawson, *Undercurrents of Power: Aquatic Culture in the African Diaspora* (Philadelphia: University of Pennsylvania Press, 2018).

27. For an early invocation for continental history, see Andrew R. L. Cayton, "Writing North American History," *Journal of the Early Republic* 22, no. 1 (2002): 105–111. For tensions between continental and Atlantic schools of thought, see Juliana Barr, "The Red Continent and the Cant of the Coastline," *WMQ* 69, no. 3 (2012): 521–526; Michael Witgen, "Rethinking Colonial History as Continental History," *WMQ* 69, no. 3 (2012): 527–530. See also Michael Witgen, *An Infinity of Nations: How the Native New World Shaped Early North America* (Philadelphia: University of Pennsylvania Press, 2012); Daniel K. Richter, *Facing East from Indian Country: A Native History of Early America* (Cambridge, Mass.: Harvard University Press, 2001). For Indigenous peoples in an Atlantic framework, its usefulness and limitations, see Jace Weaver, *The Red Atlantic: American Indigenes and the Making of the Modern World, 1000–1927* (Chapel Hill: University of North Carolina Press, 2014); Paul Cohen, "Was There an Amerindian Atlantic? Reflections on the Limits of a Historiographical Concept," *History of European Ideas* 34, no. 4 (2008): 388–410; Nathaniel F. Holly, "Transatlantic Indians in the Early Modern Era," *History Compass* 14, no. 10 (2016): 522–532. For an overview of Atlantic World frameworks, see David Armitage, "Three Concepts of Atlantic History," in *The British Atlantic World, 1500–1800,* ed. David Armitage and Michael J. Braddick (New York: Palgrave Macmillan, 2002); Bernard Bailyn, *Atlantic History: Concept and Contours* (Cambridge, Mass.: Harvard University Press, 2005); Alison Games, "Atlantic History: Definitions, Challenges, and Opportunities," *American Historical Review* 111, no. 3

(2006): 741–757. For novel works of scholarship bridging continental and Atlantic spheres, see François Furstenberg, "The Significance of the Trans-Appalachian Frontier in Atlantic History," *American Historical Review* 113, no. 3 (2008): 647–677; Catherine Cangany, *Frontier Seaport: Detroit's Transformation into an Atlantic Entrepôt* (Chicago: University of Chicago Press, 2014). My own thinking on these intertwined amphibious "networks and nodes" of contact draws on Samuel Truett, "Settler Colonialism and the Borderlands of Early America," *WMQ* 76, no. 3 (2019): 435–442, quote on 438. Literary scholars have proven more aware of the interplay between waterways and landscapes in the history of European expansion and American colonization. See, for instance, the works of John Seelye, including *Prophetic Waters: The River in Early American Life and Literature* (New York: Oxford University Press, 1977); *Beautiful Machine: Rivers and the Republican Plan, 1755–1825* (New York: Oxford University Press, 1991). Some recent examples include Stephanie LeMenager, *Manifest and Other Destinies: Territorial Fictions of the Nineteenth-Century United States* (Lincoln: University of Nebraska Press, 2004); Lowell Duckert, *For All Waters: Finding Ourselves in Early Modern Wetscapes* (Minneapolis: University of Minnesota Press, 2017); Michele Currie Navakas, *Liquid Landscape: Geography and Settlement at the Edge of Early America* (Philadelphia: University of Pennsylvania Press, 2018).

28. The Laurentian Great Lakes basin is indeed unique in world geography, both in terms of scale—the lakes span 95,160 square miles of open water—and in potability—the watershed holds roughly 21 percent of the world's fresh water. Gales on the lakes produce winds nearing hurricane speed and waves topping thirty-eight feet. Thus, the lakes offer both the benefits of interior freshwater access and sustenance while also posing the challenges of the open sea. Ecologically, the lakes also offer a distinct bounty of waterfowl, aquatic fauna, wetlands and shoreline flora, and anadromous fish species. See Reeve M. Bailey and Gerald R. Smith, "Origin and Geography of the Fish Fauna of the Laurentian Great Lakes Basin," *Canadian Journal of Fisheries and Aquatic Sciences* 38, no. 12 (1981): 1539–1561; Riley, *Once and Future Great Lakes Country*; Jamie Benidickson, "From Boundary Waters to Watersheds: Legal Change and the Geography of the Great Lakes-St. Lawrence System," *Canadian Geographer/Le Géographe Canadien* 60, no. 4 (2016): 435–445; Jerry Dennis, *The Living Great Lakes: Searching for the Heart of the Inland Seas* (New York: Thomas Dunne Books, 2003).

29. The paradoxical nature of Chicago as part-collaborative bridge, part-contested barrier fits with the widely accepted definitions of a borderland. See the characteristic description in Pekka Hämäläinen and Samuel Truett, "On Borderlands," *Journal of American History* 98, no. 2 (2011): 338–361. In his work, James Merrell has observed similar no-man's-lands in the proverbial Pennsylvanian woods between colonial settlements and Native villages; see Merrell, *American Woods*. For an earlier iteration of borderlands in U.S. context, see Jeremy Adelman and Stephen Aron, "From Borderlands to Borders: Empires, Nation-States, and the Peoples in Between in North American History," *American Historical Review* 104, no. 3 (1999): 814–841. For the most cogent recent interpretation of Chicago as a contested space, see Ann Durkin Keating, *Rising Up from Indian Country: The Battle of Fort Dearborn and the Birth of Chicago* (Chicago: University of Chicago Press, 2012), though this work almost exclusively focuses on the American period of Chicago's colonial past.

30. As the political scientist James Scott has noted in his studies of state power through world history, "The work of civilization, or more precisely the state . . . consists in the elim-

ination of mud and its replacement by its purer constituents, land and water." Scott, *Against the Grain*, 56. For other examples of the relationship between state capacity, muddy environments, and drainage efforts, see Rohan D'Souza, *Drowned and Dammed: Colonial Capitalism and Flood Control in Eastern India* (New York: Oxford University Press, 2016); Eric H. Ash, *The Draining of the Fens: Projectors, Popular Politics, and State Building in Early Modern England* (Baltimore, Md.: Johns Hopkins University Press, 2017); Karl Butzer, *Early Hydraulic Civilization in Egypt* (Chicago: Chicago University Press, 1976). See also Vittoria Di Palma, *Wasteland: A History* (New Haven, Conn.: Yale University Press, 2014); S. Max Edelson, "Clearing Swamps, Harvesting Forests: Trees and the Making of a Plantation Landscape in the Colonial South Carolina Lowcountry," *Agricultural History* 81, no. 3 (2007): 381–406. For related work on wider hydrology and state capacity, see Simon Schama, *Landscape and Memory* (New York: Knopf: distr. by Random House, 1995), 257–262; Donald Worster, *Rivers of Empire: Water, Aridity, and the Growth of the American West* (New York: Pantheon Books, 1985).

31. The idea of expansion across space as the defining characteristic of American history goes back to Frederick Jackson Turner and his infamous "Frontier Thesis," but such sweeping analyses continue to obscure the particulars of localized interactions and "preserve the blind spots of frontier history," as one scholar has put it. For the historiographic significance of Turner's thesis, see William Cronon, "Revisiting the Vanishing Frontier: The Legacy of Frederick Jackson Turner," *Western Historical Quarterly* 18, no. 2 (1987): 157–176; Martin Ridge, "Turner the Historian: A Long Shadow (A Centennial Symposium on the Significance of Frederick Jackson Turner)," *Journal of the Early Republic* 13, no. 2 (1993): 133–144; Patricia Nelson Limerick, "Turnerians All: The Dream of a Helpful History in an Intelligible World," *American Historical Review* 100, no. 3 (1995): 697–716; Frederick Jackson Turner, *Rereading Frederick Jackson Turner: "The Significance of the Frontier in American History," and Other Essays*, ed. John Mack Faragher (New Haven, Conn.: Yale University Press, 1998). For alternatives to Turnerian theories of spatial history, see David J. Weber, "Turner, the Boltonians, and the Borderlands," *American Historical Review* 91 (1986): 66–81. For critiques of settler colonial framings as repeating the expansive vagueness of Turner's frontier method, see Truett, "Settler Colonialism and the Borderlands of Early America" (quote on 437); Roney, "Containing Multitudes," 261–268.

32. For the case for localized and grounded approaches to early American history, see Peter C. Mancall, "Pigs for Historians: Changes in the Land and Beyond," *WMQ* 67, no. 2 (April 2010): 347–375; Eric Hinderaker and Rebecca Horn, "Territorial Crossings: Histories and Historiographies of the Early Americas," *WMQ* 67, no. 3 (July 2010): 395–432; Karen Halttunen, "Grounded Histories: Land and Landscape in Early America," *WMQ* 68, no. 4 (2011): 513–532. See also Christopher Grasso and Peter C. Mancall, eds., "Forum: World and Ground," *WMQ,* 74, no. 2 (April 2017): 195–202; Matthew Dennis, "Cultures of Nature: to ca. 1810," in *A Companion to American Environmental History*, ed. Douglas Cazaux Sackman (Oxford: Wiley-Blackwell, 2010), 214–245. For "collaboration" as a useful lens for studying Indigenous interactions with European empires, see Robert Michael Morrissey, *Empire by Collaboration—Indians, Colonists, and Governments in Colonial Illinois Country* (Philadelphia: University of Pennsylvania Press, 2015). For the effectiveness of community- and microlevel studies in early American Indigenous history, see Helen Hornbeck Tanner, "The Glaize in 1792: A Composite Indian Community," *Ethnohistory* 25, no. 1 (1978): 15–39; James H. Merrell,

"Shamokin, 'the very seat of the Prince of darkness': Unsettling the Early American Frontier," in Andrew R. L. Cayton and Fredrika J. Teute, *Contact Points: American Frontiers from the Mohawk Valley to the Mississippi, 1750–1830* (Chapel Hill: University of North Carolina Press, 1998), 16–59; Joshua Piker, *Okfuskee: A Creek Indian Town in Colonial America* (Cambridge, Mass.: Harvard University Press, 2004); David L. Preston, *The Texture of Contact: European and Indian Settler Communities on the Frontiers of Iroquoia, 1667–1783* (Lincoln: University of Nebraska Press, 2009); Robert Paulett, *An Empire of Small Places: Mapping the Southeastern Anglo-Indian Trade, 1732–1795* (Athens: University of Georgia Press, 2012); Joshua Piker, *The Four Deaths of Acorn Whistler: Telling Stories in Colonial America* (Cambridge, Mass.: Harvard University Press, 2013); Murphy, *Great Lakes Creoles*; Patrick Bottiger, *The Borderland of Fear: Vincennes, Prophetstown, and the Invasion of the Miami Homeland* (Lincoln: University of Nebraska Press, 2016); Bryan C. Rindfleisch, *George Galphin's Intimate Empire: The Creek Indians, Family, and Colonialism in Early America* (Tuscaloosa: University Alabama Press, 2019).

33. For recent historiographic debates about "Vast Early America," see Gordon S. Wood, "History in Context," *Washington Examiner*, February 23, 2015; Joshua Piker, "Getting Lost," Omohundro Institute of Early American History and Culture (blog), January 21, 2016, https://blog.oieahc.wm.edu/getting-lost/; Johann N. Neem, "From Polity to Exchange: The Fate of Democracy in the Changing Fields of Early American Historiography," *Modern Intellectual History* (November 2018): 1–22; Karin Wulff, "Must Early America Be Vast?" Omohundro Institute of Early American History and Culture (blog), May 2, 2019, https://blog.oieahc.wm.edu/must-early-america-be-vast/; "Forum: Situating the United States in Vast Early America," *WMQ* 78, no. 2 (April 2021): 187–280.

A Guide to the Peoples of Chicago's Portages

1. See, for example, the treaty of August 1816 in George Fay, ed., *Treaties between the Potawatomi Tribe of Indians and the United States of America, 1789–1867* (Greeley: University of Northern Colorado, 1971), 38–39.

2. For the Anishinaabeg as the "Three Fires," see James A. Clifton, *People of the Three Fires: The Ottawa, Potawatomi and Ojibway of Michigan* (Grand Rapids: Michigan Indian Press, Grand Rapids Inter-Tribal Council, 1986); Donald L. Fixico, "The Alliance of the Three Fires in Trade and War, 1630–1812," *Michigan Historical Review* 20, no. 2 (Fall 1994): 1–23. For good recent work on the Odawas during this period, see McDonnell, *Masters of Empire*. For the Ojibwe, see Witgen, *Infinity of Nations*. For the Potawatomi, see R. David Edmunds, *The Potawatomis: Keepers of the Fire* (Norman: University of Oklahoma Press, 1978); Low, *Imprints*. For Anishinaabe kinship structures and identity, see Heidi Bohaker, "'Nindoodemag': The Significance of Algonquian Kinship Networks in the Eastern Great Lakes Region, 1600–1701," *WMQ* 63, no. 1 (2006): 23–52. For Anishinaabe governance and diplomacy, see Rebecca Kugel, *To Be the Main Leaders of Our People: A History of Minnesota Ojibwe Politics, 1825–1898* (East Lansing: Michigan State University Press, 1998); Cary Miller, *Ogimaag: Anishinaabeg Leadership, 1760–1845* (Lincoln: University of Nebraska Press, 2010); Heidi Bohaker, *Doodem and Council Fire: Anishinaabe Governance through Alliance* (Toronto: University of Toronto Press, 2020).

3. For a recent work that examines the influence the Oceti Sakowin had over the history of the Great Lakes and upper Mississippi River, see Pekka Hämäläinen, *Lakota America: A New History of Indigenous Power* (New Haven, Conn.: Yale University Press, 2019). See also Gary Clayton Anderson, *Kinsmen of Another Kind: Dakota-White Relations in the Upper Mississippi Valley, 1650–1862* (Lincoln: University of Nebraska Press, 1984); Guy Gibbon, *The Sioux: The Dakota and Lakota Nations* (Malden, Mass.: Blackwell, 2003).

4. The historiography on the Haudenosaunee is extensive, but for a work related to their activities in the western Great Lakes and Illinois Country, see José António Brandão, *Your Fyre Shall Burn No More Iroquois Policy toward New France and Its Native Allies to 1701* (Lincoln: University of Nebraska Press, 1997). For more general overviews, see Daniel K. Richter, *The Ordeal of the Longhouse: The Peoples of the Iroquois League in the Era of European Colonization* (Chapel Hill: University of North Carolina Press, 1992); Jon Parmenter, "After the Mourning Wars: The Iroquois as Allies in Colonial North American Campaigns, 1676–1760," *WMQ* 64, no. 1 (2007): 39–76; Timothy J. Shannon, *Iroquois Diplomacy on the Early American Frontier* (New York: Viking, 2008); Jon Parmenter, *The Edge of the Woods: Iroquoia, 1534–1701* (East Lansing: Michigan State University Press, 2010).

5. For Ho-Chunk history, see Paul Radin, *The Winnebago Tribe* (Lincoln: Bison Books, 1990); Patty Loew, *Indian Nations of Wisconsin: Histories of Endurance and Renewal* (Madison: Wisconsin Historical Society Press, 2001). For Oneota archaeology, see William Green and David W Benn, eds., *Oneota Archaeology: Past, Present, and Future* (Iowa City: University of Iowa Press, 1995).

6. For two good recent works on the Illinois people during the colonial period, see Morrissey, *Empire by Collaboration*; Lee, *Middle Waters*. For a dated, but good, population geography of the Illinois people and their relations with the French, see Joseph Zitomersky, *French Americans—Native Americans in eighteenth-century French colonial Louisiana: The population geography of the Illinois Indians, 1670s–1760s: the form and function of French-Native settlement relations in eighteenth century Louisiana* (Lund: Lund University Press, 1994).

7. Jablow, *Illinois, Kickapoo, and Potawatomi Indians*; for an older ethnohistorical account, see Arrell Morgan Gibson, *The Kickapoos: Lords of the Middle Border* (Norman: University of Oklahoma Press, 1963). For very early ethnographic work, see Charles Christopher Trowbridge, "Kēēkarpoo Indians" (1823), C. C. Trowbridge Papers Circa 1823–1840, Box 1, BHL.

8. For an overview of the Mascouten and clarity as to their "mysterious" identity, see Ives Goddard, "Historical and Philological Evidence Regarding the Identification of the Mascouten," *Ethnohistory* 19, no. 2 (1972): 123–134; David A. Baerreis, Erminie Wheeler-Voegelin, and Remedios Wycoco-Moore, "Appendix A: The Identity of the Mascoutens, Royce Area 148[a]," in *Indians of Northeastern Illinois* (New York: Garland Publishing, 1974), 247–299.

9. The best work on the Meskwaki experience during the colonial period remains R. David Edmunds and Joseph L. Peyser, *The Fox Wars: The Mesquakie Challenge to New France* (Norman: University of Oklahoma Press, 1993).

10. For further reading on mixed-ancestry peoples in the Great Lakes and elsewhere, see Jacqueline Peterson, "Prelude to Red River: A Social Portrait of the Great Lakes Métis," *Ethnohistory* 25, no. 1 (1978): 41–67; Jacqueline Peterson and Jennifer S. H. Brown, *The New Peoples: Being and Becoming Métis in North America* (Lincoln: University of Nebraska,

1985); Patrick J. Jung, "The Mixed Race Métis of the Great Lakes Region and Early Milwaukee," *Milwaukee History: The Magazine of the Milwaukee County Historical Society* 25, no. 3–4 (Fall–Winter 2002): 45–55; Lucy Eldersveld Murphy, "Public Mothers: Native American and Métis Women as Creole Mediators in the Nineteenth Century Midwest," *Journal of Women's History* 14, no. 4 (Winter 2003): 142–166; Murphy, "Métis," in *The Encyclopedia of Chicago* (Chicago: University of Chicago Press, 2005); Anne Farrar Hyde, *Empires, Nations, and Families: A History of the North American West, 1800–1860* (Lincoln: University of Nebraska Press, 2011); Nicole St-Onge, Carolyn Podruchny, and Brenda Macdougall, eds., *Contours of a People: Métis Family, Mobility, and History* (Norman: University of Oklahoma Press, 2012); Tony Belcourt, "For the Record . . . On Métis Identity and Citizenship Within the Métis Nation," *Aboriginal Policy Studies (Edmonton, Alberta, Canada)* 2, no. 2 (2013): 128–141; Chris Andersen, *"Métis": Race, Recognition, and the Struggle for Indigenous Peoplehood* (Vancouver: UBC Press, 2014); Murphy, *Great Lakes Creoles*; Michel Hogue, *Métis and the Medicine Line: Creating a Border and Dividing a People* (Chapel Hill: University of North Carolina Press, 2015); Gerhard John Ens and Joe Sawchuk, eds., *From New Peoples to New Nations: Aspects of Métis History and Identity from the Eighteenth to the Twenty-First Centuries* (Toronto: University of Toronto Press, 2016); James LaForest, "A Métis Family in the Detroit River Region and Pays d'en Haut," *Michigan Historical Review* 42, no. 2 (2016): 67–80.

11. Charles Christopher Trowbridge, "Traditions, Manners, and Customs of the Twaatwää or Meeärmeear Indians" (1823), C. C. Trowbridge Papers Circa 1823–1840, Box 1, BHL.

12. The best source for up-to-date Miami history can be found through the Myaami Center. This academic and cultural hub, affiliated with both Miami University of Ohio and the Miami Tribe of Oklahoma, is leading the field in linguistic and historical recovery work. See "Myaamia Center—Miami University," https://miamioh.edu/myaamia-center/. Older monographs on Miami include Erminie Wheeler-Voegelin, *An Anthropological Report on the Miami, Wea, and Eel-River Indians* (New York: Garland Publishing, 1974); Stewart Rafert, *The Miami Indians of Indiana: A Persistent People, 1654–1994* (Indianapolis: Indiana Historical Society, 1996).

13. For two good monographs dealing, in part, with the history of the Osage people in their own homeland, see DuVal, *Native Ground*; Lee, *Middle Waters*.

14. For a historical overview of the Sauk, see Nancy Bonvillain and Frank W. Porter, *The Sac and Fox* (Broomall, Pa.: Chelsea House Publishing, 1995); "History | Meskwaki | Sac & Fox Tribe of the Mississippi," *Meskwaki Nation* (blog), www.meskwaki.org/history/.

15. Edmond Atkin, quoted in Sami Lakomäki, *Gathering Together: The Shawnee People through Diaspora and Nationhood, 1600–1870* (New Haven, Conn.: Yale University Press, 2014), 13.

16. For further reading on the Shawnee in colonial America, see Colin G. Calloway, *The Shawnees and the War for America* (New York: Viking, 2007); Laura Keenan Spero, "'Stout, Bold, Cunning and the Greatest Travellers in America': The Colonial Shawnee Diaspora" (PhD diss., University of Pennsylvania, 2010); Stephen Warren, *The Worlds the Shawnees Made: Migration and Violence in Early America* (Chapel Hill: University of North Carolina Press, 2014); Lakomäki, *Gathering Together*; Stephen Warren, ed., *The Eastern Shawnee Tribe of Oklahoma: Resilience Through Adversity* (Norman: University of Oklahoma Press, 2017).

17. For statistics, see "Indigenous Tribes of Chicago," American Library Association, December 2, 2019, www.ala.org/aboutala/offices/diversity/chicago-indigenous; William Scarborough, Amanda Lewis, and Angela Walden, "The Invisibility of Chicago's Native American Residents," *Chicago Tribune*, June 7, 2019, sec. Commentary, www.chicagotribune .com/opinion/commentary/ct-perspec-native-american-inequity-uic-20190605-story .html; Daniel Hautzinger, "'We're Still Here': Chicago's Native American Community," *WTTW Chicago*, November 8, 2018, https://interactive.wttw.com/playlist/2018/11/08 /native-americans-chicago. See also the American Indian Center Chicago, https:// aicchicago.org/.

18. Milo Milton Quaife, *Chicago and the Old Northwest, 1673–1835: A Study of the Evolution of the Northwestern Frontier, Together with a History of Fort Dearborn* (Chicago: University of Chicago Press, 1913), 68–69.

19. For some historical overviews on the Huron-Petun-Wendat peoples in colonial history, see Bruce G. Trigger, *The Children of Aataentsic: A History of the Huron People to 1660* (Montreal: McGill-Queen's University Press, 1987); Kathryn Magee Labelle, *Dispersed but Not Destroyed: A History of the Seventeenth-Century Wendat People* (Vancouver: UBC Press, 2013); Charles Garrad, *Petun to Wyandot: The Ontario Petun from the Sixteenth Century* (Ottawa: Canadian Museum of History and University of Ottawa Press, 2014).

Chapter One

1. Robert Cavelier, Sieur de la Salle, "Relation de La Salle Arrive aux Illinois," January 1682, reprinted in Pierre Margry, *Découvertes et Établissements Des Français Dans l'ouest et Dans Le Sud de l'Amérique Septentrionale (1614–1754),* 6 vols. (Paris: D. Jouaust, 1876), 2:164–167. For an early engineer's assessment that a colony of beavers reinforced the continental divide between the Des Plaines and Chicago Rivers in ancient times, see James Goldthwait, *Physical Features of the Des Plaines Valley* (Urbana: University of Illinois Press Publishing, 1909), 56–57.

2. Hugh C. Prince, *Wetlands of the American Midwest: A Historical Geography of Changing Attitudes* (Chicago: University of Chicago Press, 1997), 37–41.

3. Joel Greenberg, *A Natural History of the Chicago Region* (Chicago: University of Chicago Press, 2002), 118–129, quote from 118.

4. La Salle, January 1682, in Margry, *Découvertes*, 2:164–167.

5. James Allison Brown and Patricia J. O'Brien, *At the Edge of Prehistory: Huber Phase Archaeology in the Chicago Area* (Kampsville: Published for the Illinois Department of Transportation by the Center for American Archeology, 1990), 94–99, 164. For the possibility of the canoe portages being used as early as the Middle Woodland Period, see Benjamin Sells, *A History of the Chicago Portage: The Crossroads That Made Chicago and Helped Make America* (Evanston: Northwestern University Press, 2021), 22–23.

6. See map 1.1; Skinner, "Sinews of Empire," 319. For attempts to locate and relocate all the varying takeout points and portages through Chicago's wetlands, see Robert Knight and Lucius H. Zeuch, *The Location of the Chicago Portage Route of the Seventeenth Century: A Paper Read before the Chicago Historical Society, May 1, 1923, and Later Elaborated for Publication* (Chicago: Chicago Historical Society, 1928); John F. Steward, "The Chicago Portage," unpublished manuscript (Chicago, Newberry Library, March 17, 1904); Henry W.

Lee, "The Calumet Portage," *Transactions of the Illinois Historical Society* (1912); Alfred Meyer, "Circulation and Settlement Patterns of the Calumet Region of Northwest Indiana and Northeast Illinois (The First Stage of Occupance—The Pottawatomie and the Fur Trader–1830)," *Annals of the Association of American Geographers* 44 (1954): 245–274; Richard Gross, "Mapping the Chicago Portage: Seventeenth-Century Explorations by Jolliet, Marquette, La Salle, and Joutel," *Terrae Incognitae* 54, no. 2 (2022): 162–186. For the changing course of the Calumet rivers, see Kenneth J. Schoon, *Calumet Beginnings: Ancient Shorelines and Settlements at the South End of Lake Michigan* (Bloomington: Indiana University Press, 2013), 39–42. For readers interested in locating the portage geography today, the acknowledged route as determined by the National Park Service would have begun near present-day Leavitt Street. The continental divide itself lay along Kedzie Avenue. The Mud Lake wetland would have extended from Kedzie west for over five miles to Harlem Avenue, where the wetlands exited through a small creek into the Des Plaines River. The southern route, in the estimation of John Swenson and other local history advocates, would have run somewhere between Hickory and Butterfield Creeks in the vicinity of present-day Olympia Fields and Spirit Trail Park. See Irv Leavitt, "Did the Story of Chicago Begin in the South Suburbs?" *Cook County Chronicle*, March 27, 2019, https://chronicleillinois.com /news/cook-county-news/did-the-story-of-chicago-begin-in-the-south-suburbs/. All these routes have compelling elements as to their historical usage; together they serve to drive home the reality that the Chicago area wetlands of the seventeenth century was a labyrinth of waterways providing several possible overland links between the two great watersheds of the interior.

7. William J. Mitsch and James Gosselink, *Wetlands* (New York: Van Nostrand Reinhold, 1993), 70; David M. Gates, C. H. D. Clark, and James T. Harris, "Wildlife in a Changing Environment," in *The Great Lakes Forest: An Environmental and Social History*, ed. Susan Flader (Minneapolis: University of Minnesota Press, 1983), 52–80, esp. 62. For the importance of beavers as a keystone species for a range of North American environments, see Glynnis Hood, *The Beaver Manifesto: An RMB Manifesto* (Calgary: Rocky Mountain Books, 2011).

8. For the geopolitical implications of such spaces "in-between" in early America, see Michael N. McConnell, *A Country Between: The Upper Ohio Valley and Its Peoples, 1724–1774* (Lincoln: University of Nebraska Press, 1992).

9. In older histories, this violence is chalked up to Haudenosaunee aggression during the Mourning Wars. See, for instance, White, *Middle Ground*. A more compelling explanation, however, might be that violence emulated forth from centers of European encroachment far from Chicago (in places like Albany and the St. Lawrence), triggering waves of Indigenous warfare ever deeper into the continent. For a parallel example of this phenomenon, see Ned Blackhawk, *Violence Over the Land: Indians and Empires in the Early American West* (Cambridge, Mass.: Harvard University Press, 2006).

10. The nature of lower Lake Michigan also meant that Chicago's portages intersected with a number of overland trails that spanned the region. While not nearly as efficient as the waterborne movement offered by the portages and their adjacent waterways, pedestrians traveling on foot necessarily passed around lower Lake Michigan on their movements. Several ancient pathways wore along the lakeshore connecting points of land from Green Bay to the north of Chicago to the eastern woodlands and Ohio Valley to the south and east

of Chicago. In later centuries, these pathways would take on names of their respective destinations or frequent users—the Great Sauk Trail, the Detroit Road, Hubbard's Trace to Vincennes, and similar monikers described trails that dated to before the founding of many of these settlements. Their original Indigenous names are mostly lost to history. For glimpses of such routes, see Helen Hornbeck Tanner, "The Land and Water Communication Systems of the Southeastern Indians," in *Powhatan's Mantle: Indians in the Colonial Southeast*, ed. Gregory A. Waselkov, Peter H. Wood, and M. Thomas Hatley (Lincoln: University of Nebraska Press, 2006), 27–42. For more on this facet of movement through Chicago, see Milo Milton Quaife, *Chicago's Highways, Old and New* (Chicago: DF Keller, 1923); Helen Hornbeck Tanner and Adele Hast, *Atlas of Great Lakes Indian History* (Norman: University of Oklahoma Press, 1986).

11. For the terminology of "oak openings" and savannas in the context of Illinois, see John White, "How the terms *savanna, barrens,* and *oak openings* were used in early Illinois," in *Proceedings of the North American Conference on Barrens and Savannas*, eds. J. S. Fralish et al. (1994), 25–63. For the importance of ecotones in history, see Robert Michael Morrissey, "The Power of the Ecotone: Bison, Slavery, and the Rise and Fall of the Grand Village of the Kaskaskia," *Journal of American History* 102, no. 3 (2015): 667–692; Morrissey, *People of the Ecotone: Environment and Indigenous Power at the Center of Early America* (Seattle: University of Washington Press, 2022). For more on the tallgrass prairie bioregion, see John Madson, *Where the Sky Began: Land of the Tallgrass Prairie* (Iowa City: University of Iowa Press, 2004). For the contesting environmental forces shaping the Illinois River watershed as an environmental transition zone, see Morrissey, "Climate, Ecology and History in North America's Tallgrass Prairie Borderlands," *Past & Present* 245, no. 1 (2019): 39–77.

12. *JR*, 58:97. For an interesting and descriptive look at the ecology of the Oak Openings as a setting for cross-cultural interactions, see James Fenimore Cooper, *The Oak-Openings, or, The Bee Hunter* (New York: Stringer and Townsend, 1852).

13. Fr. Claude Allouez, "Narrative of a Voyage Made to the Illinois," in John Shea, *Discovery and Exploration of the Mississippi Valley with the Original Narratives of Marquette, Allouez, Membre, Hennepin, and Anastase Douay* (Clinton Hall, N.Y.: Redfield, 1852), 72–73.

14. Allouez, for instance, described how he "saw herds of stags and does drinking and feeding on the young grass" in "Narrative of a Voyage," in Shea, *Discovery and Exploration*, 73. For oak tolerance to fire, see Greenberg, *Natural History*, 81. See also Stephan Pyne, *Fire in America: A Cultural History of Wildland and Rural Fire* (Seattle: University of Washington Press, 1982).

15. The Chicago portage routes passed through several more permanent marshes— notably, the expanse of water described by La Salle and others as a lake, maintained by beaver dams. Such marshes hosted not only a thriving diversity of aquatic plant life, such as wild rice, sedges, and cattails but also water-tolerant trees like willows, sycamore, cottonwood, and bog birch. See Prince, *Wetlands*, 45. Later observers would also name this watery expanse between the watersheds "Mud Lake." See, for instance, Hubbard, *Autobiography*, 41–42. For more on the historical ecology of the adjacent Kankakee drainage, see John White, *Kankakee River Area Assessment: Early Accounts of the Ecology of the Kankakee Area*, vol. 5 (Springfield: Illinois Department of Natural Resources, 1998). Susan Sleeper-Smith has highlighted the Maumee-Wabash portage as another contender for the shortest route to get from Montreal to New Orleans, though that portage would have been longer than Chicago's

marshy pathway during seasonal cycles of high water and did not have the potential for waters to rise to the level that negated any carry at all. See Sleeper-Smith, *Indigenous Prosperity*, 57.

16. For the abundance of waterfowl, see historical accounts such as John G. Furman's letters in Henry H. Hurlbut, *Chicago Antiquities: Comprising Original Items and Relations, Letters, Extracts, and Notes, Pertaining to Early Chicago* (Chicago, 1881), 88–89, 94. For the role of wild rice in the waterfowl migrations through the area, see William Keating, "The Natural History, Indians, and Disadvantages of Chicago," excerpted in Bessie Louise Pierce, *As Others See Chicago: Impressions of Visitors, 1673–1933*, reprint ed. (Chicago: University of Chicago Press, 2004), 37. For a detailed overview of the Chicago area's historically complex riverine ecology, see Greenberg, *Natural History*, 177–220. For the important relationship between the Chicago area wetlands and Lake Michigan, see Greenberg, *Natural History*, 231–232.

17. Carole D. Goodyear et al., *Atlas of the Spawning and Nursing Areas of Great Lakes Fishes*, vol. 4, *Lake Michigan* (Washington, D.C.: U.S. Fish and Wildlife Service, 1982), 98–99.

18. Jeanne Kay, "Wisconsin Indian Hunting Patterns, 1634–1836," *Annals of the Association of American Geographers* 69, no. 3 (September 1979): 402–418, specifically Table 1, 406.

19. Brown and O'Brien, *At the Edge of Prehistory*, 275–279.

20. For initial findings at the Middle Grant Creek site, see Madeleine McLeester et al., "Detecting Prehistoric Landscape Features Using Thermal, Multispectral, and Historical Imagery Analysis at Midewin National Tallgrass Prairie, Illinois," *Journal of Archaeological Science: Reports* 21 (October 2018): 450–459.

21. For Oneota in the context of Illinois history, see Morrissey, *Empire by Collaboration*, 19. For bison migrations into the upper "Prairie Peninsula," see John White, *A Review of the American Bison in Illinois with an Emphasis on Historical Accounts* (Urbana: University of Illinois Press, 1996).

22. For the shifting balance of power between Illinois groups and the Oneota/Ho-Chunk around the Chicago area, see Lee, *Middle Waters*, 34–37. See also La Potherie's account in Emma Helen Blair, ed., *The Indians Tribes of the Upper Mississippi Valley and Region of the Great Lakes, As Described by Nicolas Perrot, French Commandant of the Northwest; Bacqueville De La Potherie, French Royal Commissioner to Canada; Morrell Marston, American Army Officer; and Thomas Forsyth, United States Agent at Fort Armstrong*, reprint (Lincoln: University of Nebraska Press, 1996), 1:293–300; Jablow, *Illinois, Kickapoo, and Potawatomi Indians*, 47.

23. Charles Christopher Trowbridge, "Traditions, Manners, and Customs of the Twaatwāā or Meeārmeear Indians" (1823), C. C. Trowbridge Papers Circa 1823–1840, Box 1, BHL. Trowbridge recorded these oral histories from Miami informants such as Le Gros and Pinšiwa in Indiana in the early nineteenth century. For *mihsooli* as the Myaamia singular for canoe, see entry in *Miami-Illinois Indigenous Language Digital Archive*, https://mc.miamioh.edu/ilda-myaamia/dictionary/entries/5048. For Miami emergence stories relating to water and earth, see also "Text 18: Eehonci Kiintoohki Pyaawaaci Myaamiaki, 'Where the Miamis First Came From,'" in David J. Costa, *As Long as the Earth Endures: Annotated Miami-Illinois Texts*, annot. ed. (Lincoln: University of Nebraska Press, 2022), 227–230. These traditions resonate with recent retellings by Miami people of their origins along the St. Joseph River. See, for example, George Ironstrack, "A Myaamia Beginning,"

Aacimotaatiiyankwi (blog), August 13, 2010, https://aacimotaatiiyankwi.org/2010/08/13/a
-myaamia-beginning/.

24. Morrissey, *Empire by Collaboration*, esp. ch. 1.

25. See Michael McCafferty, "Illinois Voices, Observations on the Miami-Illinois Language," in *Protohistory at the Grand Village of the Kaskaskia: The Illinois Country on the Eve of
Colony*, ed. Robert Mazrim et al. (Urbana: University of Illinois Press, 2015), 122–123. For
the hydrological concept of a united watercourse within the Indigenous geography of the
area, see Mccafferty, "Author Response Revisiting Chicago," *Journal of the Illinois State
Historical Society (1998–)* 98, no. ½ (April 1, 2005): 82–98, esp. 84. For more on the name of
Chicagou's association with wild alliums around the portages, see John F. Swenson, "Chica-
goua/Chicago: The Origin, Meaning, and Etymology of a Place Name," *Illinois Historical
Journal* 84, no. 4 (1991): 235–248; Michael McCafferty, "A Fresh Look at the Place Name
Chicago," *Journal of the Illinois State Historical Society* 96, no. 2 (2003): 116–129.

26. Archaeological evidence even suggests earlier trading between the Illinois and
Huron-Wendat peoples of the northern Great Lakes during the 1620s: Lucien Campeau,
"La route commerciale de l'ouest au dix-septième siècle," *Les Cahiers des dix* 49 (1994):
21–49. See also Kathleen L. Ehrhardt, *European Metals in Native Hands: Rethinking Technological
Change, 1640–1683* (Tuscaloosa: University of Alabama Press, 2005); Margaret Kimball
Brown, *The Zimmerman Site: Further Excavations at the Grand Village of Kaskaskia*, Reports
of Investigations, Illinois State Museum No. 32 (Springfield: Illinois State Museum, 1975).
Morrissey, *Empire by Collaboration*, 30. For speculation on where exactly Nicollet met the
Illinois and Ho-Chunk, see Robert L. Hall, "Relating the Big Fish and the Big Stone:
Reconsidering the Archaeological Identity and Habitat of the Winnebago in 1634," in
Green, *Oneota Archeology*, 19–30; Hall, "Rethinking Jean Nicolet's Rout to the Ho-Chunks
in 1934," in *Theory, Method, and Practice in Modern Archaeology*, ed. Robert J. Jeske and
Douglas K. Charles (Westport, Conn.: Praeger, 2003), 238–251; Robert Mazrim and Duane
Esarey, "Rethinking the Dawn of History: The Schedule, Signature, and Agency of Euro-
pean Goods in Protohistoric Illinois," *Midcontinental Journal of Archaeology* 32, no. 2 (2007):
145–200, esp. 165–185. For a different interpretation, see Michael Mccafferty, "Where Did
Nicollet Meet the Winnebago in 1634? A Critique of Robert L. Hall's 'Rethinking Jean
Nicollet's Route to the Ho-Chunks in 1634,'" *Ontario History* 96, no. 2 (2004): 170–183.

27. La Potherie, in Blair, *Upper Mississippi Valley*, 1:321.

28. For the best recent overview of Indigenous-Jesuit relations, with particular emphasis
on the Illinois, see Tracy Neal Leavelle, *The Catholic Calumet: Colonial Conversions in
French and Indian North America* (Philadelphia: University of Pennsylvania Press, 2012).

29. *JR*, 54:167.

30. *JR*, 54:191.

31. For more on the Indigenous slave trade in the Great Lakes and New France, see Brett
Rushforth, *Bonds of Alliance: Indigenous and Atlantic Slaveries in New France* (Chapel Hill:
University of North Carolina Press, 2012), esp. 163–174; Lee, *Middle Waters*, 44–46. See also
Alan Gallay, ed., *Indian Slavery in Colonial America* (Lincoln: University of Nebraska Press,
2009).

32. For Potawatomi trading ventures through Chicago, see "Unfinished Journal of Father
Jacques Marquette," in Louise Phelps Kellogg, *Early Narratives of the Northwest, 1634–1699*
(New York: C. Scribners's Sons, 1917), 263–265.

33. Louis Jolliet and Jacques Marquette were the first two visitors to write about their journey through Chicago, in 1673. Neither related anything regarding settlements anywhere upriver from the towns of the Kaskaskias. Marquette wintered at Chicago in 1674–1675 and recorded the nearest Illinois settlement as twenty leagues (approximately sixty miles) away. Both Tonti and La Salle passed over Chicago's portage in the early 1680s with no reference to Native settlements, though they both note settlements farther downriver on the Illinois and east along the St. Joseph.

34. For an overview of the overlapping Indigenous activities around the Chicago portage corridor, see Wheeler-Voegelin and Blasingham, *Indians of Western Illinois and Southern Wisconsin.*

35. For wetlands serving as multitribal crossroads and spaces of resource acquisition in the Native Northeast, see DeLucia, *Memory Lands,* 164–200, esp. 192.

36. For evidence of the Indigenous exchange of geographic knowledge in the Great Lakes region, see G. Malcolm Lewis, "Indicators of Unacknowledged Assimilations from Amerindian 'Maps' on Euro-American Maps of North America: Some General Principles Arising from a Study of La Vérendrye's Composite Map, 1728–29," *Imago Mundi* 38 (1986): 9–34; G. Malcolm Lewis, ed., *Cartographic Encounters: Perspectives on Native American Mapmaking and Map Use* (Chicago: University of Chicago Press, 1998); G. Malcolm Lewis, "Intracultural Mapmaking by First Nations Peoples in the Great Lakes Region: A Historical Review," *Michigan Historical Review* 32, no. 1 (Spring 2006): 1–17. For the ancient trade links between Cahokia and the Great Lakes, see John E. Kelley, "Cahokia and Its Role as a Gateway Center in Interregional Exchange," in *Cahokia and the Hinterlands: Middle Mississippian Cultures of the Midwest,* ed. Thomas E. Emerson and R. Barry Lewis (Urbana: University of Illinois Press, 1991), 61–80; Timothy Pauketat, *Cahokia: Ancient America's Great City on the Mississippi* (New York: Penguin Books, 2009).

37. The geographic importance of these portages has been underemphasized in modern scholarship. For much earlier work on the significance of such sites, see Hulburt, *Keys to the Continent*; John M. Lamb, *The Chicago and Kankakee Portages: A Comparison, 1673–1848* (Lockport: Illinois Canal Society, 1982). For more on the St. Joseph–Kankakee Route as a preferrable alternative to Chicago, see George Albert Baker, *The St. Joseph-Kankakee Portage: Its Location and Use by Marquette, La Salle and the French Voyageurs* (South Bend: Northern Indiana Historical Society, 1899).

38. *JR,* 59:86. See also "General Memoir on the State of Canada," Frontenac to Colbert, Quebec, November 14, 1674, in *NYCD,* 9:121.

39. Frs. Jacques Marquette and Claude Dablon, "Of the First Voyage Made by Father Marquette Toward New Mexico," in Kellogg, *Early Narratives of the Northwest,* 229.

40. Marquette, "First Voyage" in Kellogg, *Narratives,* 235–236.

41. *JR,* 59:159; see also *JR,* 58: 97–99.

42. *JR,* 58:95.

43. Marquette, "First Voyage," in Kellogg, *Narratives,* 257.

44. Marquette, "First Voyage," in Kellogg, *Narratives,* 257.

45. *JR,* 58:99. For the entire relation of the expedition, written by Jacques Marquette and edited by his superior, Father Claude Dablon, see "Le premier Voÿage qu'a fait ie P. Marquette vers le nouueau Mexique," in *JR,* 59:85–163, taken and translated from the original manuscripts held at St. Mary's College, Montreal.

46. *JR*, 58:91.

47. *JR*, 58: 101–103.

48. "General Memoir on the State of Canada," Frontenac to Colbert, Quebec, November 14, 1674, in *NYCD*, 9:121; see figure 1.1. Jolliet's so-called X Map, which he sketched from memory in 1674 in Montreal, has been lost, but a map thought to be hand copied from it is held in the John Carter Brown Library and republished in Sara Jones Tucker, *Indian Villages of the Illinois Country, Part 1, Atlas*, vol. 2, Scientific Papers, Illinois State Museum (Springfield: State of Illinois, 1942), Plate IV. For description and provenance, see Tucker, *Illinois Country*, 2.

49. Such a project was feasible—this was the era when Louis XIV's France undertook a series of modern canal projects across the home nation, flexing the new state capacity of such efforts. See Chandra Mukerji, *Impossible Engineering: Technology and Territoriality on the Canal Du Midi* (Princeton, N.J.: Princeton University Press, 2009).

50. *JR*, 58:101.

51. *JR*, 58:103.

52. *JR*, 59:171–172.

53. Marquette, "Unfinished Journal," in Kellogg, *Early Narratives*, 267.

54. Marquette, "Unfinished Journal," in Kellogg, *Early Narratives*, 268.

55. Marquette, "Unfinished Journal," in Kellogg, *Early Narratives*, 263. For this linguistically accurate spelling of "Chachagwessiou" see Michael Mccafferty, *Native American Place-Names of Indiana* (Urbana and Chicago: University of Illinois Press, 2008), 24. For Chachag8essi8's later career as a Peoria leader, see Lee, *Middle Waters*, 79–80.

56. Marquette, "Unfinished Journal," in Kellogg, *Early Narratives*, 265–266.

57. Marquette, "Unfinished Journal," in Kellogg, *Early Narratives*, 266.

58. For the coureurs de bois as a destabilizing force in the region, from the perspective of colonial officials, see Morrissey, *Empire by Collaboration*, 93–94. For an expansive view of the coureurs de bois in North American history, see Gilles Havard, *Histoire Des Coureurs De Bois: Amérique du Nord, 1600–1840* (Paris: Indes Savantes, 2016).

59. Marquette, "Unfinished Journal," in Kellogg, *Early Narratives*, 269.

60. Marquette, "Unfinished Journal," in Kellogg, *Early Narratives*, 268.

61. *JR*, 59:190–193. The fact that the party went on a new route underscores the myriad ways one could portage between the two watersheds, depending on localized circumstances such as time of year and water levels.

62. For Illinois authority over Chicago's route, see Lee, *Middle Waters*, 35–39.

63. *JR*, 60:157–159.

64. *JR*, 60:159. For population estimates and calculations based on Marquette and Allouez's accounts, see Erminie Wheeler-Voegelin and Emily J. Blasingham, "An Anthropological Report on the Indian Occupancy of Royce Areas 77 and 78," in *Indians of Western Illinois and Southern Wisconsin* (New York: Garland Publishing, 1974), xvii, 8.

65. The well-armed and well-organized Haudenosaunee raided to seize captives, to replace family members lost to disease and warfare in the east. Historians have long researched and debated how to best interpret the Haudenosaunee experience and these costly wars. For Haudenosaunee motives in the Mourning Wars, earlier mischaracterized as the "Beaver Wars," see Daniel Richter, "War and Culture: The Iroquois Experience," *WMQ* 40, no. 4 (1983): 528–559; Richter, *Ordeal of the Longhouse*; Brandão, *Your Fyre Shall Burn*

No More; Parmenter, *Edge of the Woods*. For a review of more recent trends in Haudeno-saunee history, see Edward Countryman, "Toward a Different Iroquois History," *WMQ* 69, no. 2 (2012): 347–360.

66. Perrot's Memoir in Blair, *Upper Mississippi Valley*, 1:153–157.

67. In Richard White's *Middle Ground*, for instance, Algonquian peoples of the lakes moved "toward a larger unity" and a "common identity as children of Onontio [New France's governor]" in the wake of Haudenosaunee campaigns. These disparate Algon-quian groups, White argues, formed "a collective sense of themselves" in refugee settle-ments at places like Green Bay and the upper Illinois River, ultimately settling into joint, mediated alliance under the auspices of the French. White, *Middle Ground*, 57, xi, 29. For correctives to this Iroquois-Algonquian dichotomy, see James Merrell, "Indian History during the English Colonial Era," in *A Companion to Colonial America*, ed. Daniel Vickers (Malden, Mass.: Blackwell, 2003); Brett Rushforth, "Slavery, the Fox Wars, and the Limits of Alliance," *WMQ* 63, no. 1 (January 2006): 53–80, esp. 80.

68. *JR*, 59:227.

69. *JR*, 54:219.

70. *JR*, 60:199.

71. Marquette, "Unfinished Journal," in Kellogg, *Early Narratives*, 267.

72. Robert Cavelier, Sieur de la Salle, *Relation of the Discoveries and Voyages of Cavelier De La Salle from 1679 to 1681, the Official Narrative*, trans. Melville Best Anderson (Chicago: Caxton Club, 1901), 193.

73. Letter of La Salle to La Barre, June 4, 1683, in Margry, *Découvertes*, 2:319–320.

74. Perrot in Blair, *Upper Mississippi Valley*, 1:227. For chronology of this event, see Jablow, *Illinois, Kickapoo, and Potawatomi*, 79.

75. La Salle to the Abbé Bernou, August 12, 1682; La Salle to the Abbé Bernou, August 22, 1682, both in Margry, *Découvertes*, 2:215–216; see also "Anthropological Report on the Chippewa, Ottawa, and Potawatomi Indians of Royce Area 148," in David A. Baerreis et al., *Indians of Northeastern Illinois*, American Indian Ethnohistory (New York: Garland Publishing, 1974), 38.

76. La Potherie, in Blair, *Upper Mississippi Valley*, 1:349–350.

77. La Potherie, in Blair, *Upper Mississippi Valley*, 1:350.

78. La Potherie, in Blair, *Upper Mississippi Valley*, 1:349.

79. La Salle, *Official Narrative*, 195.

80. La Salle, *Official Narrative*, 197.

81. See Henri de Tonti, "Memoire," in Kellogg, *Early Narratives*, 291–294. For Haudeno-saunee brutality in the valley, see *Official Narrative*, 211–215.

82. La Salle, *Official Narrative*, 193.

83. Jacques Duchesneau, "Memoire on the Western Indians," in *NYCD*, 9:162.

84. Membre, in Shea, *Discovery and Exploration*, 158.

85. La Salle, *Official Narrative*, 209; Henri de Tonti, "Memoire," in Kellogg, *Early Narra-tives*, 294.

86. La Salle, *Official Narrative*, 263.

87. La Salle, *Official Narrative*, 257–259. For a good overview of the 1680 campaign and the Illinois' rebounding strength along the Illinois River, see Morrissey, *Empire by Collabo-ration*, 39–40, 49.

88. La Salle, "Relation of La Salle on Rivers," in Margry, *Découvertes*, 2:201.

89. René-Robert Cavelier, Sieur de La Salle, "Enclosed with the letter of Monsieur de Frontenac November 9, 1680," in *CISHL*, 1:3.

90. La Salle, "Relation of La Salle on Rivers," in Margry, *Découvertes*, 2:201. For population estimates based on La Salle's "huts," see Wheeler-Voegelin and Blasingham, "Anthropological Report," in *Indians of Western Illinois and Southern Wisconsin*, 20. Jean Baptiste Louis Franquelin, the French hydrographer of New France, sketched a map of North America the following year that demonstrated La Salle's efforts in the Illinois Country. The map denoted the multiethnic resettlements taking place near the Chicago portages as more Native peoples strategically moved down the Illinois River to Starved Rock, Fort Saint-Louis, and Kaskaskia. See Copy of Jean Baptiste Louis Franquelin, *Carte de la Louisiane ou des voyages du Sr. De La Salle*, 1896, in Library of Congress Geography and Map Division, https://lccn.loc.gov/2001620469. See also https://www.loc.gov/resource/g3300.ct000656/?r=-0.643,-0.011,2.286,0.853,0. The original 1684 copy was housed in the Archives de la Marine, Paris, before it was lost.

91. For a long-standing debate on how to categorize this small French outpost at Chicago, see Milo M. Quaife, "Was There a French Fort at Chicago?" in *Transactions of the Illinois State Historical Society for the Year 1908* (Springfield: Illinois State Historical Society, 1909), 115–121; Sells, *History of the Chicago Portage*, 42, 46–50.

92. Voyageur Contract Between Henri de Tonty and Two Voyageurs, September 27, 1684, Quebec, Ayer MSS. For the merchant society of New France at this time, see Louise Dechêne, *Habitants and Merchants in Seventeenth-Century Montreal* (Montreal: McGill-Queen's University Press, 1992).

93. For a translated firsthand account from one of the traders, Mathieu Brunet dit Letang, see Timothy J. Kent, *Phantoms of the French Fur Trade: Twenty Men Who Worked in the Trade between 1618 and 1758*, 3 vols. (Ossineke: Silver Fox Enterprises, 2015), 2:1153–1157. See also Cadwallader Colden, *The History of the Five Indian Nations of Canada* (1747), 60–62; Peter Wraxall, *An Abridgment of the Indian Affairs Contained in Four Folio Volumes, Transacted in the Colony of New York from the Year 1678 to the Year 1751* (Cambridge, Mass.: Harvard University Press, 1915), 12–13; Statistics of the raid included in Appendix D of Brandão, *Your Fyre Shall Burn No More*.

94. La Potherie, in Blair, *Upper Mississippi Valley*, 2:58.

95. Louis Hennepin, *A New Discovery of a Vast Country in America [. . .]*, 2 vols., ed. Victor Hugo Paltsits (Chicago: A. C. McClurg, 1903), 2:627.

96. Margry, *Découvertes*, 2:338; de Tonti's Memoir in Kellogg, *Early Narratives*, 305; Fr. Enjalran to Lefevre de la Barre, Michilimackinac, August 26, 1683, in *WHC*, 16:110–113.

97. Henri Joutel, *A Journal of La Salle's Last Voyage* (New York: Corinth Books, 1962), 163–164.

98. Joutel, *Journal*, 165–166.

99. Pierre-Charles DeLiette, "Memorial Concerning the Illinois Country," ca. 1693, Ayer MSS; a translated edition can be found in Milo Milton Quaife, *The Western Country in the 17th Century* (Chicago: Lakeside Press, 1947), 168.

100. La Potherie, in Blair, *Upper Mississippi Valley*, 2:13–14; Emily J. Blasingham, "Indians of the Chicago Area," in *Chicago Area Archaeology*, Bulletin no. 3, Illinois Archaeological Survey (Carbondale: Southern Illinois University, 1961), 164. For more on the hunting

techniques of the fur trade, and French aversion to it, see Claiborne A. Skinner, *The Upper Country: French Enterprise in the Colonial Great Lakes* (Baltimore, Md.: Johns Hopkins University Press, 2008).

101. Prince, *Wetlands*, 97–98.

102. DeLiette, in *Western Country*, 91.

103. La Potherie, in Blair, *Upper Mississippi Valley*, 1:332.

104. DeLiette, in *Western Country*, 168.

105. Fr. Jacques Gravier to Monseigneur de Laval, Ville-Marie, September 17, 1697, *JR*, 65:51–57; Gravier to Laval, Ville-Marie, September 20, 1698, *JR*, 65:57–63.

Chapter Two

1. Jean François Buisson de St. Cosme wrote a letter in the winter of 1699 describing the expedition's entire voyage from Michilimackinac to the Arkansas Country. A copy can be found as "Jean François de St. Cosme Buisson Letter," January 2, 1699, CHM, transcribed from the original in Laval University Archives, Quebec. An English language translation is included in Kellogg, *Early Narratives*, 342–361. These "seminarians" came from a new religious order started in Quebec in 1663, known simply as the Séminaire de Québec, or sometimes, the Seminary of Foreign Missions (Missions Étrangères). For a comparative look at how they operated alongside, and in competition with, the Jesuits in the Illinois Country, see Michaela Kleber, "A Tale of Two Missions: Illinois Choices and Conversions at the Turn of the Eighteenth Century," *Church History* 91, no. 2 (2022): 286–305.

2. St. Cosme Letter, CHM. See also Kellogg, *Early Narratives*, 347.

3. St. Cosme Letter, CHM. See also Kellogg, *Early Narratives*, 347.

4. Portions of this argument and chapter first appeared in John William Nelson, "The Ecology of Travel on the Great Lakes Frontier: Native Knowledge, European Dependence, and the Environmental Specifics of Contact," *Michigan Historical Review* 45, no. 1 (2019): 1–26. Reprinted with permission from the Historical Society of Michigan. For the wider maritime strategy and logic of the French empire, see Kenneth J. Banks, *Chasing Empire across the Sea—Communications and the State in the French Atlantic, 1713–1763* (Montreal: McGill-Queen's University Press, 2008). For France as an Atlantic empire, Christopher Hodson and Brett Rushforth, "Absolutely Atlantic: Colonialism and the Early Modern French State in Recent Historiography," *History Compass* 8, no. 1 (2010): 101–117. See also Christian Ayne Crouch, *Nobility Lost: French and Canadian Martial Cultures, Indians, and the End of New France* (Ithaca, N.Y.: Cornell University Press, 2014); James S. Pritchard, *In Search of Empire: The French in the Americas, 1670–1730* (Cambridge: Cambridge University Press, 2004).

5. For "middle ground" as both a cultural and geographic phenomenon, see White, *Middle Ground*. Downriver, in the Illinois Country proper, sustained collaboration persisted beyond the reach of French imperial wishes. See Morrissey, *Empire by Collaboration*. Richard White himself made the case that the middle ground was at its weakest when French officials pursued retaliatory campaigns such as those in the Fox Wars. See *Middle Ground*, 182–183. For the wider historiographic implications of the concept in histories of contact, imperialism, and Native America, see Richard White, "Creative Misunderstandings and New Understandings," *WMQ* 63, no. 1 (2006): 9–14. Scholars such as Gilles

Havard have argued that a so-called Pax-Gallica defined the region after the diplomatic triumph of 1701, portraying French power in the Great Lakes. Such French hegemony never penetrated Chicago's portages. See Gilles Havard, *The Great Peace of Montreal of 1701: French-Native Diplomacy in the Seventeenth Century*, trans. Phyllis Aronoff and Howard Scott (Montreal: McGill-Queen's University Press, 2001); Gilles Havard, *Empire et métissages: Indiens et Français dans le Pays d'en Haut, 1660–1715* (Sillery, Québec: Septentrion, 2003). Other works have questioned the reach of either a middle ground or a "Pax-Gallica" in a continent still defined by Indigenous power and kin structures. For explicit examples, see Bohaker, "'Nindoodemag'"; DuVal, *Native Ground*; Michael Witgen, "The Rituals of Possession: Native Identity and the Invention of Empire in Seventeenth-Century Western North America," *Ethnohistory* 54, no. 4 (2007): 639–668; Michael McDonnell, "Rethinking the Middle Ground: French Colonialism and Indigenous Identities in the Pays d'en Haut," in *Native Diasporas: Indigenous Identities and Settler Colonialism in North America*, ed. Gregory D. Smithers and Brooke N. Newman (Lincoln: University of Nebraska Press, 2014), 79–108.

6. For this argument, see Rushforth, "Slavery, the Fox Wars, and the Limits of Alliance"; Morrissey, *Empire by Collaboration*; McDonnell, *Masters of Empire*.

7. Wayne C. Temple, *Indian Villages of the Illinois Country, Historic Tribes*, 2nd ed., vol. 2, 2 vols., Scientific Papers, Illinois State Museum, Part Two (Springfield: Illinois State Museum, 1966), 33. The dictionary was only rediscovered and identified in 1999 by Michael McCafferty in the Jesuit Archive of St-Jerome, Quebec. See Michael McCafferty, "The Latest Miami-Illinois Dictionary and Its Author," *Papers of the 36th Algonquian Conference* (Winnipeg: University of Manitoba, 2005): 271–286. The dictionary was compiled by four different authors, including Father Pinet, Father Mermet, Father Marest, and the lay brother and former Chicago fur trader, Jacques Largillier. See Michael McCafferty, "Jacques Largillier: French Trader, Jesuit Brother, and Jesuit Scribe 'Par Excellence,'" *Journal of the Illinois State Historical Society (1998–)* 104, no. 3 (2011): 188–198. For the cultural and religious implications of better translation and communication between the Jesuits and Illinois Indians, see also Robert Michael Morrissey, "'I Speak It Well': Language, Cultural Understanding, and the End of a Missionary Middle Ground in Illinois Country, 1673–1712," *Early American Studies* 9, no. 3 (2011): 617–648.

8. DeLiette in *Western Country*, 166. Miami and Illinois peoples were culturally and linguistically similar, and the main differences seem to be political structures and settlement patterns. Miami-speakers like the Wea tended toward more hierarchical social organizations, with sagamores at the head of political and diplomatic decisions. The Miami also remained along the river valleys of the lower Great Lakes, whereas the Illinois began to grow more powerful as they exploited the resources of the tallgrass prairie—namely, bison. They settled in larger towns in the lower Illinois Country. French observers believed the Illinois to be allied into a loose confederacy, whereas they reported noticeable political divisions between the Miami subgroups of Wea, Piankeshaw, Pepikokia, Kilatika, and others throughout the lower Great Lakes. For more on the Miami political structure, see Trowbridge, "Traditions, Manners, and Customs," Trowbridge Papers; Kinietz, *Indians of the Western Great Lakes*, 161–211; Bottiger, *Borderland of Fear*, 13–19. For older tribal histories of the Miami, see Rafert, *Miami Indians of Indiana*; Bert Anson, *The Miami Indians* (Norman: University of Oklahoma Press, 1970). See also Sleeper-Smith, *Indigenous Prosperity*.

9. For corroborating reports of these Wea settlements and Chicago as their center, see Temple, *Indian Villages*, 60.

10. DeLiette, in *Western Country*, 125–126. For the role of Native women in agricultural production, especially among the Miami, see Sleeper-Smith, *Indigenous Prosperity*. See also Sleeper-Smith, *Indian Women and French Men*.

11. Jacques-Charles de Sabrevois, "Memoir on the Savages of Canada as Far as the Mississippi River, Describing Their Customs and Trade," 1718, *WHC*, 16:375.

12. DeLiette, in *Western Country*, 125–126.

13. DeLiette, in *Western Country*, 127.

14. DeLiette, in *Western Country*, 91, 130.

15. DeLiette, in *Western Country*, 122. For more context and modern linguistic analysis, see "apahkwaya" entry at https://mc.miamioh.edu/ilda-myaamia/dictionary/entries/3103. See also "apac8ana" and "apac8eia(ki)" entrees in Father Gravier's Illinois-Miami Dictionary, circa 1690s, republished as Carl Masthay, ed. *Kaskaskia Illinois-to-French Dictionary* (St. Louis, 2002), 64. The original dictionary, circa 1690s, is now held in Watkinson Library Special Collections, Trinity College Library, Hartford, Connecticut. While the document is labeled as "Dictionary of the Algonquin Illinois Language" and attributed to Father Jacques Gravier, Michael McCafferty has made a compelling case that credit for the manuscript should go primarily to Jacques Largillier, the voyageur-turned-lay-brother and linguist who served as scribe for a succession of Jesuit missionaries in the Illinois Country. See McCafferty, "Jacques Largillier"; McCafferty, "Largillier," entry, *Miami-Illinois Indigenous Language Digital Archive*, https://mc.miamioh.edu/ilda-myaamia/documents/14.

16. As Iberville related in a 1702 memorandum, "the Miamis, who have withdrawn from the banks of the Mississippi" had "gone to Chicago for the convenience of beaver hunting." See Margry, *Découvertes*, 4:662. For the importance of wetlands across the region in serving as Native American "breadbaskets," see Sleeper-Smith, *Indigenous Prosperity*, quote on 39. For more on Indigenous appreciation of wetland ecosystems, see Ann Vileisis, *Discovering the Unknown Landscape: A History of America's Wetlands* (Washington, D.C.: Island Press, 1999), 20–27.

17. DeLiette, in *Western Country*, 87–88.

18. For similar Miami activities at the Maumee-Wabash portage, see Bottiger, *Borderland of Fear*, 16, fn. 10.

19. Cadillac reported as early as 1697 that de Tonti and la Forest "have now a warehouse at Chicagou in the country of the Miamis." He also shared that "Sr. de Liette, a subordinate officer of the troops, cousin of the Sieurs de Tonty, commands at the post of Chicagou, in the country of the Miamis. All have but one and the same mind and but one interest, tending solely to trading." See Cadillac Papers, *MPHC*, 33:75.

20. It should be noted that in 1696, King Louis XIV had ordered a moratorium on the fur trade in the west to contend with a surplus of peltry, closing the posts at Michilimackinac, St. Joseph, and La Baye. La Forest and de Tonti were able to negotiate an exemption to the law from their political ally, Louis de Buade de Frontenac, the governor of New France. Hence, while the flow of legal European goods and North American furs slowed elsewhere across the lakes, the route through Chicago remained open to the Illinois posts. See Sleeper-Smith, *Indigenous Prosperity*, 87–88; E. B. Osler "Henri de Tonty," in *Dictionary of Canadian Biography*, online ed. (Toronto: University of Toronto Press, 1979–2016). See

also W. J. Eccles, *The Canadian Frontier, 1534–1760*, rev. ed. (Albuquerque: University of New Mexico Press, 1983).

21. Pierre-François-Xavier de Charlevoix, *History and General Description of New France*, 6 vols. (1870), 5:142.

22. Alain Beaulieu, ed., "Copie du traité de paix de 1701," *La Grande Paix, Chronique d'une saga diplomatique* (Montréal: Éditions Libre Expression, 2001), 3. For the treaty, see also "Ratification of the Peace concluded in the month of September last between the colony of Canada, its Indigenous allies, and the Iroquois, in a General Meeting of the Chiefs of each of these Nations convoked by Chevalier de Callieres Governor and Lieutenant-General for the King in New France. At Montreal, the fourth of August One thousand seven hundred and One," *NYCD*, 9:722–725; Havard, *Great Peace*. For the signatures specifically, and the possibility that the Wea signature represented a place as well as a profession, see Yann Guillaud, Denys Delâge, and Mathieu d'Avignon, "Les signatures amérindiennes: essai d'interprétation des traités de paix de Montréal de 1700 et 1701," *Reserches amérindiennes au Québec* 31, no. 2 (2001): 21–41, esp. 37.

23. Cadillac, in *Western Country*, 5–6.

24. Cadillac, in *Western Country*, 76.

25. Cadillac, in *Western Country*, 74.

26. Cadillac, in *Western Country*, 69, 75–77. Cadillac's larger work is riddled with such arguments of a maritime, freshwater space within the American interior. For further articulation of Chicago's role in Cadillac's freshwater network of trade, see "Account of Detroit," in Cadillac Papers, *MPHC*, 33:131. For Cadillac's role as promoter and embellisher of empire in the interior, see also Richard Weyhing, "'Gascon Exaggerations': The Rise of Antoine Laumet dit de Lamothe, Sieur de Cadillac, the Foundation of Colonial Detroit, and the Origins of the Fox Wars," in Robert Englebert and Guillaume Teasdale, *French and Indians in the Heart of North America, 1630–1815* (East Lansing: Michigan State University Press, 2013), 77–112.

27. The western Great Lakes remained only one of many regions where French visions of empire never quite measured up to reality. See Scott Berthelette, "'Frères et Enfants Du Même Père': The French Illusion of Empire West of the Great Lakes, 1731–1743," *Early American Studies* 14, no. 1 (2016): 174–198. See also G. Herbert Smith, *The Explorations of the La Verendryes on the Northern Plains* (Lincoln: University of Nebraska Press, 1980), 11–42. For the historiography of the French maritime empire more broadly, see Jordan Kellman, "Beyond Center and Periphery: New Currents in French and Francophone Atlantic Studies," *Atlantic Studies: Literary, Cultural, and Historical Perspectives* 10, no. 1 (2013): 1–11; Dale Miquelon, "Envisioning the French Empire: Utrecht, 1711–1713," *French Historical Studies* 24, no. 4 (2001): 653–677; Pritchard, *In Search of Empire*; and especially Banks, *Chasing Empire*. For "imaginative geography," see also West, *Contested Plains*.

28. See figure 2.1. Jean Baptiste Louis Franquelin, *Carte de l'Amerique Septentrionnale: depuis le 25, jusqu'au 65° deg. de latt. & environ 140, & 235 deg. de longitude* (1688).

29. Quaife, "Was There a French Fort at Chicago?" Interestingly, cartographers of both the French and their European rivals perpetuated the myth of a French fort at Chicago. For British examples, see Robert Morden and Christopher Browne, *A New Map of the English Empire in America viz Virginia, New York, Maryland, New Jersey, Carolina, New England, Pennsylvania, New England, Pennsylvania, New Foundland, New France and c.* (1695); revised

(1719) and reissued in John Senex, *A New General Atlas: Containing a Geographical and Historical Account of All the Empires, Kingdoms, and Other Dominions of the World, with the Natural History and Trade of Each Country* (London, 1721), Baskes Collection, Newberry Library; Henry Popple, *A Map of the British Empire in America with the French and Spanish Settlements Adjacent Thereto* (London, 1733). For French examples, see Louis Hennepin, *Amerique Septentrionalis—Carte d'un tres grand Pays entre le Nouveau Mexique et la Mer Glaciace* (1698); Rigobert Bonne, *Partie occidentale du Canada, contenant les cinq Grands Lacs, avec les pays circonvoisins* (reprint: Genève, 1780); Jacques Nicolas Bellin, *Carte de la Louisiane et des pays voisins* (Paris, 1755). Sometimes mapmakers labeled this post "Fort Miamis" or other variations. The conflation of Fort Miami with Fort Checagou originated with Louis Hennepin's map of 1698, which indicated a fort at the mouth of a river along Lake Michigan's southern shore. Hennepin likely meant to show La Salle's early fortifications at the mouth of the St. Joseph River, but the poor representation of the lake and river allowed European mapmakers to conflate reports of Chicago's importance as a portage route with Hennepin's misplaced geography. For a contemporaneous example clearly distinguishing between Chicago's portages and Fort Miami on the St. Joseph River, see Baron Lahonton, *Carte de la riviere Longue: et de quelques autres, qui se dechargent dans le grand fleuve de Missisipi* (La Haye, Netherlands, 1703).

30. DeLiette, in *Western Country*, 121. See also Jablow, *Illinois, Kickapoo, and Potawatomi Indians*, 118.

31. For more on the Indigenous slave trade between the Illinois and French, see Brett Rushforth, "'A Little Flesh We Offer You': The Origins of Indian Slavery in New France," *WMQ* 60, no. 4 (2003): 777–808; Rushforth, *Bonds of Alliance*.

32. For a detailed firsthand account of birchbark canoe construction, see Joseph-François Lafitau, *Customs of the American Indians Compared with the Customs of Primitive Times*. Translated by William N. Fenton and Elizabeth Moore. 2 vols. (Toronto: Champlain Society, 1974), 2:124–126; Thomas McKenney, *Sketches of a Tour to the Lakes: Of the Character and Customs of the Chippeway Indians, and of Incidents Connected with the Treaty of Fond Du Lac* (Fielding Lucas: June 1827), 319–320. For secondary assessments, see Edwin Tappan Adney, *The Bark Canoes and Skin Boats of North America*, Bulletin (Washington, D.C.: Smithsonian Institution, 1964); Timothy J. Kent, *Birchbark Canoes of the Fur Trade*, 2 vols. (Ossineke, Mich.: Silver Fox Enterprises, 1997); John Jennings, *The Canoe: A Living Tradition* (Toronto/ Buffalo: Firefly Books, in collaboration with the Canadian Canoe Museum, 2002); Mark Neuzil and Norman Sims, *Canoes: A Natural History in North America* (Minneapolis: University of Minnesota Press, 2016); Bruce Erickson and Sarah Wylie Krotz, eds., *The Politics of the Canoe* (Winnipeg: University of Manitoba Press, 2021).

33. Quoted in Skinner, "Sinews of Empire," 206–207.

34. Lafitau, *Customs*, 2:124.

35. For the role of birchbark canoes in Great Lakes travel, see Nelson, "Ecology of Travel."

36. For French unwillingness to invest in European-style shipping on the Great Lakes during the eighteenth century, see Skinner, "Sinews of Empire," 128–130. See also "Abstract of Dispatches from Beauharnois and Hocquart to the French Minister," January 25, 1729, in *NYCD*, 9:1014.

37. Capturing the particulars of this early era in the fur trade has proved a challenge to scholars, but social historians have produced masterful accounts of later eighteenth- and

early nineteenth-century voyageur culture, where more detailed records survive. As many of the elements of the fur trade continued from early French practices into the mid-nineteenth century, this literature can still be helpful for understanding earlier fur trade history. For later fur trade dynamics, see Carolyn Podruchny, *Making the Voyageur World: Travelers and Traders in the North American Fur Trade* (Lincoln: University of Nebraska Press, 2006); Robert Englebert, "Merchant Representatives and the French River World, 1763–1803," *Michigan Historical Review* 34, no. 1 (2008): 63–82. For one of the best histories of early French trade networks in the Great Lakes, see Skinner, "Sinews of Empire."

38. Cadillac in *Western Country*, 16–18.

39. For more on the rigors of portaging, see Skinner, "Sinews of Empire," 222–225. For "kinetic" empires as a potential lens for understanding Indigenous networks of mobility in North America, see Pekka Hämäläinen, "What's in a Concept? The Kinetic Empire of the Comanches," *History and Theory* 52, no. 1 (2013): 81–90. Whereas Hämäläinen's horse cultures used animal power to enhance mobility, peoples of the lakes utilized their own muscle power and watercourses for a similar effect. See also George Colpitts, *Pemmican Empire: Food, Trade, and the Last Bison Hunts in the North American Plains, 1780–1882* (New York: Cambridge University Press, 2015).

40. Account republished in Raymond Phineas Stearns, "Joseph Kellogg's Observations on Senex's Map of North America (1710)," *Mississippi Valley Historical Review* 23, no. 3 (December 1936): 345–354, quotes from 352–353. Kellogg would eventually escape back to Britain's American colonies, where he would provide officials with one of the first eyewitness accounts of the western Great Lakes in the English language. For more on the circumstances that put Kellogg into the French fur trade orbit as an English-speaking voyageur, see John Demos, *The Unredeemed Captive: A Family Story from Early America* (New York: Vintage, 1995).

41. Claude Maugue, "Fur trade contract between François Francoeur and four voyageurs for transport of goods and purchase of beaver pelts in Michilimackinac and Chicagou, Before the Clerk Claude Maugue in Ville-Marie, Québec," September 15, 1692, Ruggles Collection, Newberry Library.

42. Maugue, "Fur trade contract," Ruggles Collection, Newberry Library.

43. Pierre-François-Xavier de Charlevoix, *A Voyage to North-America: Undertaken by Command of the Present King of France; Containing the Geographical Description and Natural History of Canada and Louisiana; with the Customs, Manners, Trade and Religion of the Inhabitants; a Description of the Lakes and Rivers, with Their Navigation and Manner of Passing the Great Cataracts*, 2 vols. (Dublin: J. Exshaw, and J. Potts, 1766), 2:2.

44. See figure 2.2. Nicolas de Fer, *Le Cours de Missisipi, Ou de St. Louis Dressée Sur Les Relations et Mémoires Du Pere Hannepin et de Mrs. de La Salle, Tonti, Laotan, Ioustel, Des Hayes, Joliet, et Le Maire* (Paris: Chez J. F. Benard, 1718), Newberry Library.

45. Havard, *Coureurs de bois*. For the exclusionary nature of the conge system and motives for becoming an illegal trader, especially by the eighteenth century, see Skinner, "Sinews of Empire," 117. See also Jean Lunn, "The Illegal Fur Trade out of New France, 1713–60," *Report of the Annual Meeting* 18, no. 1 (1939): 61–76.

46. For a study of France's Louisiana colony, see Sophie White, *Wild Frenchmen and Frenchified Indians: Material Culture and Race in Colonial Louisiana* (Philadelphia: University of Pennsylvania Press, 2012).

47. For such schemes, see Louis XIV to Callières, Versailles, May 3, 1702 *NYCD*, 9:735–736.

48. "Petition that beavers not be carried to the Mississippi," in Margry, *Découvertes*, 4:610.

49. Callières to Pontchartrain, October 16, 1700, in *WHC*, 16:201–202.

50. Callières to Pontchartrain, October 16, 1700, in *WHC*, 16:201–202. See also Callières and Champigny to Minister, October 5, 1701, in Margry, *Découvertes*, 5:356–360.

51. Jablow, *Illinois, Kickapoo, and Potawatomi Indians*, 129, quoting Iberville's Memorandum in Margry, *Découvertes*, 4:653.

52. Jablow, *Illinois, Kickapoo, and Potawatomi Indians*, 128.

53. Quoted in Jablow, *Illinois, Kickapoo, and Potawatomi Indians*, 127.

54. Governor Callières, in New France, seems to have also endorsed this ecological division, suggesting to the French minister that Louisiana officials "forbid those who will settle in that country receiving any beaver either directly or indirectly, or going to trade for any to the Indian nations, permitting them only to trade in Buffalo skins and other articles that can be procured on that continent," while they preserve the fur trade for Canadian merchants. See Callières to Pontchartrain, October 16, 1700, in *WHC*, 16:203.

55. Jablow, *Illinois, Kickapoo, and Potawatomi Indians*, 129–130; Margry, *Découvertes*, 4:660.

56. Louis XIV to Callières, Versailles, May 3, 1702 *NYCD*, 9:735–736; Memoir of the King to Callieres and Champigny, Versailles, May 31, 1701, *NYCD*, 9:721.

57. French officials often used watersheds as boundaries of dominion and tried to impose similar boundaries on their colonial rivals. See, for instance, Ramezay's insistence that northern New York "ought for this reason to be considered as being in territory under French domination, if the boundaries of the upper country were determined by the height of Lands [meaning, the divides between watersheds]," in "Extracts from Ramezay and Begon to the French Minister," September 13 and 16, 1715, *WHC*, 16:316–317. See also Lauren A. Benton, *A Search for Sovereignty: Law and Geography in European Empires, 1400–1900* (Cambridge: Cambridge University Press, 2010), esp. 54–59. For a visual representation of this hydrographical logic of sovereignty, see Guillaume Delisle, *Carte du Mexique et de la Florida des Terres Angloises et des Isles Antilles du Cours et des Environs de la Riviere de Mississipi* (Paris, Chez l'Auteur sur le Quai de l'Horloge, 1703).

58. St. Cosme Letter, CHM; also in Kellogg, *Early Narratives*, 346.

59. Charlevoix, *Voyages*, 2:33.

60. Charlevoix, *Voyages*, 2:71–72.

61. "Narrative of Jean Perrault," in *MPHC*, 37:539–541.

62. St. Cosme Letter; Kellogg, *Early Narratives*, 346.

63. Charlevoix, *Voyages*, 2:72.

64. Shea, ed., "Letter of Mr. Thaumur de la Source," *Early Voyages*, 84–85.

65. Cadillac, in *Western Country*, 18.

66. For similar activities by Miamis on the Wabash, see Bienville to Pontchartrain, October 27, 1711, in A. G. Sanders and Dunbar Rowland, eds., *Mississippi Provincial Archives*, 3 vols. (Jackson: Press of the Mississippi Department of Archives and History, 1927), 3:162; Cadillac to Pontchartrain, October 26, 1713, in *Mississippi Provincial Archives*, 2:171. For Miamis exacting tolls, see Bottiger, *Borderland of Fear*, 16, fn. 10.

67. Jablow, *Illinois, Kickapoo, and Potawatomi Indians,* 126.

68. Jablow, *Illinois, Kickapoo, and Potawatomi Indians,* 126. These Dakota raids into the upper Illinois River valley would continue into the next decade. See, for example, Marest to Germon, November 9, 1712, in *JR,* 66:269, 289.

69. Callieres to Pontchartrain, November 4, 1702, *NYCD,* 9:737.

70. For more reports of Dakota-Miami hostilities, see Callieres to Pontchartrain, October 16, 1700, in *WHC,* 16:203, and Jablow, *Illinois, Kickapoo, and Potawatomi Indians,* 125–126. For an explanation of Dakota motivations for offensives during this period, see Hämäläinen, *Lakota America,* 42–49.

71. For tensions between Anishinaabeg and Weas at this time, see Cadillac, "Memorandum of M. de la Mothe Cadillac Concerning the Establishment of Detroit, From Quebec," November 19, 1704, *MPHC,* 33:234.

72. Sabrevois, "Memoir," 1718, *WHC,* 16:373. See editor's note on "canoe people." According to Iberville, there were still two settlements of Weas at Chicago in 1702, but it seems migrations were already underway to the Wabash. By the time Sabrevois wrote his memoir in 1718, the Wea had been gone for many years. For Iberville references, see Margry, *Découvertes,* 4:661–662. For other references to the Weas' flight, see *WHC,* 16:313, 322; Frances Kraskopf, ed., *Ouiatanon Documents* (Indianapolis: Indiana Historical Society, 1955), 164, 169; *MPHC,* 19:6. See also Wheeler-Voeglin and Blasingham, *Indians of Western Illinois and Southern Wisconsin,* 110–115.

73. Father Jean Mermet to Lamothe Cadillac, River St. Joseph, April 19, 1702, in Margry, *Découvertes,* 5:219.

74. Jean François de St. Cosme Buisson, "Jean François de St. Cosme Buisson Letter," January 2, 1699, CHM, transcribed from original in Laval University Archives, Quebec. For an English translation, see Kellogg, *Early Narratives,* 344.

75. While rumors of Meskwaki violence detoured French travelers along the Lake Michigan routes, to the west open war defined the hazards of Meskwaki blockades. In August 1700, a large party of Meskwakis and their allies waylaid five French voyageurs on the Mississippi, stripping them of their clothes and goods and leaving one man "dangerously wounded." In 1702, French adventurers farther north on the Mississippi suffered an attack from a combined force of Meskwakis and Mascoutens, leaving three French dead. For these early indications of Meskwaki resistance, see "Account of Le Sueur's Voyage up the Mississippi," 1700, in *WHC,* 16:181, and Penicault's Relation in Margry, *Découvertes,* 5:425–426. See also *WHC,* 16:199–200; La Potherie, in Blair, *Upper Mississippi Valley,* 1:322–323; Bernard de la Harpe, "Historical Journal of the Establishment of Louisiana by France," *WHC,* 16:181–183; Report on Detroit, *MPHC,* 33:173–176; Memoire De La Forest, *MPHC,* 33:512; *WHC,* 16:221–227. For Meskwaki strategy involving the Fox-Wisconsin portage route to the west, see the anthropological work of Jeffrey A. Behm, "The Meskwaki in Eastern Wisconsin: Ethnohistory and Archaeology," *Wisconsin Archaeologist* 89, nos. 1–2 (January–December 2008): 7–85.

76. For firsthand reports of the incident at Detroit, see Dubuisson to Vaudreuil, June 15, 1712, *WHC,* 16:267–287; "Another Account of the Siege of Detroit," *WHC,* 16:293–295. For the best secondary account, see Edmunds and Peyser, *Fox Wars,* 62–74. For the role of kin networks, see Karen L. Marrero, *Detroit's Hidden Channels: The Power of French-Indigenous Families in the Eighteenth Century* (East Lansing: Michigan State University Press, 2020).

77. Le Tonnerre (the Thunderer)'s war party of Meskwakis threatened strategic portages as far afield as Niagara according to one report. See Vaudreuil to Pontchartrain, September 6, 1712, *MPHC*, 33:561.

78. Marest to Vaudreuil, July 21, 1712, *WHC*, 16:288–292, quotes on 289. See also Dubuisson to Vaudreuil, June 15, 1712, *WHC*, 16:267–287.

79. Dubuisson to Vaudreuil, June 15, 1712, *WHC*, 16:267; attacks along the Illinois routes and elsewhere continued from 1712 onward. In 1713, Governor Vaudreuil reported that Meskwakis near Green Bay had killed a French trader named l'Epine, and closer to Detroit, other raids left eight more French dead. See Vaudreuil to Minister, November 15, 1713, *WHC*, 16:298–299.

80. Extracts from Letters of Ramezay and Bégon to French Minister, October 23 and November 12, 1714, *WHC*, 16:311.

81. Ramezay and Bégon to the French Minister, September 13 and 16, 1715, *WHC*, 16:311–312.

82. For the Pimiteoui raid, see Edmunds and Peyser, *Fox Wars*, 76. For quote, see Ramezay to Minister, September 18, 1714, *WHC*, 16:301.

83. Ramezay to the Minister, November 3, 1715, *WHC*, 16:323. For more on this planned rendezvous and the reasons for selecting Chicago, see Extracts from Ramezay and Begon to the French Minister, September 13 and 16, 1715, *WHC*, 16:311–320.

84. For the failed French rendezvous at Chicago, see Ramezay to the Minister, November 3, 1715, *WHC*, 16:322–326. For attack against the Kickapoos and Mascoutens, see Vaudreuil to the Council of Marine, October 14, 1716, *WHC*, 16:341. See also Edmunds and Peyser, *Fox Wars*, 81.

85. Ramezay and Bégon to the French Minister, November 7, 1715, *WHC*, 16:332; original document in Archives Nationales de la Marine, series B1, 8:274. For more on this dangerous settlement of Illinois Indians and renegade French in the Illinois River bottom, see Morrissey, *Empire by Collaboration*, ch. 4.

86. Ramezay and Bégon to the Minister, November 7, 1715, *WHC*, 16:332.

87. Ramezay to the Minister, November 3, 1715, *WHC*, 16:325–326. In this same letter, Ramezay argued that illicit coureurs de bois "may become even more lawless than they are, which can only be prevented by having troops to restrain them," and worried over how "to prevent them [the Weas] from trading with their English neighbors."

88. Ramezay to the Minister, November 3, 1715, *WHC*, 16:326.

89. For more on this logic, see Sabrevois, "Memoir," 1718, *WHC*, 16:373.

90. Extracts from Ramezay and Bégon to the French Minister, September 13 and 16, 1715, *WHC*, 16:313, 319–320.

91. "Order to unite and incorporate the Country of the Savages of the Illinois under the government of Louisiana," September 27, 1717, discussed in Morrissey, *Empire by Collaboration*, 104–105. For Canadian reactions, see Vaudreuil to Council, November 4, 1720, Fox Subseries, GLOVE; Vaudreuil to Minister, October 2, 1723, *WHC*, 16:428–431; Vaudreuil to Minister, October 11, 1723, *WHC*, 16:433–441.

92. For Anishinaabe instigation of the Meskwaki conflict, see McDonnell, *Masters of Empire*. For Illinois aggression, see Morrissey, *Empire by Collaboration*. For the role of captive taking and slave trading, see Rushforth, "Slavery, the Fox Wars, and the Limits of Alliance."

93. Robert Morrissey has made similar observations for how these new colonial boundaries disrupted and defined the Illinois Country downriver from Chicago. See Morrissey, *Empire by Collaboration*, ch. 4, esp. 88–90.

94. Brett Rushforth has demonstrated the blatant hypocrisy of French officials in New France, who half-heartedly tried to end the war with the Meskwakis while they continued to benefit from the flow of Indigenous slaves generated by the conflict. Rushforth, "Limits of Alliance," 53–80. Meanwhile, French officials in the Illinois Country actively sided with the Illinois in their refusals to cease hostilities. See, for example, Pierre Dugué de Boisbriand's promises to the Illinois people when he arrived in Kaskaskia in 1717 in Morrissey, *Empire by Collaboration*, 119. Another Illinois commander, DeLiette, was sending propositions to exterminate the Meskwakis even on the brink of peace negotiations in "Memoire of the French King to Monsieurs Beauharnois and Dupuy," April 29, 1727, *WHC*, 3:160–161. Whether such officials were actually hamstrung by administrative divisions or simply unwilling, the muddled border between New France and Louisiana gave them a plausible excuse for their inability to broker peace.

95. Vaudreuil to Council, October 28, 1719, in *WHC*, 16:380–381.

96. For raids and counterraids, see Vaudreuil to Council, October 28, 1719, *WHC*, 16:380–383; Vaudreuil to Council, October 22, 1720, *WHC*, 16:392–395; Stanley Faye, ed., "A Search for Copper on the Illinois River: The Journal of Le Gardeur Delisle, 1722," *Journal of the Illinois State Historical Society* 38 (March 1945): 52–53; Nicolas Michael Chassin to Fr. Bobe, July 1, 1722, in *Mississippi Provincial Archives*, 2:274–279.

97. For the death toll of these raids, see "Anakapita and Massauga and Jonachin, recorded by duTisné," January 14, 1725, *WHC*, 16:458–463. At Chicago, "Kikagwa" specifically, at least one Frenchman named LeSueur was killed while several other convoys under Lafleur and Alphonse de Tonti (or possibly Charles Henri Joseph DeLiette de Tonti) were pillaged by either Meskwakis or their allies. See "Anakapita and Massauga and Jonachin, recorded by duTisné," January 14, 1725, *WHC*, 16:459. Allies of New France, such as the Anishinaabeg, also suffered from the bloodshed around Chicago. In 1720, Meskwakis captured two Potawatomis, including the son of a prominent headman. See Vaudreuil to Council of the Marine, October 22, 1720, *WHC*, 16:392–393.

98. duTisné to Vaudreuil, Fort de Chartres, January 14, 1725, *WHC*, 16:450.

99. Boullenger, Kereben, and la Source to duTisné, Mission at Kaskaskia, January 10, 1725, *WHC*, 16:455.

100. For the importance of Louvigny's peace of 1716 in reopening illegal trade to the west, see Skinner, "Sinews of Empire," 102.

101. DuTisné to Vaudreuil, Fort de Chartres, January 14, 1725, *WHC*, 16:450–452.

102. Boullenger, Kereben, and la Source to duTisné, Mission at Kaskaskia, January 10, 1725, *WHC*, 16:453.

103. "Anakapita and Massauga and Jonachin, recorded by duTisné," January 14, 1725, *WHC*, 16:459–462.

104. Other Illinois Indians took their complaints about French inaction against the Meskwakis directly to the king. In 1725, a delegation of Illinois headmen traveled to France. In his audience with King Louis, Chikagou, a leader of the Metchigamea Illinois, called the French to action, promising, "I, Chicagou, will myself lead the way and teach them how to cut off a scalp." Quoted in Morrissey, *Empire by Collaboration*, 133.

105. For Beauharnois's new policy, see Edmunds and Peyser, *Fox Wars*, 117.

106. Beauharnois to DeSiette [*sic*], August 20, 1727, *WHC*, 3:163–164.

107. For DeLiette's skirmish at Chicago and Lignery's campaign in Wisconsin, see Edmunds and Peyser, *Fox Wars*, 111–115.

108. See Edmunds and Peyser, *Fox Wars*, 119–133.

109. Hocquart to the Minister, January 15, 1731, *WHC*, 17:129.

110. Hocquart to the Minister, January 15, 1731, *WHC*, 17:129–130.

111. Report by Beauharnois and Hocquart, October 12, 1731, *WHC*, 17:142–147. See also Beauharnois and Hocquart to Minister, November 2, 1730, *WHC*, 5:107–108. For firsthand accounts of the siege and subsequent massacre, see Letter by Unknown, September 1730, *WHC*, 17:109–113; Villiers to Beauharnois, September 23, 1730, *WHC*, 17:113–118. See also Edmunds and Peyser, 137–157; Joseph L. Peyser, "The 1730 Siege of the Foxes: Two Maps by Canadian Participants Provide Additional Information on the Fort and Its Location," *Illinois Historical Journal* 80, no. 3 (1987): 147–154.

112. Hocquart to the Minister, January 15, 1731, *WHC*, 17:130.

113. This disagreement, and the varying reports of the location, have perplexed historians and archaeologists as they have tried to locate the historical battlefield. See, for instance, Peyser, "The 1730 Siege of the Foxes"; Temple, *Indian Villages*, 42; Stanley Faye, "The Foxes' Fort: 1730," *Journal of the Illinois State Historical Society (1908–1984)* 28, no. 3 (1935): 123–163; Lenville J. Stelle, "History and Archaeology: The 1730 Mesquakie Fort," in John A. Walthall and Thomas E. Emerson, *Calumet & Fleur-De-Lys: Archaeology of Indian and French Contact in the Midcontinent* (Washington, D.C.: Smithsonian Institution Press, 1992), 265–307. See also Morrissey, *People of the Ecotone*.

114. Quoted in Edmunds and Peyser, *Fox Wars*, 157.

115. Beauharnois and Hocquart to Minister, November 2, 1730, *WHC*, 5:107–108.

116. The French hunted the Meskwakis far and wide after 1730 in efforts to exterminate the People of the Red Earth, ranging as far west as Iowa. Beauharnois, in his own words, wished to demonstrate French willingness to "seek the enemy at the extremity of the world," which he hoped would induce France's Native allies "to exterminate the race." See Beauharnois to the French Minister, October 9, 1735, *WHC*, 17:216–221. Despite Beauharnois optimistic outlook on these genocidal efforts, the French made more enemies than allies in their reckless campaigns of extermination. See, for example, "Copy of Relation of the Journey of the Sieur de Noyelle Commanding the War-Party against the Renards and Sakis, sent to Monsieur the Marquis Beauharnois," Enclosed in Beauharnois to the French Minister, October 9, 1735, *WHC*, 17:221–229.

117. An anonymous French census of the Native population of the region in 1736 listed only one small settlement of Peoria Illinois at Starved Rock and no Kickapoo, Mascouten, or Meskwaki settlements around Chicago. By the 1730s, many Kickapoos and Mascoutens had relocated either farther west or with the Weas along the Wabash River. By 1740, some Meskwakis and Sauk began relocating to Chicago's portages and away from French garrisons. See *NYCD*, 9:1057.

118. For conflicts with the Chickasaws, see James R. Atkinson, *Splendid Land, Splendid People: The Chickasaw Indians to Removal* (Tuscaloosa: University of Alabama Press, 2004), esp. 36–87. For waning French investment in its western regions at this time, see Morrissey, *Empire by Collaboration*, 180.

119. The French maintained garrisons at St. Joseph from 1717 onward, controlling the St. Joseph–Kankakee route; at La Baye (Green Bay) from 1717 onward, controlling the Fox-Wisconsin route; at Sault Ste. Marie from 1725 onward, controlling the St. Mary's River route into Lake Superior; at Kekionga and Ouiatenon from 1715 and 1717, respectively, controlling the Maumee-Wabash route; and at Fort Kaministiquia beginning in 1717, on Lake Superior's Thunder Bay, controlling the northern portage route into Lake Nipigon. Fort des Chartres, well down the Mississippi, remained the closest garrisoned post to guarding the Illinois River route. It lay over three hundred miles from Chicago's portages by water. For an overview of French posts and fortifications throughout the *Pays d'en Haut*, see René Chartrand, *The Forts of New France: The Great Lakes, the Plaines, and the Gulf Coast 1600–1763* (Oxford: Osprey Publishing, 2010).

120. Ramezay to Minister, November 3, 1715, *WHC*, 16:325.

121. Beauharnois to Minister, September 26, 1741, *WHC*, 17:362–366.

122. Beauharnois to Minister, September 26, 1741, *WHC*, 17:363.

123. Beauharnois to Minister, September 26, 1741, *WHC*, 17:363, 365.

124. Beauharnois to Minister, September 26, 1741, *WHC*, 17:336–337, 362, 366. For a summary of the back and forth raiding between the Dakotas, Meskwakis, Sauks, and Peorias along the upper Illinois River, see Wheeler-Voeglin and Blasingham, *Indians of Western Illinois and Southern Wisconsin*, 130–131.

125. *WHC*, 17:362–363, 365, 404, 437; 18:39–40. See also Temple, *Indian Villages*, 92.

126. For parallel examples of swamps as sites of resistance and imperial consternation across early America, see Lisa Brooks, "'Every Swamp Is a Castle': Navigating Native Spaces in the Connecticut River Valley, Winter 1675–1677 and 2005–2015," *Northeastern Naturalist* 24, no. sp7 (2017): 45–80; DeLucia, *Memory Lands*, 121–124; Sleeper-Smith, *Indigenous Prosperity*, 251–254, 320; Daniel O. Sayers, *A Desolate Place for a Defiant People: The Archaeology of Maroons, Indigenous Americans, and Enslaved Laborers in the Great Dismal Swamp* (Gainesville: University Press of Florida, 2014); J. Brent Morris, *Dismal Freedom: A History of the Maroons of the Great Dismal Swamp* (Chapel Hill: University of North Carolina Press, 2022).

Chapter Three

1. Macarty to Vaudreuil, Kaskaskia, September 2, 1752, in *CISHL*, 29:678. For the importance of the Calumet ceremony in the diplomacy of the Illinois Country, and its ancient context along the borderlands, see Donald J. Blakeslee, "The Origin and Spread of the Calumet Ceremony," *American Antiquity* 46, no. 4 (1981): 759–768, and Leavelle, *Catholic Calumet*, 2–8. For the Illinois adoption of the calumet ceremony a century earlier under similar circumstances, see Morrissey, *Empire by Collaboration*, 11–12.

2. La Jonquiere to Rouille, October 15, 1750, *CISHL*, 29:239–240.

3. The Anishinaabe peoples (plural Anishinaabeg) represent a culturally and linguistically coherent network of tribal entities, including the Ojibwes (Chippewa), Odawas (Ottawa), and Potawatomis, sometimes referred to as the "Three Fires Confederacy." As members from all three of these collective Anishinaabe groups migrated west of Lake Michigan in the eighteenth century, tribal distinctions became blurry through intertribal marriage, multiethnic settlements, and intricate kinship networks. For context of the collective

Anishinaabeg, see Fixico, "Alliance of the Three Fires"; Clifton, *People of the Three Fires*; Bohaker, "'Nindoodemag.'"

4. Portions of this chapter were originally published in John William Nelson, "Sigenauk's War of Independence: Anishinaabe Resurgence and the Making of Indigenous Authority in the Borderlands of Revolution," *WMQ* 78, no. 4 (October 2021): 653–686. My thanks to the *William and Mary Quarterly* for generously allowing this material to be incorporated into the book. Sabrevois, "Memoir," 1718, *WHC*, 16:373. See editor's note on "canoe people." See also McDonnell, *Masters of Empire*, 24–25. For other examples of the eighteenth century as a time of revival and resurgence for Native power in the Great Lakes region and beyond, see Sleeper-Smith, *Indigenous Prosperity*, esp. 10; McDonnell, *Masters of Empire*; Claudio Saunt, *West of the Revolution: An Uncommon History of 1776* (New York: Norton, 2014), 148–168; Pekka Hämäläinen, *The Comanche Empire* (New Haven, Conn.: Yale University Press, 2008). Parmenter, "After the Mourning Wars"; Dowd, *Spirited Resistance*; Richard Aquila, *The Iroquois Restoration: Iroquois Diplomacy on the Colonial Frontier, 1701–1754* (Detroit, Mich.: Wayne State University Press, 1983).

5. For earlier examples of Anishinaabe presence along the region's trade routes, see "Unfinished Journal of Father Jacques Marquette," in Kellogg, *Early Narratives*, 263–265. For hunting trips in the region prior to 1743, see "Proceedings of Council regarding letter (dated October 6, 1721) of Governor Vaudreuil," December 2, 1721, in *WHC*, 16:397–398. The Nassauakueton clan of Odawa—another Anishinaabe group from the Michilimackinac region—had long maintained commercial ties in the Chicago area. For Nassauakueton connections to southwestern Lake Michigan, dating to the seventeenth century, see McDonnell, *Masters of Empire*, 99–101, 107–109, 117, 245–247.

6. Part of the initial draw to the region for Anishinaabe migrants may have been preexisting kinship ties with the Chicago-area Sauk, who maintained towns in the region through the early 1740s against French wishes. As seventeenth-century diplomat Nicolas Perrot observed to the Potawatomis, "Thou, Pouteouatamis, thy tribe is half Sakis; the Sakis are in part Renards [Fox]; thy cousins and thy brothers-in-law are Renards [Fox] and Sakis." See Perrot in Blair, *Upper Mississippi Valley*, 1:270. For Sauk-Meskwaki settlements at Milwaukee and Chicago, see Marquis de Beauharnois to Minister, September 26, 1741, *WHC*, 17:362–365. For direct Odawa-Sauk kinship ties dating to the mid-seventeenth century, see Perrot, in Blair, *Upper Mississippi Valley*, 1:188; McDonnell, *Masters of Empire*, 98–100, 117–118. For parallel Potawatomi-Mascouten connections dating as early as the 1640s, see *JR*, 18:230.

7. For the specificity of this date as attested by Anishinaabeg themselves, see Auguste Chouteau, "Notes on Boundaries of Various Indian Nations," in *Glimpses of the Past*, Missouri Historical Society, 10 vols. (St. Louis: Missouri Historical Society, 1933), 7:125–132.

8. Scholars have traced Anishinaabe expansion elsewhere in the Great Lakes, but the extension of Anishinaabe settlements southwest of Lake Michigan remains opaque. See Witgen, *Infinity of Nations*; McDonnell, *Masters of Empire*; Fixico, "Alliance of the Three Fires."

9. The work of geographers like Jeanne Kay has highlighted the importance of conflict zones to animal populations' recovery during the fur trade era. See especially Kay, "Wisconsin Indian Hunting Patterns, 1634–1836," *Annals of the Association of American Geographers* 69, no. 3 (1979): 412; Jeanne Kay, "Native Americans in the Fur Trade and Wildlife Depletion," *Environmental Review* 9, no. 2 (1985): 119–122. For earlier work on this theory, see Harold Hickerson, "The Virginia Deer and Intertribal Buffer Zones in the Upper Mis-

sissippi Valley," in Anthony Leeds, *Man, Culture, and Animals: The Role of Animals in Human Ecological Adjustments* (American Association for the Advancement of Science, 1965). For the diversity of wildlife still in abundance in the Chicago area by midcentury, see "Minutes of Mr. Hamburgh's Journal, 1763," in Newton Dennison Mereness, ed., *Travels in the American Colonies* (New York: Macmillan, 1916), 362–363. For the larger importance of wetlands ecological abundance to state-resistant and anticolonial societies, see Scott, *Against the Grain*, 47–57.

10. Bohaker, "'Nindoodemag'"; Kinietz, *Indians of the Western Great Lakes*, 240. The Anishinaabeg believe many of the skills necessary to flourish in the Great Lakes were endowed to them by Nanabush near the sacred Mackinac Island, including canoe building and netting fish.

11. Sabrevois, "Memoir," 1718, *WHC*, 16:369. For more on Sabrevois as an observer of Native lifeways, see Sleeper-Smith, *Indigenous Prosperity*, 52–57.

12. Antoine de Lamothe Cadillac, "Relation Du Sieur de La Motte Cadillac" (1718), Ayer MSS. See also "Minutes of Mr. Hamburgh's Journal, 1763" in Mereness, ed., *Travels*, 361.

13. Chicago's marshland ecology would have been similar to habitats in Green Bay and the marshlands southwest of Lake Superior where Anishinaabeg had prospered for decades before the 1740s migrations. For wild rice in the Chicago area, see Hubbard, *Autobiography*, 41; Keating, "Disadvantages of Chicago," in Pierce, *As Others See Chicago*, 37; Greenberg, *Natural History*, 123. For its role in the Illinois Country, see Morgan, *Land of Big Rivers*, 162–164. For the importance of wild rice to Anishinaabeg, see Charles Christopher Trowbridge, "Response to Lewis Cass's Questionnaire Concerning the Chippewa, Ottawa, and Winnebago Indians, Mackinac," 1823, Trowbridge Papers Circa 1823–1840, Box 1, BHL; Thomas Vennum, *Wild Rice and the Ojibway People* (St. Paul: Minnesota Historical Society, 1988). For the wider importance of wild rice as a commodity in the political ecology of the Great Lakes fur trade, see Nelson, "Ecology of Travel," 19–20.

14. For more on Anishinaabeg adoption of bison hunting in the tallgrass prairie, see William Duncan Strong and Berthold Laufer, *The Indian Tribes of the Chicago Region, with Special Reference to the Illinois and the Potawatomi*, Anthropology Leaflet; No. 24 (Chicago: Field Museum of Natural History, 1926), 22–23.

15. Louis Armand de Lom d'Arce baron de Lahontan, *New Voyages to North-America*, 2 vols., ed. Reuben Gold Thwaites (Chicago: A. C. McClurg & Company, 1905), 1:113.

16. Lahontan, *Voyages*, 2:481–483; 1:210; 1:319. See also Kinietz, *Indians of the Western Great Lakes*, 237. Such sustainable hunting methods changed over time as economic pressures mounted, and we should be cautious in essentializing Indigenous environmental consciousness. For more on this, see Shepard Krech, *The Ecological Indian: Myth and History* (New York: Norton, 2000); Gregory D. Smithers, "Beyond the 'Ecological Indian': Environmental Politics and Traditional Ecological Knowledge in Modern North America," *Environmental History* 20, no. 1 (2015): 83–111.

17. This version of the Anishinaabe tradition of a Great Flood is taken from Perrot's account of the Odawa creation story, found in Blair, *Upper Mississippi Valley*, 1:35–37. For similar relations in other sources, see Charlevoix's *Histoire de la Nouvelle France*, 3:344. For canoe version, see Charles E. Cleland, *Rites of Conquest: The History and Culture of Michigan's Native Americans* (Ann Arbor: University of Michigan Press, 1992), 39. For further context, see Bohaker, "'Nindoodemag,'" 32.

18. McDonnell, *Masters of Empire*.

19. For examples of Anishinaabe tolls in other parts of the Great Lakes, see Trigger, *Children of Aataentsic*, 268; 341.

20. For Anishinaabe expertise in canoe building and boat sales, see "Mr. Hamburgh's Journal, 1763," in Mereness, ed., *Travels*, 361; Kinietz, *Indians of the Western Great Lakes*, 245. For a more detailed overview of Indigenous canoe construction and its environmental context, see Nelson, "Ecology of Travel," 1–26, esp. 8–12. For the spiritual significance of canoes in Anishinaabe cosmology, see Erickson and Wylie Krotz, eds., *Politics of the Canoe*, 3.

21. For the many uses of birchbark in Anishinaabeg traditions, see Neuzil and Sims, *Canoes*, 50–53. See also Wendy Djinn Geniusz, *Our Knowledge Is Not Primitive: Decolonizing Botanical Anishinaabe Teachings* (Syracuse, N.Y.: Syracuse University Press, 2009). For a primary account of the presence of birch groves in the upper reaches of the Illinois River country, see "Mr. Patrick Kennedy's Journal of an Expedition undertaken by himself and several coureurs de bois in the year 1773, from Kaskaskias village in the Illinois Country, to the head waters of the Illinois River," in Gilbert Imlay, *A Topographical Description of the Western Territory of North America*, 3rd ed. (London: J. Debrett, 1797), 506–511.

22. For the geographically determined transition between birchbark canoes and heavier dugout pirogues in the respective watersheds of the Great Lakes and Mississippi, see Nelson, "Ecology of Travel," 9–10; Peter Wood, "Beyond Birchbark: How Lahontan's Images of Unfamiliar Canoes Confirm His Remarkable Western Expedition in 1688," in *Politics of the Canoe*, ed. Erickson and Wylie Krotz, 154–175. For more on the importance of dugouts, or pirogues, in the Mississippi watershed, see Peter H. Wood, "Missing the Boat: Ancient Dugout Canoes in the Mississippi-Missouri Watershed," *Early American Studies* 16, no. 2 (2018): 197–254. For the lack of birchbark canoes available in the lower Illinois Country, see Thomas Gage to the Earl of Hillsborough, New York, November 10, 1770, GP, vol. 19, English Series (hereafter ES). For descriptions of this exchange of boats at Chicago, see "Journal of a Voyage made by Mr. Hugh Heward to the Ilinois Country," in Milo M. Quaife, ed. *The John Askins Papers*, 2 vols. (Detroit, 1931) 2:339–359.

23. It is worth noting that expansion remained the dominant theme in the self-recorded written and oral histories of the Anishinaabeg for the period covering the eighteenth century. See D. Macleod, "The Anishinabeg Point of View: The History of the Great Lakes Region to 1800 in Nineteenth-Century Mississauga, Odawa and Ojibway Historiography," *Canadian Historical Review* 73, no. 2 (1992): 194–210. See also McDonnell, *Masters of Empire*, 91–92, 245–247; Witgen, *Infinity of Nations*.

24. "Speech of the Marquis de Beauharnois to the Ottawas of Mackinac," July 8, 1741, *WHC*, 17:351.

25. "Speech of the Marquis de Beauharnois to the Ottawas of Mackinac," July 8, 1741, *WHC*, 17:352.

26. Kinietz, *Indians of the Western Great Lakes*, 230; McDonnell, *Masters of Empire*, 132–136.

27. "Journal of whatever occurred of interest at Quebec in regard to the operations of the war, and the various intelligences received there since the sailing of the ships in November 1747, sent by the governor and intendant of New France to the French Minister," November 1747, Quebec, *NYCD*, 10:137–145, quotes from 137–138.

28. La Jonquiere to Rouille, September 17, 1751, *CISHL*, 29:347–351. For further evidence of Anishinaabe-Illinois conflict at this time, see La Jonquiere to Rouille, October 15, 1750, *CISHL*, 29:239–240; La Jonquiere to Rouille, September 25, 1751, *CISHL*, 29:359–360; Macarty to Vaudreuil, Kaskaskia, September 2, 1752, *CISHL*, 29:664–665.

29. For the most devastating of these campaigns, see Macarty to Vaudreuil, Kaskaskia, September 2, 1752, in *CISHL*, 29:654–655. See also Raymond E. Hauser, "The Fox Raid of 1752: Defensive Warfare and the Decline of the Illinois Indian Tribe," *Illinois Historical Journal* 86, no. 4 (1993): 210–224.

30. Macarty to Vaudreuil, Kaskaskia, March 27, 1752, in *CISHL*, 29:541. For the abortive hunt, see Macarty to Vaudreuil, Kaskaskia, September 2, 1752, *CISHL*, 29:677.

31. DuQuesne to the French Minister, October 14, 1754, in *WHC*, 18:141.

32. "Mr. Hamburgh's Journal, 1763," in Mereness, ed., *Travels*, 364. The Illinois withdrawal in the face of Anishinaabe expansion would lead later generations of Potawatomi and Odawa to claim the region "by right of conquest" according to Auguste Chouteau. Chouteau, a trader and U.S. Indian agent, seems to have gathered his detailed information of the Anishinaabe conquest directly from Indigenous informants. Chouteau, "Notes," in *Glimpses of the Past*, 7:125–132. For another source on this same expansion, but from the Sauk perspective, see Donald Jackson, ed., *Black Hawk: An Autobiography* (Champaign: University of Illinois Press, 1990), 46. Intermittent conflict between the newly settled Anishinaabeg and lingering Illinois Indians during the 1750s and 1760s were likely the source of a persistent myth recorded in the nineteenth century, that the Odawa, Potawatomi, and Ojibwa fought the Illinois to near extinction in a climatic siege at Starved Rock in 1769. A Potawatomi informant, named Méachelle, from the Chicago area told John D. Caton in 1833 about the supposed siege at Starved Rock in 1769. See John Dean Caton, *The Last of the Illinois and a Sketch of the Pottawatomies* (Chicago: Rand McNally & Co. Printers, 1870), 14–19. Subsequent scholars have found no evidence of such a siege, and indeed, Illinois Indians were alive and well downriver at Kaskaskia, Cahokia, and in settlements across the Mississippi long after the 1760s. For the persistence of the myth, see Mark Walczynski, "The Starved Rock Massacre of 1769: Fact or Fiction," *Journal of the Illinois State Historical Society* 100, no. 3 (October 2007): 215–236; Mark Walczynski, *1769: The Search for the Origin of the Legend of Starved Rock* (St. Louis: Center for French Colonial Studies, 2013). For the survival of the Illinois Indians, see Morrissey, *Empire by Collaboration*. For Illinois withdrawal from the upper Illinois River, see also Wheeler-Voeglin and Blasingham, *Indians of Western Illinois and Southern Wisconsin*, 149.

33. "Mr. Hamburgh's Journal," in Mereness, ed., *Travels*, 362.

34. "Mr. Hamburgh's Journal," in Mereness, ed., *Travels*, 363.

35. For the Anishinaabe attack on Pickawillany as the opening salvo in what would become a global conflict between France and Britain, see Michael A. McDonnell, *Masters of Empire: Great Lakes Indians and the Making of America* (New York: Hill and Wang, 2015), 152–166. For the best overview of the Seven Years' War more generally, see Fred Anderson, *Crucible of War: The Seven Years' War and the Fate of Empire in British North America, 1754–1766* (New York: Vintage Books, 2001).

36. Extract of a letter from D'Abbadie, New Orleans, August 9, 1764, in *WHC*, 18:221–222. For his intended route through Chicago, see "Intelligence," Enclosed in Amherst to Johnson, August 12, 1761, *JP*, 4:415. Beaujeu estimated these trade goods were worth 65,000 livres.

See David A. Armour, "Louis Liénard de Beaujeu," in *Dictionary of Canadian Biography*, online ed. (Toronto: University of Toronto Press, 1979–2016).

37. Louise P. Kellogg, ed., "La Chapelle's Remarkable Retreat Through the Mississippi Valley, 1760–61," *Mississippi Valley Historical Review* 22, no. 1 (1935): 63–81; Joseph Gagné, "'Fidèle à Dieu, à la France, et au roi'—Les retraites militaires de La Chapelle et de Beaujeu vers la Louisiane après la perte du Canada 1760–1762" (master's thesis, Université Laval, 2014); Joseph Gagné, *Inconquis: Deux Retraites Françaises Vers La Louisiane Après 1760* (Quebec: Septentrion, 2016).

38. Alexander Henry, *Alexander Henry's Travels and Adventures in the Years 1760–1776*, Milo Milton Quaife, ed. (Chicago: Lakeside Press, 1921), 44.

39. See "Journal and Report of Thomas Hutchins, 4 April–24 September 1762," enclosed in Johnson to Amherst, November 12, 1762, *JP*, 10:521–529. The closest the British had to an accurate account of the route through Chicago was an obscure reference in a report generated in 1721 to the Board of Trade. See "Representation of the Lords Commissioners for Trade and Plantations to the King upon the State of His Majesties Colonies & Plantations on the Continent of North America, dated September 8th, 1721," in *NYCD*, 5:622.

40. See figure 3.1. Detail from Thomas Hutchins map (1762), original Map HM 1091. Courtesy of the Huntington Library.

41. "Intelligence from Michilimackinac," *JP*, 4:169 ("Puttewatimes and Ottawas"). It is worth noting that George Croghan reports a similar rumor—almost verbatim—nearly two years later. Given that both are said to have originated from a Wea Indian, they may have a common origin. See Croghan to William Johnson, Ouiatenon, July 12, 1765, *JP*, 11:838. For the persistence of such rumors and conspiracies, and their significance, in the British-occupied Great Lakes, see Gregory Evans Dowd, "The French King Wakes up in Detroit: 'Pontiac's War' in Rumor and History (French-Indian Relations, 1763–65)," *Ethnohistory* 37, no. 3 (1990): 254–278. See also Gregory Evans Dowd, *Groundless: Rumors, Legends, and Hoaxes on the Early American Frontier* (Baltimore, Md.: Johns Hopkins University Press, 2015). For Minweweh's town at Chicago, see map 13, "Indian Villages and Tribal Distribution c. 1768," in Tanner and Hast, *Atlas of Great Lakes Indian History*, 58–59 ("Grand Saulteur"). Composite Indigenous towns were common throughout the Great Lakes, often comprised of multiple coresident ethnicities and not necessarily represented by shared leadership. Groups coalesced around kinship and shared interests. See Tanner, "Glaize." Within Anishinaabewaki, see Kugel, *Main Leaders* 21, no. 2; Witgen, *Infinity of Nations*, 64–68. For Pontiac's War as a region-wide Indigenous resistance movement, see Gregory Evans Dowd, *War under Heaven: Pontiac, the Indian Nations & the British Empire* (Baltimore, Md.: Johns Hopkins University Press, 2002); Keith R. Widder, *Beyond Pontiac's Shadow: Michilimackinac and the Anglo-Indian War of 1763* (Mackinac Island and Lansing: Mackinac State Historic Parks; Michigan State University Press, 2013).

42. Croghan describes the Indians accompanying Minweweh in these talks as "Ottawas and Chippewas from Chicago" in his notes. See "Croghan's Official Journal," May 15–September 25, 1765, in *CISHL*, 11:49. Even the St. Joseph Potawatomi, often viewed as some of the most hostile to British overtures following Pontiac's War, preceded the southwestern Anishinaabe in affirming the peace by January 1765. See Gregory Evans Dowd, "Indigenous Catholicism and St. Joseph Potawatomi Resistance in 'Pontiac's War,' 1763–1766," *Ethnohistory* 63, no. 1 (2016): 156.

43. Historians have overstated the hostility of the St. Joseph Potawatomi in the years after Pontiac's War. While young men from the St. Joseph were responsible for several killings in the Detroit area, they proved cooperative in comparison to the Anishinaabe towns west of the lake. The local trader Louis Chevalier Sr. played a key role in reconciling the St. Joseph Potawatomi to the British peace. See, for instance, William Johnson to Thomas Gage, Johnson Hall, December 12, 1766, in *JP*, 12:228. There remains no evidence of such local traders advocating for peace with the British in the southwestern towns. For more on the St. Joseph Potawatomi between Pontiac's War and the American Revolution, see Dowd, "Indigenous Catholicism and St. Joseph Potawatomi Resistance."

44. "Declaration of Hugh Crawford concerning the unfavourable disposition of the Indians at the Illinois," July 22, 1765, *CISHL*, 10:483–484.

45. "Journal of William Howard," Enclosed in Johnson to Gage, July 25, 1765, GP, vol. 40, American Series (hereafter AS).

46. "James Grant's Report," enclosed in Henry Gladwin to Thomas Gage, Detroit, June 9, 1764, GP, vol. 20, AS.

47. George Turnbull to Gage, Michilimackinac, May 29, 1771, in *JP*, 8:117.

48. For "ill-disposed," see Johnson to Gage, Johnson Hall, December 12, 1766, *JP*, 12: 227–228. For presents, see Johnson to Robert Adems, Johnson Hall, June 26, 1766, *JP*, 12:113.

49. For Minweweh's activities, see Johnson to Gage, December 21, 1765, in *JP*, 11:981–984, esp. 982; Johnson to Gage, December 12, 1766, *JP*, 12:227–230, esp. 227–28; Gage to Johnson, September 10, 1771, *JP*, 8:251–253, esp. 252.

50. Johnson to Gage, Johnson Hall, December 21, 1765, *JP*, 11:982.

51. See, for examples, Benjamin Roberts to Johnson, Michilimackinak September 31 [Oct. 1], 1767, *JP*, 5:714; Jonathan Carver, *Travels through the Interior Parts of North-America in the Years 1766, 1767, and 1768*, ed. John Coakley Lettsom (London: J. Walter and S. Crowder, 1778): 90–95.

52. For Anishinaabe offensives in the Illinois Country, see Gage to Johnson, New York, September 10, 1771, *JP*, 8:252. The Indian agent Daniel Claus reported that Minweweh led between fifteen and twenty warriors as he raided up and down the Illinois River during the late 1760s. See Journal of Daniel Claus, 1770, in *JP*, 7:947–961, esp. 949; George Turnbull to Gage, February 22, 1768, in GP, vol. 74, AS. See also Lee, *Middle Waters*, 154.

53. Turnbull to Gage, February 23, 1768, GP, vol. 74, AS; Edmunds, *Potawatomis*, 98. See also Howard H. Peckham, ed., *George Croghan's Journal of His Trip to Detroit in 1767* (Ann Arbor, Mich.: University of Michigan Press, 1939), 40–45.

54. Edmunds, *Potawatomis*, 96–98.

55. For Anishinaabe settlements in the region, and Minweweh's [Grand Saulteur's] town at Chicago, see map 13, "Indian Villages and Tribal Distribution c. 1768," in Tanner and Hast, *Atlas of Great Lakes Indian History*, 58–59.

56. Claus to Johnson, La Chine, August 3, 1771, *JP*, 8:213; Turnbull to Gage, May 29, 1771, June 27, 1771, GP, AS. While many secondary sources incorrectly place the location of Bruce's attack near Michilimackinac, the primary sources all indicated Minweweh died in the Anishinaabe territory west or south of Lake Michigan. This misinformation originated from the useful but not entirely reliable source Carver, *Travels*. For more on Carver's *Travels* as a source, see Edward Gaylord Bourne, "The Travels of Jonathan Carver," *American Historical Review* 11, no. 2 (January 1, 1906): 287–302.

57. Gage to Johnson, New York, September 10, 1771, *JP*, 8:252.

58. Gage to the Earl of Hillsborough, New York, November 10, 1770, GP, vol. 19, ES. As if to prove Gage's point, the following year southwestern Anishinaabeg hunting south along the Illinois River killed a British soldier near Fort Chartres in the spring of 1772. See Gage to the Earl of Hillsborough, New York, July 1, 1772, GP, vol. 22, ES. See also Edmunds, *Potawatomis*, 98. For more on Gage's gathering of information about the region, see "Extract of a Deposition of a Trader Taken at St. Joseph's and Carried to the Illinois," n.d., GP, vol. 139b, AS.

59. As historian R. David Edmunds has observed, during the 1760s, British traders and officials continued to incorrectly refer to these dispersed and often hostile Anishinaabeg as "St. Joseph Potawatomis," acknowledging the migrants' origins and kinship ties to the St. Joseph Potawatomi even as they eschewed their affiliations to newly settled towns to the west of Lake Michigan. See Edmunds, *Potawatomis*, 96.

60. "Journal of William Howard," Enclosed in Johnson to Gage, July 25, 1765, Gage Papers, vol. 40, AS.

61. John Swenson, "Jean Baptiste Point de Sable: A Documented History," rev. ed. (2006), unpublished manuscript.

62. As late as the 1820s, some residents of the Chicago area referred to the north branch of the Chicago as the river Guarie, or Gary, after this early trader's short-lived post. For the fleeting information we have on Guillory, see Quaife, *Chicago and the Old Northwest*, 138. For Guillory's continued association with the North Branch, see Hubbard, *Autobiography*, 40–41; William H. Keating, *Narrative of an Expedition to the Source of St. Peter's River, Lake Winnepeek, Lake of the Woods, &c., &c. Performed in the Year 1823, by Order of the Hon. J.c. Calhoun, Secretary of War, Under the Command of Stephen H. Long, Major* (Philadelphia: H. C. Carey & I. Lea, 1824), 1:172; John G. Furman's letters in Hurlbut, *Chicago Antiquities*, 86.

63. Judith A. Franke, *French Peoria and the Illinois Country: 1673–1846* (Springfield: Illinois State Museum, 1995), 35–41. See also Ulrich Danckers, "French Chicago During the 18th Century," *Baybury Review* 1, no. 1 (1997).

64. For 1763 as a watershed moment, see Jane T. Merritt, *At the Crossroads: Indians and Empires on a Mid-Atlantic Frontier, 1700–1763* (Chapel Hill: University of North Carolina Press, 2003); Colin G. Calloway, *The Scratch of a Pen: 1763 and the Transformation of North America* (Oxford: Oxford University Press, 2006). For more on the Spanish as a new counterweight after French withdrawal, see Kathleen DuVal, *Independence Lost: Lives on the Edge of the American Revolution* (New York: Random House, 2015). For Spain's acceptance of French Louisiana and the geopolitical rationale behind it, see David J. Weber, *The Spanish Frontier in North America* (New Haven, Conn.: Yale University Press, 1992), 198–203; Paul W. Mapp, *The Elusive West and the Contest for Empire, 1713–1763* (Chapel Hill: University of North Carolina Press, 2011), 387–412. For enduring modes of Indigenous-colonial diplomacy in the Illinois Country after 1763, see Morrissey, *Empire by Collaboration*. For the growth of St. Louis as a regional hub, see Carl J. Ekberg and Sharon K. Person, *St. Louis Rising: The French Regime of Louis St. Ange De Bellerive* (Urbana: University of Illinois Press, 2015); Stephen Aron, *American Confluence: The Missouri Frontier from Borderland to Border State* (Bloomington: Indiana University Press, 2006); Jay Gitlin, Robert Michael Morrissey, and Peter J. Kastor, eds., *French St. Louis: Landscape, Contexts, and Legacy* (Lincoln: University of Nebraska Press, 2021).

65. Known as "La Fourche," this important site, just downriver from Chicago's portage, housed mainly Mascoutens by the 1760s–1770s, important local allies of the incoming An-ishinaabe. See Wheeler-Voeglin and Blasingham, *Indians of Western Illinois and Southern Wisconsin*, 159; Blasingham, "Indians of the Chicago Area," 167.

66. "Contemporary Copy: Copy of a Speech Made by Machioquisse a Chief of the Put-tawattamees of This Place," Detroit, August 14, 1768, *JP*, 12:585–586.

67. "Report of Francisco Rui. . . . ," May 2, 1769, *WHC*, 18:299–304.

68. For Sigenauk (Letourneau/El Heturno) as the leader of Le Petit Fort, see "The Illinois Indians to Captain Abner Prior, 1794," *American Historical Review* 4, no. 1 (1898): 107–111. For the location of Petit Fort, see Jacqueline Peterson, "Prelude to Red River: A Social Portrait of the Great Lakes Métis," *Ethnohistory* 25, no. 1 (1978): 44. For more con-text on Nakiowin and Sigenauk, their home villages, and their origins, see Guillaume La Mothe to Haldimand, April 24, 1782, Detroit, *MPHC*, 10:569–571; Wheeler-Voeglin and Blas-ingham, *Indians of Western Illinois and Southern Wisconsin*, 170–173; Nelson, "Sigenauk's War of Independence."

69. Francisco Cruzat to Bernardo de Galvez, November 15, 1777, "Summary of the Indian Tribes of the Misuri River, Who Generally Come to Receive Presents at This Post, and the Number of Their Warriors; the Name of the Principal Chief of Each Tribe; the Districts Where They Are Located; Their Distances and Directions from This Village; in What Each One Is Occupied; the Profit or Harm Derived from Each Tribe in the Past; and the Enemies of Each One," in Louis Houck, ed., *The Spanish Regime in Missouri: A Collection of Papers and Documents Relating to Upper Louisiana Principally Within the Present Limits of Missouri During the Dominion of Spain, from the Archives of the Indies at Seville, Etc., Trans-lated from the Original Spanish Into English, and Including Also Some Papers Concerning the Supposed Grant to Col. George Morgan at the Mouth of the Ohio, Found in the Congressional Library*, 2 vols. (Chicago: R. R. Donnelley & Sons Company, 1909), 1:141–148. This census from Spanish St. Louis described the Odawa as "especially well affected toward this dis-trict; and as we have heard are not very well satisfied with the English." They recorded Sigenauk's name as Leturno, a variation on his French appellation L'étourneau, the Starling, or Black Bird. Interestingly, they locate him as a principal chief with the Ojibwe of Lake Huron, hinting at the kinship networks still connecting the southwestern Anishinaabe with their eastern progenitors.

70. "Summary of the Indian Tribes," in *Spanish Regime in Missouri*, 1:141–148.

71. Arendt Schuyler De Peyster to Sir Guy Carleton, Michilimackinac, June 6, 1777, *WHC*, 7:405–407; De Peyster to Carleton, Michilimackinac, June 17, 1777, *WHC*, 7: 407–408. See also De Peyster to Unknown, Michilimackinac, June 4, 1777, *MPHC*, 10:275.

72. Clark's Memoir in *CISHL*, 8:252–253.

73. Clark's Memoir in *CISHL*, 8:253 (quotations). As Richard White succinctly puts it, "George Rogers Clark hated Indians" (White, *Middle Ground*, 368).

74. Clark's Memoir, in *CISHL*, 8:252–255 ("a Polite Gen[tleman]") 253; "perfectly Right," "he long suspected," "pains too tell," "inquisitive Indian," 254). Sigenauk remains the only Native leader Clark and Henry record by name in the records from Virginia's Illinois expe-dition. See Patrick Henry to the Virginia Delegates in Congress, November 16, 1778, *CISHL*, 8:72–73 ("Great Blackbird," 72). Several surviving letters from Clark and Joseph Bowman indicate an ongoing correspondence between the southwestern Anishinaabe towns and

Virginian forces. See Joseph Bowman to Mech Kigie, Chef du village de Chicagou, April 20, 1779, and Clark to Nanaloibi, Chef des Pout, April 20, 1779, both in Clark's Memoir, in *CISHL*, 8:311–313. For the relationship between an ogimaa's authority and the redistribution of gifts and resources within Anishinaabe society, see White, *Middle Ground*, 128–132; Witgen, *Infinity of Nations*, 5, 183–184; McDonnell, *Masters of Empire*, 101. See also David Murray, *Indian Giving: Economies of Power in Indian-White Exchanges* (Amherst: University of Massachusetts Press, 2000), 33–39.

75. Clark's Memoir, *CISHL*, 8:255.

76. Joseph Bowman to Mech Kigie Chef du village de Chicagou, April 20, 1779, and Clark to Nanaloibi, Chef des Pout, April 20, 1779, in *CISHL*, 8:311–315. Unfortunately, the letter Bowman refers to from November 12, 1778, from Mech Kigie, has not been located. R. David Edmunds asserts that "Nanaloibi" is a misrendering of the prominent Potawatomi chief Nanaquiba from the St. Joseph River, but it is unclear how he makes this assertion. The letter does not name Nanaloibi's location and no other scholars who have used this letter have made the connection to Nanaquiba. See Edmunds, *Potawatomis*, 105, fn. 26. See also Frederick Jackson Turner, ed., "Intercepted Letters and Journal of George Rogers Clark, 1778, 1779," *American Historical Review* 1, no. 1 (1895): 90–96; Clark to Nanaloibi, *CISHL*, 1:430.

77. Bowman to Clark, Cahokia, May 28, 1779, *CISHL*, 2:610.

78. In general, the eastern Anishinaabe maintained a lukewarm allegiance to the British throughout the war. For the best assessment of the L'Arbre Croche Anishinaabe's strategy along these lines, see McDonnell, *Masters of Empire*, 272–379, esp. 292–298. The St. Joseph Potawatomi are an anomaly in their varying positions toward the British before and throughout the war. At times they engaged in campaigns willingly as British allies while at other times they threatened British officers and looted traders. For such threats and robberies, see "Lieut. Bennett's Report," in *WHC*, 18:398–401; "Council between Bennett and the Potawatomis," August 3, 1779, *MPHC*, 10:348–353, in Reuben Gold Thwaites, *Early Western Travels, 1748–1846: A Series of Annotated Reprints of Some of the Best and Rarest Contemporary Volumes of Travel, Descriptive of the Aborigines and Social and Economic Conditions in the Middle and Far West, During the Period of Early American Settlement*, 32 vols. (Cleveland: A. H. Clark Company, 1904), 2:181–184; "Memorial and Petition by Mathew Lessy and others," 1779, *MPHC*, 10:367. For the prewar stance of the St. Joseph Potawatomis, see Dowd, "Indigenous Catholicism and St. Joseph Potawatomi Resistance." For eastern Anishinaabe activity in support of the British, see Hamilton to Haldimand, Petite Riviere, November 1, 1778, *WHC*, 11:178.

79. This is not to suggest that the eastern Anishinaabeg always acted as a unified body or wholeheartedly served British interests. Eastern Anishinaabeg pursued their own set of divergent strategies to navigate the revolutionary conflict. See McDonnell, *Masters of Empire*, 272–309, esp. 292–298. My use of *faction* here highlights a well-established diplomatic and political strategy within Anishinaabe society. For more on factionalism as a political tool in Indigenous-colonial relations, see Rebecca Kugel, "Factional Alignment among the Minnesota Ojibwe, 1850–1880," *American Indian Culture and Research Journal* 9, no. 4 (January 1985): 23–48; Kugel, *To Be the Main Leaders*, esp. 6–8, 83–86, 180–188. For Anishinaabe settlements on both sides of Lake Michigan, refer to map 3.1.

80. Mason Bolton to Haldimand, Niagara, May 20, 1779, *MPHC*, 19:415–416.

81. De Peyster to Haldimand, Michilimackinac, May 2, 1779, Canadian Archives, Series B, vol. 96. Also in *CISHL*, 1:433.

82. Mason Bolton to Haldimand, Niagara, May 20, 1779, *MPHC*, 19:415–416.

83. See Brehm to Haldimand, Niagara, May 8, 1779, in *MPHC*, 19:404–405. Quote from Brehm to Haldimand, Niagara, May 12, 1779, in same. For Brehm's earlier reconnaissance of the southwestern Great Lakes, including Chicago, see "Lieut. [Diederick] Brehm's Report to his Excellency General Amherst of a Scout Going from Montreal by La Galette—round part of the North Shore of Lake Ontario to Niagara, from thence round the South Shore of Lake Erie to Detroit, up Lake St. Claire and part of Lake Huron, returning by land to Fort Pitt," in *New England Historical and Genealogical Register* 37 (1883): 22–26, and "Thompson Maxwell's Narrative—1760–1763," *WHC*, 11:213–217. See also Keith Widder, "The Cartography of Dietrich Brehm and Thomas Hutchins and the Establishment of British Authority in the Western Great Lakes Region, 1760–1763," *Cartographica: International Journal for Geographic Information and Geovisualization* 36, no. 1 (1999): 1–23.

84. De Peyster to Haldimand, May 13, 1779, in *MPHC*, 9:397 (quotation); De Peyster to Bolton, July 6, 1779, *MPHC*, 19:448–449. For the *Welcome*'s subsequent service on Lake Huron from the fall of 1779 to 1781, see Alexander Harrow, "Log-Book of the Welcome, 1779–1782," Alexander Harrow Family Papers, 1775–1930, Burton Historical Collection, Detroit Public Library, Michigan. For more on Linctot as a regional player, see George A. Brennan, "De Linctot, Guardian of the Frontier," *Journal of the Illinois State Historical Society (1908–1984)* 10, no. 3 (1917): 323–366.

85. The stanzas discussing the threats from Chicago follow: "To Detroit, Linctot bends his way/ I therefore turn you from the Pey/ To intercept the chevalier/ At Fort St. Joseph's and O Post/ Go—lay in ambush for his host/ While I send round Lake Mitchigan/ To raise the warriors—to a man—/ Who, on their way to get to you/ Shall take a peep at—at *Eschickagou*/ [. . .] Those runagates at Milwakie / Must now *per* force with you agree, / Sly Siggenaak and Naakewoin, / Must with Langlade their forces join; / Or, he will send them *tout au diable*, / As he did *Baptist Point de Saible*." Arent Schuyler De Peyster, *Miscellanies, by an Officer: (Colonel Arent Schuyler De Peyster, B. A.), 1774–1813*, ed. J. Watts De Peyster, 2nd ed. (New York: A. E. Chasmar, 1888), 10. The Earl W. and Florence C. De La Vergne Collection, BHL.

86. For the log of the entire voyage, see Samuel Roberts, "Remarks on Board his Majesty's Sloop Felicity," October 1779, *WHC*, 11:203–212, quote from 210. De Peyster later claimed that he also attempted to buy back the loyalty of Sigenauk and Nakiowin directly, though it remains unclear whether this bribery was part of the *Felicity*'s 1779 mission or constituted a separate attempt. See "Indian Council," March, 11, 1781, Detroit, in *MPHC*, 10:454–455.

87. "Documents Relating to the Attack Upon St. Louis 1780," *Missouri Historical Society Collections*, 5 vols. 2:41–50. Carolyn Gilman, "L'Annee du Coup: The Battle of St. Louis, 1780," *Missouri Historical Review* 103, no. 3 (April 2009): 133–147.

88. Patrick Sinclair to Haldimand, July 8, 1780, in *MPHC*, 9:558–560 ("Band of Indians," 558); Cruzat to Gálvez, November 13, 1780, in Kinnaird, *Spain in the Mississippi Valley*, 2: 396–397 ("courageous resistance," 397); "John Montgomery to the Hon. Board of Commissioners for the Settlement of Western Accounts," February 22, 1783, in William Pitt Palmer, ed., *Calendar of Virginia State Papers and Other Manuscripts Preserved in the Capitol at Richmond*, 11 vols. (Richmond: R.F. Walker, 1875), 3:443.

89. Cruzat to Gálvez, St. Louis, November 13, 1780, in Lawrence Kinnaird, ed. *Annual Report of the American Historical Association for the Year 1945: Spain in the Mississippi Valley, 1765–1794*, 4 vols. (Washington D.C.: Government Printing Office, 1949), 2:398.

90. Cruzat to Gálvez, St. Louis, January 10, 1781, in Kinnaird, ed. *Spain in the Mississippi Valley*, 2:415. De Peyster to Brig. Gen. Powell, January 8, 1781, in *MPHC*, 19:591–592. See also Fernando de Leyba to the governor, June 20, 1780, in A. P. Nasatir, "St. Louis during the British Attack of 1780," in *New Spain and the Anglo-American West: Historical Contributions Presented to Herbert Eugene Bolton. . . .*, ed. Charles W. Hackett et al. (Los Angeles, 1932), 1:253–254.

91. Cruzat to Gálvez, St. Louis, January 10, 1781, in Kinnaird, ed. *Spain in the Mississippi Valley*, 2:416. See also Cruzat to Gálvez, November 13, 1780, in Kinnaird, ed. *Spain in the Mississippi Valley*, 2:396–397.

92. For the significance of Chevalier's participation in the raid and his important kin network, see Cruzat to Esteban Miró, August 6, 1781, in Kinnaird, *Spain in the Mississippi Valley*, 2:431–434; Susan Sleeper-Smith, "'Ignorant bigots and rebels': The American Revolution in the Western Great Lakes," in *The Sixty Years' War for the Great Lakes*, ed. Skaggs and Nelson, 145–165. For the full report of the expedition, as well as details of La Gesse's negotiations, see Cruzat to Miró, St. Louis, August 6, 1781, in Kinnaird, *Spain in the Mississippi Valley*, 2:431–434.

93. Eugene Pouré, "Spanish Act of Possession for the Valleys of the St. Joseph and Illinois Rivers," February 12, 1781, in Kinnaird, *Spain in the Mississippi Valley*, 2:418–419. Such acts of possession had a long precedent in the Great Lakes and Mississippi valley. For early French ceremonies, see Tracy Leavelle, "Geographies of Encounter: Religion and Contested Spaces in Colonial North America," *American Quarterly* 56, no. 4 (2004): 913–943. For similar Spanish actions during the Revolution, see Balthazar de Villiers, Act of Possession of East Bank of the Mississippi River North of the District of Natchez," November 22, 1780, in Kinnaird, *Spain in the Mississippi Valley*, 2:400.

94. Cruzat to Miró, St. Louis, August 6, 1781, in Kinnaird, *Spain in the Mississippi Valley*, 2:432.

95. Sigenauk's warriors killed two of the fur traders as they tried to escape. Cruzat to Miró, St. Louis, August 6, 1781, in Kinnaird, *Spain in the Mississippi Valley*, 2:433. For limited violence and targeted objectives as tenants of early modern Indigenous warfare, see Wayne E. Lee, "Peace Chiefs and Blood Revenge: Patterns of Restraint in Native American Warfare, 1500–1800," *Journal of Military History* 71, no. 3 (July 2007): 701–741.

96. See, for instance, the postwar activities of Sigenauk and La Gesse in Nelson, "Sigenauk's War of Independence," esp. 680–683.

97. Abraham P. Nasatir, *Spanish War Vessels on the Mississippi, 1792–1796* (New Haven, Conn.: Yale University Press, 1968), 139–140.

98. "Cahokia Records, 1778–1790," in *CISHL*, 2:152–153.

99. For a description of Maillet's Peoria settlement from 1790, see "Journal of a Voyage made by Mr. Hugh Heward to the Ilinois Country," in Milo M. Quaife, ed. *The John Askins Papers*, 2 vols. (Detroit, Detroit Library Commission, 1931), 2:359. See also Franke, *French Peoria*, 35–41; Danckers, "French Chicago During the 18th Century"; Swenson, "Point de Sable." For further evidence of French families' presence at Chicago after the war, see Philippe de Rocheblave to Haldimand, Quebec, November 3, 1783, in *CISHL*, 5:352–353.

100. William Burnett to Hand, St. Josephs, February 2, 1790, in Hurlbut, *Chicago Antiquities*, 56–57. The following year, Burnett also reported that "the Pouwatamies, at Chicago,

has killed a Frenchman," likely another hapless trader. See Burnett to Hand, St. Josephs, February 6, 1791, in Hurlbut, *Chicago Antiquities*, 59.

101. "Journal of a Voyage," *Askins Papers*, 2:357–358. For other examples along Heward's voyage, see 2:350. For corroborating accounts of Indigenous tolls along the Chicago route at this time, see Hubbard, *Autobiography*, 43–44.

102. For the importance of local support, labor, and expertise, see "Journal of a Voyage," *Askins Papers*, 2:341, 343, 344, 352, 356–358.

103. For the expedition's time at Chicago, see "Journal of a Voyage," *Askins Papers*, 2: 356–358. The occupation of Indigenous porters was not unique to Anishinaabe Chicago. For Haudenosaunee parallels at the Oneida Carrying Place, see Gail D. MacLeitch, "'Red' Labor: Iroquois Participation in the Atlantic Economy," *Labor* 1, no. 4 (2004): 69–90; Gail D. MacLeitch, *Imperial Entanglements: Iroquois Change and Persistence on the Frontiers of Empire* (Philadelphia: University of Pennsylvania Press, 2011), 198–199.

104. For the full journal, plus some editorial context on Heward and his unusual route, see "Journal of a Voyage," *Askins Papers*, 2:339–359. See also 342, fn. 7.

105. For examples of navigation and landmarks, see "Journal of a Voyage," *Askins Papers*, 2:341, 348, 350, 357.

106. For other examples of exchanges along the route, see "Journal of a Voyage," *Askins Papers*, 2:344–346, 350, 352, 357–359.

107. "Journal of a Voyage," *Askins Papers*, 2:354.

108. "Journal of a Voyage," *Askins Papers*, 2:356.

Chapter Four

1. For two compelling studies of this violence, with somewhat contradictory explanations, see Griffin, *American Leviathan*; Rob Harper, *Unsettling the West: Violence and State Building in the Ohio Valley* (Philadelphia: University of Pennsylvania Press, 2017).

2. Michael Witgen has argued that U.S. colonial policy in the far Northwest Territory indeed remained vastly different than settler colonial approaches elsewhere. See Michael Witgen, "Seeing Red: Race, Citizenship, and Indigeneity in the Old Northwest," *Journal of the Early Republic* 38, no. 4 (2018): 581–611. For more familiar patterns of settlement policy in the Old Northwest, see Andrew R. L. Cayton, *The Frontier Republic: Ideology and Politics in the Ohio Country, 1780–1825* (Kent, Ohio: Kent State University Press, 1986).

3. See Henry Knox to Josiah Harmar, December 1789, Harmar Papers, 11:118, William Clements Library, Ann Arbor, Mich. As quoted in Colton Storm, "Lieutenant John Armstrong's Map of the Illinois River, 1790," *Journal of the Illinois State Historical Society (1908–1984)* 37, no. 1 (1944): 48. See also Colton Storm, "Lieutenant Armstrong's Expedition to the Missouri River 1790," in *Mid-America* 25, no. 3 (July 1943): 180–188.

4. For the full original report, see John Armstrong to Josiah Harmar, Fort Washington, June 2, 1790, in Harmar Papers, 12:122a, William Clements Library, Ann Arbor, Mich.

5. For map and annotation, see John Armstrong, *Illinois River*, 1790, manuscript map, Harmar Papers, William Clements Library, Ann Arbor, Mich.

6. Maillet was a knowledgeable informant, having traded along the Illinois River route for years. Maillet had accompanied La Gesse and Sigenauk east in their 1781 raid on

St. Joseph and maintained close relationships with many of the Anishinaabe headmen in the region. For more on the St. Joseph Raid, see Nelson, "Sigenauk's War of Independence." For more on Maillet, see Franke, *French Peoria*, 41–44.

7. Winthrop Sargent, "Informations from Mr. Maya" (1790), The Winthrop Sargent Papers, Box 3, Folder 9 and MIC 96/Sargent Roll 3, Ohio Historical Society, Ohio History Connection.

8. For a recent study of the important riverine and kinship connections in this region, see Lee, *Middle Waters*.

9. Arthur St. Clair to George Washington, "Report of Official Proceedings in the Illinois Country from March 5th to June 11th, 1790," in William Henry Smith, *The St. Clair Papers: The Life and Public Services of Arthur St. Clair: Soldier of the Revolutionary War, President of the Continental Congress; and Governor of the North-Western Territory: With His Correspondence and Other Papers* (Cincinnati: R. Clarke & Company, 1882), 174.

10. St. Clair to Washington, *St. Clair Papers*, 174.

11. Robert Buntin to Winthrop Sargent, Vincennes, June 21, 1793, Potawatomi Subseries, Box 6511, GLOVE.

12. St. Clair to Washington, 176. Given Sargent's extensive interviews with Jean Baptiste Maillet, a prominent figure in the trading village of Peoria, it is not surprising that St. Clair highlighted this location as a potential site for later fortifications.

13. N. Mitchell, "A Letter on the Illinois Country to Alexander Hamilton, 1792," in "Notes and Documents," *Mississippi Valley Historical Review* 8, no. 3 (December 1921): 265–266.

14. See Colin G. Calloway, *The Victory with No Name: The Native American Defeat of the First American Army* (Oxford: Oxford University Press, 2015).

15. For a recent biography of Wayne, see Mary Stockwell, *Unlikely General: "Mad" Anthony Wayne and the Battle for America* (New Haven, Conn.: Yale University Press, 2018).

16. His paranoia about British and Spanish intrigues reflected common sentiments among U.S. officials and American citizens at the time, as Wayne's friend and confidant, Sharp Delany, relayed from Philadelphia: "Rumours are passing here that Spain as well as G. Britain encourage the Savages against us." Sharp Delany to Anthony Wayne, Philadelphia, November 2, 1792, Box 7, WP. For Wayne's own fears of British influence in the west, see Anthony Wayne to Sharp Delany, Legion Ville, April 27, 1793, Box 7, WP.

17. It remains unclear whether Collins was working in concert with Wilkinson and the Spanish, on behalf of the British, or was, indeed, a legitimate spy working on behalf of the United States' interests. For Wayne's interrogation of Collins, and his doubts, see Anthony Wayne, "Collins, Joseph E., Deposition," February 16, 1793, Potawatomi Collection, Box 6511, GLOVE; Wayne to Henry Knox, Greenville Headquarters, July 7, 1794, Letterbook, WP. For more on Wilkinson's dubious connections, see Andro Linklater, *An Artist in Treason: The Extraordinary Double Life of General James Wilkinson* (New York: Walker, 2009); David E. Narrett, "Geopolitics and Intrigue: James Wilkinson, the Spanish Borderlands, and Mexican Independence," *WMQ* 69, no. 1 (2012): 101–146.

18. Wayne, "Collins, Joseph E., Deposition," February 16, 1793, Potawatomi Collection, Box 6511, GLOVE.

19. Underlined in the original letter. Wayne to Knox, Greenville Headquarters, July 7, 1794, Letterbook, WP. For more evidence of Wayne's convictions that Indians, "Spaniards and British" were all engaged in "dark intrigues & Nefarious attempts" together against the

United States, even after his victory at Fallen Timbers, see Wayne to Knox, Miami Villages, September 20, 1794, Letterbook, WP. For Wayne's ongoing premonitions leading up to Fallen Timbers, see Wayne to Delany, Headquarters, Greenville, July 10, 1794, Box 8, WP; Wayne to Isaac Wayne, Headquarters, Greenville, July 14, 1794, and "Items Related to the Last Will of A. Wayne," July 14, 1794, Box 8, WP.

20. For a recent history of this campaign, see William Hogeland, *Autumn of the Black Snake: George Washington, Mad Anthony Wayne, and the Invasion That Opened the West* (New York: Farrar, Straus and Giroux, 2018).

21. "Examination of Antoine Lassell, captured in the action of the 20th instant," Camp Grand Glaize, August 28, 1794, in Walter Lowrie and Matthew St. Clair Clarke, eds., *American State Papers: Documents, Legislative and Executive, of the Congress of the United States*, 38 vols. (Washington, D.C.: Gales and Seaton, 1832), *Indian Affairs* 2:494.

22. The southwestern Anishinaabeg around Chicago remained distant to the struggles between Native American and U.S. forces to the east, and, indeed, by late August 1794, it seems many Anishinaabeg had generally disengaged with the Indigenous confederacy opposed to Wayne's army. See Wayne, "Examination of a Patawatomi Warrior," July 23, 1794, Potawatomi Subseries, Box 6511, GLOVE; Hogeland, *Autumn of the Black Snake*, 320–323. For Blue Jacket's recruitment trip to the Illinois River Anishinaabeg in 1792, see John Sugden, *Blue Jacket, Warrior of the Shawnees* (Lincoln: University of Nebraska Press, 2000), 128–130.

23. Timothy Pickering, "Pickering's Report to the House of Representatives Regarding Number of Troops Necessary to Protect the Frontiers of the United States," January 26, 1795, Box 9, WP.

24. Quotes from Wayne to William R. Atlee, Greenville, February 22, 1795, Anthony Wayne Folder, CHM. See also Wayne to Knox, Greenville, *confidential*, December 14, 1794, Box 9, WP.

25. Wayne to Knox, Greenville Headquarters, May 26, 1794, Letterbook, WP. Wayne's fears of Spanish naval activities on the Mississippi and Ohio Rivers were not wholly unfounded. For more on this, see Nasatir, *Spanish War Vessels on the Mississippi*.

26. Wayne to Knox, Greenville Headquarters, May 26, 1794, Letterbook, WP.

27. Wayne to John Francis Hamtramck, Greenville Headquarters, March 13, 1795, Box 9, WP.

28. Much of this raiding began at the prompting of the Spanish, who were at war with the Little Osage in the 1790s. Spanish involvement only added to U.S. officials' fears over foreign machinations along the region's waterways. See Edmunds, *Potawatomis*, 153–156; Keating, *Rising Up*, 41, fn. 19.

29. Knox to Wayne, "Instructions to Major General Anthony Wayne, Relatively to a Proposed Treaty with the Indians Northwest of the Ohio," April 4, 1794, Folder One: Oversize Manuscripts, WP; quote from Pickering to Wayne, April 8 and 14, 1795; draft of the proposed treaty, Northwest Territory Collection, William Henry Smith Memorial Library of the Indiana Historical Society, Indianapolis. See also Dwight L. Smith, "Wayne and the Treaty of Greenville," *Ohio State Archeological and Historical Quarterly* 63, no. 1 (January 1954).

30. This was directly against the advice of Timothy Pickering, the secretary of war, who doubted the "expediency" of pressing for Chicago in the treaty negotiations. Pickering to

Wayne, War Office, April 15, 1795, in Richard C. Knopf, ed., *Anthony Wayne, a Name in Arms: Soldier, Diplomat, Defender of Expansion Westward of a Nation; the Wayne-Knox-Pickering-Mchenry Correspondence* (Pittsburgh: University of Pittsburgh Press, 1960), 32–36. See also Pickering to Wayne, War Office, April 8, 1795," in Knopf, *Correspondence*, 19–32.

31. Gomo, from Peoria, and Sigenauk, from the Chicago area, served as the only identifiable representatives for the southwestern Anishinaabeg at the Treaty of Greenville. In all, twenty-three Potawatomis signed the treaty, mostly from the St. Joseph and Huron Rivers. Wheeler-Voegelin and Blasingham, *Indians of Western Illinois and Southern Wisconsin*, 112. See also Edmunds, *Potawatomis*, 135; Wayne to Pickering, June 17, 1795, in Knopf, *Correspondence*, 427–428; John Mills to Sargent, July 16, 1795, Sargent Papers, Massachusetts Historical Society.

32. *American State Papers, Indian Affairs* I, 576.

33. "A Treaty of Peace between the United States of America and the Tribes of Indians Called the Wyandots, Delawares, Shawanoes, Ottawas, Chipewas, Puttawatimies, Miamis, Eel River, Weea's and Kickapoo's," Fort Greenville, August 3, 1795, Treaty of Greenville with appendix, manuscript copy, Anthony Wayne Folder, CHM.

34. Wayne also secured a six-mile-square area at Peoria and a twelve-mile-square area at the mouth of the Illinois River. Together, these three strategic sites represented Wayne's ambitions to fortify the route connecting lower Lake Michigan with the Mississippi River. See Treaty of Greenville.

35. For the Jay Treaty and its impact on Great Lakes Indian policy, see Willig, *Restoring the Chain of Friendship*, esp. 248–249; Lawrence B. A. Hatter, "The Jay Charter: Rethinking the American National State in the West, 1796–1819," *Diplomatic History* 37, no. 4 (2013): 693–726.

36. Wayne to the Secretary of Treasury (Oliver Wolcott), September 4, 1796, *TP*, 2: 569–572. In general, Wayne seemed to have a fixation on securing the water routes connecting British Canada with Spanish Louisiana. See his similar understanding of the role Fort Wayne, on the Maumee-Wabash portage, would play in American security on the frontier in Wayne to Hamtramck, Miami Villages, October 22, 1794, Box 8, WP.

37. "Speech of the Putawatimes Chiefs of Chikago, in Council," Greenville, July 17, 1796, Potawatomi Subseries, Box 6512, GLOVE.

38. Wayne to James McHenry, Greenville, July 22, 1796, in Knopf, *Correspondence*, 30.

39. "Speech of the Putawatimes Chiefs of Chikago, in Council," Greenville, July 17, 1796, Potawatomi Subseries, Box 6512, GLOVE. Since Anishinaabe notions of sovereignty extended not only to the lands around Chicago but also its waterways specifically, it is doubtful these leaders saw the American cession as the transformative territorial shift that Wayne hoped it would be. For notions of Anishinaabe sovereignty over water, see Victor P. Lytwyn, "Waterworld: The Aquatic Territory of the Great Lakes First Nations," in Dale Standen and David McNab, *Gin Das Winan: Documenting Aboriginal History in Ontario* (Toronto: University of Toronto Press, 1994), 14–28.

40. Refer back to chapter 3, "Crossroads," and Nelson, "Sigenauk's War of Independence."

41. Zénon Trudeau, "Noticia de los parages donde invernan las Canoas del Comerico Yngles de Michilimaquina para el trato conlos Yndios y consume particular de cada Nacion," San Luis de Yllinois, May 18, 1793, reprinted in Abraham P. Nasatir, *Before Lewis and Clark; Documents Illustrating the History of the Missouri, 1785–1804*, 2 vols. (St. Louis:

St. Louis Historical Documents Foundation, 1952), 1:179–180. For comparison, during this same time British traders only brought ten canoes through the Potawatomi towns at St. Joseph and three canoes to the Milwaukee towns. Further confirmation of the amount of foreign commerce moving through U.S. territory comes from the British themselves. Robert Hamilton, a Canadian merchant, estimated as late as 1798 that the illicit trade passing through American territory by way of the Great Lakes totaled upward of £90,000 sterling. See Robert Hamilton, "Memorandum on Trade and Commerce," September 24, 1798, in Russell Papers, 2:266–268, quoted in Hatter, "Jay Charter," 705.

42. For Spanish agents visiting the Chicago portage as late as 1797, see Nasatir, *Spanish War Vessels on the Mississippi*, 139–140. For Anishinaabe envoys to St. Louis, see Kinnaird, ed., "Northern Indians to Zenon Trudeau, [1793?]," in *Spain in the Mississippi Valley*, 3:110–111.

43. Thu-pe-ne-naw-ac, "Speech to Col. Hamtramck," Fort Wayne, March 21, 1796, Potawatomi Subseries, Box 6512, GLOVE.

44. Keating, *Rising Up*, 40–41.

45. Wabinewa was possibly the same individual ("Wabenaneto") who acted as spokesman for the Kankakee towns when the Chicago headmen visited Wayne in 1796. See "Speech of the Putawatimes Chiefs of Chikago."

46. See Keating, *Rising Up*, 41. Migrations of anti-American militants also led to intertribal tensions along the Illinois River corridor. The Anishinaabe settlements downriver, at Peoria, remained on good relations with the Americans under the leadership of Gomo. The Illinois River Anishinaabeg had continued to maintain diplomatic correspondence with the United States from the American Revolution forward, and Gomo had been part of the delegation that treated for peace during Putnam's Treaty of Vincennes in 1792 and traveled east to meet with George Washington the following year. Gomo also traveled east to Philadelphia for further diplomatic meetings after the Greenville Treaty. Such a stark divide between the Illinois River towns and those farther north along the Chicago, Kankakee, Calumet, and Milwaukee Rivers would continue through the War of 1812. For more on Gomo and the Peoria-area Anishinaabeg, see R. David Edmunds, "The Illinois River Potawatomi in the War of 1812," *Journal of the Illinois State Historical Society (1908–1984)* 62, no. 4 (1969): 341–362.

47. "Journal of a Voyage," in *Askins Papers*, 2:356–358. For a detailed biographical sketch of Point de Sable from his early business ventures in Cahokia, through the tumultuous years of the Revolution and into his time at Chicago, see Swenson, "Point de Sable."

48. Antoine Ouilmette to John H. Kinzie, Racine, June 1, 1839, Ayer MSS. For a brief biographical sketch of Ouilmette, see Frank R. Grover, *Antoine Ouilmette: A Resident of Chicago A.D. 1790–1826. The First Settler of Evanston and Wilmette (1826–1838) with a Brief History of His Family and the Ouilmette Reservation* (Evanston, Ill.: Evanston Historical Society, 1908).

49. For these early French habitants at Chicago, see Keating, *Rising Up*, 30–32. See also "Mrs. Whistler's Relation," in Hurlbut, *Chicago Antiquities*, 25; Danckers, "French Chicago During the 18th Century."

50. For more on the Burnett trading business, and Kakima's key role, see Sleeper-Smith, *Indian Women and French Men*.

51. William Burnett to John and James McGriger, July 29, 1802, St. Josephs, *WB*, 149–150.

52. By 1786, Burnett had at least nine engagés in his employ, dispersed at various trading locations along the waterways of lower Lake Michigan. See Burnett to John Sayer, June 26,

1786, St. Josephs, *WB*, 11–14. See, for instance, his employee Ducharme's activities along the Illinois River, Burnett to Sayer, May 14, 1786, St. Josephs, *WB*, 2–3.

53. Burnett to William Hand, May 25, 1786, St. Josephs, *WB*, 3–7. For Burnett's seasonal movements and winters on the Kankakee, see Burnett to William Hand, February 6, 1791, St. Josephs, *WB*, 47–49; Burnett to William Hand, February 2, 1790, St. Josephs, *WB*, 31–34.

54. William Burnett to Messrs. Meldrum and Park, May 25, 1786, St. Josephs, *WB*, 8–10.

55. Keating, *Rising Up*, 46.

56. "Inventory Sale of Property Sold by Point Sable to Jean Lalime," May 7, 1800, Collections About Jean Baptiste Pointe DuSable, Folder 1, CHM. For more context, see Milo Milton Quaife, "Property of Jean Baptiste Point Sable," *Mississippi Valley Historical Review* 15, no. 1 (June 1928): 89–92.

57. For Kinzie's vacillating citizenship, and the divided loyalties of his Detroit associates after Jay's Treaty, see Keating, *Rising Up*, 44.

58. "Inventory Sale of Property," May 7, 1800, Collections About Jean Baptiste Pointe DuSable, Folder 1, CHM. Interestingly, given Kinzie and LaLime's future rivalry, LaLime was the official buyer on the deed of sale. Kinzie and Burnett were listed as witnesses. Point de Sable may have proven further inclined to sell *because* of the incoming Americans and his own mixed affiliations with the Americans, Spanish, and British. During the American Revolution he had initially been captured by the British and suspected as an American sympathizer, only to be promoted by Patrick Sinclair to a prominent position managing the outpost at the Pinery, near Detroit. He worked on behalf of the British for at least the remainder of the war, but in the decades after, he cooperated with both Americans and Spanish to convey letters, host messengers, serve as an interpreter, and facilitate reconnaissance of the Chicago area. It remains difficult to tell his motivations for leaving Chicago on the eve of U.S. arrival, but several recent historians have speculated Chicago's rise to prominence may have worried someone with so many disparate ties to competing regional powers. See Keating, *Rising Up*, 30–31, 46–47; Swenson, "Point de Sable."

59. Burnett to Messrs Parker, Gerrard, & John OGilvy, August 24, 1798, St. Josephs, *WB*, 107–108.

60. As John Rice Jones wrote from Kaskaskia in 1798, he continued to hope "a garrison will shortly be erected on the Illinois River to protect us from the Ravages of the Indians living thereon; Sure I am that 'till that is the case we shall never be free from their Depredations." John Rice Jones to Winthrop Sargent, Kaskaskia, May 9, 1798, Potawatomi Subseries, Box 6512, GLOVE.

61. For Main Poc's role in these raids, and his rise to prominence in the early nineteenth century, see William Franz, "'To Live By Depredations' Main Poc's Strategic Use of Violence," *Journal of the Illinois State Historical Society (1998–)* 102, no. 3/4 (2009): 238–247; Mark J. Wagner, "'He Is Worst Than the Shawnee Prophet': The Archaeology of Nativism Among the Early Nineteenth Century Potawatomi of Illinois," *Midcontinental Journal of Archaeology* 31, no. 1 (January 1, 2006): 89–116; R. David Edmunds, "Main Poc: Potawatomi Wabeno," *American Indian Quarterly* 9, no. 3 (July 1, 1985): 259–272. For Black Bird's ascension to leadership, and his skepticism toward American encroachments, see "Indian Council," Amherstberg, June 8, 1805, in *MPHC*, 23:41.

62. As early as 1781, Jefferson himself had gleaned sufficient information about Chicago's portage to realize its importance. See Thomas Jefferson, *Notes on the State of Virginia* (Boston

and Walpole, NH: Printed by David Carlisle, for Thomas & Andrews, 1801), 14. For more on Jefferson's views regarding western lands and Indigenous peoples, see Anthony F. C. Wallace, *Jefferson and the Indians: The Tragic Fate of the First Americans* (Cambridge, Mass.: Harvard University Press, 1999); Robert M. Owens, *Mr. Jefferson's Hammer: William Henry Harrison and the Origins of American Indian Policy* (Norman: University of Oklahoma Press, 2011).

63. Dearborn to Hamtramck, as quoted in Arthur Frazier, "The Military Frontier: Fort Dearborn," *Chicago History* (Summer 1980): 80–85, quote on 82.

64. For a biography of Wells, who spent time as both a Native warrior opposing the United States and a federally employed Indian agent, see William Heath, *William Wells and the Struggle for the Old Northwest* (Norman: University of Oklahoma Press, 2015).

65. Dearborn to William Wells, May 26, 1803, in *TP*, 7:115, and Dearborn to Wells, May 27, 1803, in *TP*, 7:187.

66. Dearborn to Hamtramck, as quoted in Frazier, "Military Frontier," 82.

67. For the ideology behind settler expansion and the growth of a yeoman farmer class, see Drew R. McCoy, *The Elusive Republic: Political Economy in Jeffersonian America* (Chapel Hill: University of North Carolina Press, 1980). See also Gordon Wood, *Empire of Liberty: A History of the Early Republic, 1789–1815* (Oxford: Oxford University Press, 2009).

68. See figure 4.1. John Whistler, *Map of Fort Dearborn in January 1808 by Captain John Whistler, Commandant of the Fort* (January 25, 1808). Courtesy of the Chicago History Museum.

69. For Harrison's infamous role in managing Indian affairs and divesting Native peoples of their lands in the Northwest under the Jefferson administration, see Owens, *Mr. Jefferson's Hammer*. For a more localized study of Harrison's dealings in Vincennes, Tippecanoe, and Fort Wayne, see Bottiger, *Borderland of Fear*.

70. Jouett's first wife, Eliza, would die within two years of moving to Chicago. Charles then remarried in 1809, bringing his bride, Susan Randolph Allen, and Joseph Battle, an enslaved Black man, from Kentucky to Chicago. See Hurlburt, *Chicago Antiquities*, 102–107; Keating, *Rising Up*, 62–63.

71. Keating, *Rising Up*, 54–65.

72. For an overview of the factory system in U.S. Indian policy, see David Andrew Nichols, *Engines of Diplomacy: Indian Trading Factories and the Negotiation of American Empire* (Chapel Hill: University of North Carolina Press, 2016).

73. The frailty of American sovereignty in the region was highlighted in 1804, when Patterson had a Chicago-bound shipment of liquor waylaid by the Potawatomis at the St. Joseph River. Patterson blamed the incident on Chadonai, a métis clerk present at the time. British Indian agents in Canada sent several strings of threatening wampum to the St. Joseph Potawatomis, and Patterson dared to arrest Chadonai on behalf of British Canada, well within U.S. territory. For Patterson's entry into the Lake Michigan trade, see Burnett to Messrs Parker, Gerrard, & John Ogilvy, December 20, 1798, St. Josephs, *WB*, 110–112. For the incident with Chadonai, see Burnett to Wells, March 2, 1806, St. Josephs, *WB*, 177–186.

74. Certainly, Burnett's main complaint revolved around Harrison's broken promises to reward him with a monopoly on trade in the region. Burnett saw the easily purchased trade licenses given out to British citizens as a direct threat to his own business at Chicago and elsewhere. Burnett to Harrison, September 10, 1803, *WB*, 156–163. William Wells had sworn

Burnett in as justice of the peace for Knox County, organized under the Indian Territory (which both Chicago and St. Joseph belonged to at the time), on August 29, 1801.

75. John Whistler to Jacob Kingsbury, Fort Dearborn, July 27, 1804, Letter Book No. 1, Folder 1, KP.

76. Whistler to Kingsbury, Fort Dearborn, July 26, 1804, Letter Book No. 1 Folder 1, KP.

77. Whistler to Kingsbury, Fort Dearborn, November 3, 1804, Letter Book No. 2, Folder 2, KP.

78. Whistler to Kingsbury, Fort Dearborn, August 14, 1804, Box 381, John Whistler Folder, CHM.

79. Whistler to Kingsbury, Fort Dearborn, July 12, 1805, Letter Book No. 2, Folder 2, KP; Whistler to Kingsbury, Fort Dearborn, 1 September 1, 1805, Letter Book No. 2, Folder 2, KP.

80. Whistler himself was very ill in November 1804. See Whistler to Kingsbury, Fort Dearborn, November 3, 1804, Letter Book No. 2, Folder 2, KP. For similar problems with malaria at other wetlands garrisons in the Northwest Territory at the same time, see Whipple to Kingsbury, Fort Wayne, September 1, 1804; Clemson to Kingsbury, Detroit, October 27, 1804; Rhea to Kingsbury, Maumee, July 31, 1804; Rhea to Kingsbury, Maumee, August 31, 1804; Rhea to Kingsbury, Maumee, September 8, 1804, all in Letter Books 1 and 2, KP.

81. As late as the 1830s, John Kinzie's daughter-in-law, Juliette Magill Kinzie, related that, "owing to the badness of the roads a greater part of the year, the usual mode of communication between the fort and the Point [location of Kinzie's trading post] was by boat rowed up the river, or by a canoe paddled by a skillful hand. By the latter means too, an intercourse was kept up between the residents of the fort and the Agency house." Juliette Augusta (Magill) Kinzie, *Wau-Bun, the "Early Day" of the North-West* (Chicago: Caxton Club, 1901), 149.

82. Kingsbury to Whistler, June 12, 1804, KP.

83. Even the eventual message to evacuate Fort Dearborn, in 1812, ironically came by way of Winamac, a Potawatomi messenger. See Ninian Edwards to William Eustis, July 29, 1812, in *TP*, 16:247.

84. Burnett to James May, January 20, 1804, St. Josephs, *WB*, 165.

85. For rumors of Whistler's extensive private farming, see Heald to Kingsbury, Fort Dearborn, May 31, 1810, and Kingsbury to Heald, June 11, 1810, Folder 4, KP. For more generalized complaints of Whistler's personal and professional abuses of power, see Seth Thompson to Kingsbury, Fort Dearborn, May 1, 1810, and Matthew Irwin to Kingsbury, April 29, 1810, Folder 4, KP.

86. Dr. John Cooper, as quoted in an oral history. See James Grant Wilson, "Chicago from 1802–1812," unpublished manuscript, CHM.

87. For garrison subsistence in a later era, see Francis Paul Prucha, *Broadax and Bayonet: The Role of the United States Army in the Development of the Northwest, 1815–1860* (Madison: State Historical Society of Wisconsin, 1953), esp. 120–130. For similar on-the-ground adjustments and subsistence patterns when British garrisons occupied posts in the Great Lakes after the Seven Years' War, see Michael N. McConnell, *Army and Empire: British Soldiers on the American Frontier, 1758–1775* (Lincoln: University of Nebraska Press, 2004).

88. Kinzie had formerly partnered with Whistler's son to supply the garrison, beginning in 1807, but the partnership ended in August 1809. See "Contract of Indenture, John Kinzie and John Whistler Junior," September 12, 1810, John Kinzie Collection, Folder 1, CHM; William Barry, "Transcript of John Kinzie's Four Account Books; St. Josephs 1803–1804,

Chicago, 1804–1822," John Kinzie Collection, Folder 2, CHM. The original account books were lost in the Chicago fire of 1871.

89. Seth Thompson to Jacob Kingsbury, Fort Dearborn, May 1, 1810, Folder 4, Jacob Kingsbury Papers, CHM. The trading feud became so acrimonious that Whistler's son-in-law, Lt. Hamilton, challenged John Kinzie to a duel, which the latter declined to accept. See James Grant Wilson, "Chicago from 1802–1812," unpublished manuscript, CHM. For more on this trade rivalry, see Quaife, *Chicago and the Old Northwest*, 171–173.

90. Matthew Irwin to Jacob Kingsbury, Chicago, April 29, 1810, Folder 4, KP; quotes from Seth Thompson to Kingsbury, Fort Dearborn, May 1, 1810, Folder 4, KP. See also Irwin to Eustis, December 30, 1809, in Carter, *TP*, 14:66–68.

91. Whether all these charges were accurate or merely trumped up by the various rival factions within the garrison and local trading community remains unclear. Whistler was never charged for any improprieties, and after his recall, many government officials and agents repudiated the charges made against him by Thompson and Irwin. For Whistler's own defense of his conduct, see Whistler to Kingsbury, May 27, 1810, Folder 4, KP. See also Heald to Kingsbury, May 31, 1810, Folder 4, KP.

92. Frazier, "Military Frontier," 84; Keating, *Rising Up*, 87–88.

93. Nathan Heald to Kingsbury, Fort Dearborn, June 8, 1810, Folder 4, KP.

94. Philip Ostrander to Kingsbury, Fort Dearborn, September 25, 1810, Folder 4, KP.

95. James Leigh to Kingsbury, Chicago, March 30, 1811, Folder 5, KP.

96. Jean LaLime to John Johnson, Chicago, July 7, 1811, Potawatomi Subseries, Box 6516, GLOVE.

97. Heald to William Eustis (Secretary of War), Fort Dearborn, July 9, 1811, Potawatomi Subseries, Box 6516, GLOVE.

98. LaLime to Governor Howard, February 4, 1812, in *TP*, 14:536–537.

99. Gillum Ferguson, *Illinois in the War of 1812* (Urbana: University of Illinois Press, 2012), 57.

100. Heald to William Wells, Fort Dearborn, April 15, 1812, in *TP*, 16:47. See also Irwin to Eustis, Fort Dearborn, May 15, 1812, in *TP*, 16:219–222; Kinzie, *Wau-Bun*, 239–240. While the local Anishinaabeg insisted that the perpetrators were some Ho-Chunks (Winnebagos) passing through the portage, it remains unclear who actually committed the violence.

101. Irwin to Eustis, April 16, 1812, in *TP*, 16:219–222. See Keating, *Rising Up*, 95–100, for a full account of Irwin's activities at Chicago. For ongoing suspicions by Irwin against Kinzie, even after the Battle of Fort Dearborn, see Sells, *Chicago Portage*, 81–84.

102. Matthew Irwin to Eustis, May 15, 1812, in *TP*, 16:221–222.

103. For "stabed . . . to the heart," see Heald to Irwin, April 12, 1812, in Richard C. Knopf, *Document Transcriptions of the War of 1812 in the Northwest*, 10 vols. (Columbus: Anthony Wayne Parkway Board, Ohio State Museum, 1957), 6:126. For further details, see Irwin to Secretary of War, May 15, 1812, *TP*, 16:221–222; Isaac Van Voorhis to Secretary of War, June 30, 1812, in Knopf, *War of 1812*, 6:67–68; Irwin to Secretary of War, July 1812, in Knopf, *War of 1812*, 69–72. For "fell down together" as reported in a later account of the murder, see Gurdon Hubbard to Arthur Kinzie, Chicago, February 6, 1884, in Linai Taliaferro Helm, *The Fort Dearborn Massacre, Written in 1814 by Lieutenant Linai T. Helm*, ed. Nelly Kinzie Gordon (Chicago: Rand McNally & Co., 1912), 94–95. See also Paul Dailing, "The Long Death of Jean Lalime," *Chicago Reader*, October 18, 2018.

104. Keating, *Rising Up*, 116–123.

105. White, *Middle Ground*, 501–518; Dowd, *Spirited Resistance*, 181–185; Ferguson, *Illinois*, 39–54.

106. Linai Helm to Augustus Woodward, Flemington, New Jersey, June 6, 1814, in *The Fort Dearborn Massacre*, ed. Nelly Kinzie Gordon, 15.

107. Ferguson, *Illinois*, 55.

108. Thomas Douglas, *A Sketch of the British Fur Trade in North America with Observations Relative to the North-West Company of Montreal by the Earl of Selkirk* (New York: Clayton & Kingsland, 1818): 30–31.

109. Keating, *Rising Up*, 114–115.

110. Hull to Secretary of War, July 29, 1812, quoted in Quaife, *Chicago and the Old Northwest*, 125.

111. For "Unusually large," see Copy of a letter from Capt. Heald Late Commandant at Fort Chicago, Pittsburgh, October 23, 1812, Fort Dearborn Collection, Folder 1, CHM. For other corroborating sources of the days leading up to the evacuation, see Nathan Heald, "Nathan Heald's Journal (1801–1822)," in Quaife, *Chicago and the Old Northwest*, 402–405; Linai Helm to Augustus Woodward, Flemington, New Jersey, June 6, 1814, in *The Fort Dearborn Massacre*. For a more dubious account, see *Wau-Bun*. For a recent and well-researched secondary account of the events at Fort Dearborn in August 1812, see Keating, *Rising Up*, 130–161.

112. Kinzie, *Wau-Bun*, 165. Other local Anishinaabe leaders warned Heald that "there was serious trouble ahead." See Simon Pokagon, "The Massacre of Fort Dearborn at Chicago, Gathered from the Traditions of the Indians Engaged in the Massacre, and from the Published Accounts," *Harper's New Monthly Magazine* 98 (1898): 651.

113. Kinzie, *Wau-Bun*, 174.

114. Keating, *Rising Up*, 133–143; Kinzie, *Massacre*, 29. For Wells (Eepiihkaanita)'s participation in the battle as both an agent of the United States and as a Miami warrior, see also Heath, *William Wells*; George Ironstrack, "Aacimwinki Niimihki Šikaakonki," *Aacimotaatiiyankwi* (blog), June 7, 2013, https://aacimotaatiiyankwi.org/2013/06/07/aacimwinki-niimi hki-sikaakonki/.

115. Kinzie related his recollections of the battle to Jacob Varnum in 1816; see "Journal of Jacob B. Varnum of Petersburg, VA.: Incidents in the Life of Jacob Butler Varnum, 1809 to 1822" (Journal, 1864), Box 362, Jacob Butler Varnum Folder, CHM.

116. Walter Jordan to "Betcy," Fort Wayne, October 12, 1812, and Walter Jordan to Joseph Hunter, Fort Wayne, December 17, 1812. For over a century after the battle, historians dismissed Jordan's eyewitness account as a fabrication, but in the mid-twentieth century, a second letter to his wife, found in a private collection, corroborated much of the information in Jordan's original letter, giving further credence to Jordan's account. Transcriptions of both letters, along with the argument for Jordan's authenticity, can be found in John D. Barnhart, "A New Letter About the Massacre at Fort Dearborn," *Indiana Magazine of History* 41, no. 2 (June 1945): 187–199. For a prominent example of earlier historians who dismissed the account, see Quaife, *Chicago and the Old Northwest*, 395–396.

117. This description of the battle is drawn primarily from the eyewitness accounts of Linai Helm to Augustus Woodward, Flemington, New Jersey, June 6, 1814, in *Fort Dearborn Massacre*; copy of a letter from Capt. Heald Late Commandant at Fort Chicago, Pittsburgh,

October 23, 1812, Fort Dearborn Collection, Folder 1, CHM; "Statement of James Corben of the County of Culpeper and State of Virginia, Respecting his Enlistment in the Army of the United States and his Service and Suffering in the Same," July 8, 1826, republished in M. M. Quaife, "The Story of James Corbin, A Soldier of Fort Dearborn," *Mississippi Valley Historical Review* 3, no. 2 (1916): 221–226; and secondhand relations of Sgt. Griffith in Robert B. McAfee, *History of the Late War in the Western Country* (Bowling Green, Ohio, 1816), 113–116, and John Kinzie, recorded in "Journal of Jacob B. Varnum," CHM. Kinzie also related a similar account later to a U.S. expedition led by Lewis Cass in 1820. See also Mentor L. Williams, "John Kinzie's Narrative of the Fort Dearborn Massacre," *Journal of the Illinois State Historical Society (1908–1984)* 46, no. 4 (1953): 343–362.

118. For two captivity accounts resulting from the battle, see "Statement of James Corben," and Jordan to "Betcy," Fort Wayne, October 12, 1812. Potawatomis and Kickapoos took Corben down the Illinois River, while Jordan escaped from a town farther upstream on the Des Plaines River.

119. Williams, "Kinzie Narrative," 350.

120. See entries for August 18, 1812, in Barry, "Transcript of John Kinzie's Four Account Books," John Kinzie Collection, Folder 2, CHM.

121. Thomas Forsyth to Nathan Heald, January 2, 1813, quoted in Quaife, *Chicago and the Old Northwest*, 246; Thomas Forsyth to William Wolden, St. Louis, December 9, 1826, reprinted in Quaife, "James Corbin," 227–228. Forsyth and Kinzie were able to negotiate the release of Nathan Heald and his wife, sending the couple east to William Burnett's post on the St. Joseph River. See "Copy of a Letter from Capt. Heald."

122. Ferguson, *Illinois*, 74, 125.

Chapter Five

1. John Kinzie to Lewis Cass, July 15, 1815, in *TP*, 16:200–201.

2. Kinzie to Cass, July 15, 1815, in *TP*, 16:200–201.

3. Anthony Butler to John Armstrong (Secretary of War), Detroit, January 23, 1814, Potawatomi Subseries, Box 6518, GLOVE. See also Butler to Armstrong, Detroit, February 3, 1814, National Archives, RG 107, B-420.

4. "Council between Clark and the Potawatomi Chief, Black Partridge, and Eight Warriors," St. Louis, January 2, 1814, Potawatomi Subseries, Box 6518, GLOVE.

5. Black Partridge and Petchaho, "Speech to Governor William Clark," Fort Clark, March 4, 1815, and William Clark, Ninian Edwards, and Auguste Chouteau to the Secretary of War, St. Louis, May 22, 1815, in Potawatomi Subseries, Box 6518, GLOVE.

6. For more on the struggle for a national policy of infrastructure in the early republic and the significance of government institutions, even without it, see Larson, "'Bind the Republic Together.'" See also Harry N. Scheiber, "On the New Economic History—and Its Limitations: A Review Essay," *Agricultural History* 41, no. 4 (October 1967): 383–396; Richard R. John, "Government Institutions as Agents of Change: Rethinking American Political Development in the Early Republic, 1787–1835," *Studies in American Political Development* 11, no. 2 (Fall 1997): 347–380; William J. Novak, "The Myth of the 'Weak' American State," *American Historical Review* 113, no. 3 (June 2008): 752–772; Gautham Rao, "The New Historiography of the Early Federal Government: Institutions, Contexts, and the Imperial

State," *WMQ* 77, no. 1 (2020): 97–128. See also Rachel St. John, "State Power in the West in the Early American Republic," *Journal of the Early Republic* 38, no. 1 (2018): 87–94.

7. For such internal improvements in Illinois and the Chicago area, see Bessie Louise Pierce, *A History of Chicago: The Beginning of a City, 1673–1848*, 3 vols. (New York: Knopf, 1937), vol. 1; John Henry Krenkel, *Illinois Internal Improvements, 1818–1848* (Cedar Rapids, Iowa: Torch Press, 1958).

8. The major work on Chicago's nineteenth-century development as a city is William Cronon, *Nature's Metropolis*, but most of the book focuses on the city's influence on an expanding hinterland. The local environment and the impact of Chicago's transformation on the people in the immediate area—Native or otherwise—received little attention. For a critical review highlighting these missing locals, see Peter A. Coclanis, "Urbs in Horto— Nature's Metropolis: Chicago and the Great West by William Cronon," *Reviews in American History* 20, no. 1 (1992): 14–20. For the importance of eastern investors and land speculators in Chicago's development, see John Denis Haeger, *The Investment Frontier: New York Businessmen and the Economic Development of the Northwest* (Albany: State University of New York Press, 1981).

9. See, for example, Larson, *Internal Improvement*; Larson, "'Bind the Republic Together'"; Howe, *What Hath God Wrought*; David Davidson, "Republic of Risk: Canals and Commercial Infrastructure Planning in the United States, 1783–1808" (PhD diss., Evanston: Northwestern University ProQuest Dissertations Publishing, 2012). For "transportation revolution," see George Rogers Taylor, *The Transportation Revolution, 1815–1860* (New York: Rinehart, 1951). For market revolution, see John Lauritz Larson, *The Market Revolution in America: Liberty, Ambition, and the Eclipse of the Common Good* (New York: Cambridge University Press, 2010); Charles Sellers, *The Market Revolution: Jacksonian America, 1815–1846* (New York: Oxford University Press, 1991). For the role of the state in such processes, see Goodrich, *Government Promotion*; Frymer, *Building an American Empire*; Brian Balogh, *A Government Out of Sight: The Mystery of National Authority in Nineteenth-Century America* (New York: Cambridge University Press, 2009); Max M. Edling, *A Hercules in the Cradle: War, Money, and the American State, 1783–1867* (Chicago: University of Chicago Press, 2014); Ablavsky, *Federal Ground*.

10. Recent examples of such studies include Saler, *Settlers' Empire*; Saunt, *Unworthy Republic*; Witgen, *Seeing Red*.

11. For a notable exception, see Aley Ginette, "Bringing About the Dawn: Agriculture, Internal Improvements, Indian Policy, and Euro-American Hegemony in the Old Northwest, 1800–1846," in *The Boundaries between Us: Natives and Newcomers along the Frontiers of the Old Northwest Territory, 1750–1850*, ed. Daniel Barr (Kent, Ohio: Kent State University Press, 2006), 196–218. For adjacent developments in Wisconsin slightly later, see Witgen, *Seeing Red*, especially 300–305. John Larson's recent book, while not dealing with Indigenous dispossession, does examine the environmental implications of expanding settler capitalism in the interior. See Larson, *Laid Waste!* For the significant implications of taking such a holistic view of the early republic, see Snyder, *Great Crossings*, 16–17.

12. Bradley to Parker, Chicago, August 3, 1816, in Drennan Papers, CHM; Frazier, "Military Frontier," 84.

13. Prucha, *Broadax and Bayonet*, 149–171.

14. Jacob Varnum, the government factor who arrived in 1816, recounted how one of the first tasks the American contingent had to complete when they arrived at Chicago entailed collecting "the bleached bones of about 70" casualties from the Battle of Fort Dearborn and interring them. Such tasks, plus the burned-out buildings of the old fort, went far in reiterating Anishinaabe dominance at the site, even after the return of U.S. forces. See "Journal of Jacob B. Varnum," CHM. See also Quaife, *Chicago and the Old Northwest,* 262–267.

15. Jouett arrived that fall to begin his renewed work as an Indian agent. See Jouett to Cass, Chicago, October 24, 1816, Potawatomi Subseries, Box 6519, GLOVE; Edmunds, *Potawatomis,* 217.

16. Jouett to William Crawford, Chicago, December 2, 1816, Potawatomi Subseries, Box 6519, GLOVE.

17. *Niles Weekly Register* (Baltimore, August 6, 1814).

18. See Albert Gallatin "Roads and Canals, communicated to the Senate," Treasury Department, April 4, 1808, in *ASP: Miscellaneous,* 1:724–741. See also Peter B. Porter's proposal two years later, under "Canals and Roads," February 8, 1810, *Annals of Congress,* 11th Cong., 2nd sess., II:1387–1393.

19. Duncan MacArthur to the Secretary of War, Chillicothe, May 15, 1815, Potawatomi Subseries, Box 6518, GLOVE.

20. William Clark to Thomas A. Smith, St. Louis, March 28, 1815, Potawatomi Subseries, Box 6519, GLOVE.

21. George Graham (Chief Clerk, Dept of War) to Brig. General Thomas A. Smith, War Department, November 6, 1815, Potawatomi Subseries, Box 6519, GLOVE. At the same time, George Graham directed Governor Ninian Edwards to disburse $2,000 worth of goods to the Illinois Country Potawatomis to placate them about the survey of these lands. See Graham to Edwards, War Dept., November 6, 1815, Potawatomi Subseries, Box 6519, GLOVE.

22. Forsyth to Edwards, St. Louis, March 31, 1816, in *WHC,* 11:347.

23. Benjamin Parke to the Secretary of War, Vincennes, April 3, 1816, Potawatomi Subseries, Box 6519, GLOVE.

24. For surveying and mapping as tools of the state, see Sarah S. Hughes, *Surveyors and Statesmen: Land Measuring in Colonial Virginia* (Richmond: Virginia Surveyors Foundation, 1979); Gregory H. Nobles, "Straight Lines and Stability: Mapping the Political Order of the Anglo-American Frontier," *Journal of American History* 80, no. 1 (1993): 9–35; Marcus Gallo, "Land Surveying in Early Pennsylvania: A Case Study in a Global Context," *Journal of Early American History* 6, no. 1 (2016): 9–39. This view of surveying as a state tool for dispossession anticipates just a fragment of the good work being done by Michael Borsk, a PhD candidate at Queen's University. See also Scott, *Seeing Like a State,* 11–52.

25. For official plans for military outposts and land cessions through these successive St. Louis treaties, see A. J. Dallas to William Henry Harrison, Duncan MacArthur, and John Graham, War Department, June 9, 1815, Potawatomi Subseries, Box 6518, GLOVE. See also "Treaty with the Sauk and Foxes, 1804," in Charles Jones Kappler, ed., *Indian Affairs: Laws and Treaties,* 7 vols. (Washington, D.C.: U.S. Government Printing Office: 1904), 2:74–77. Digitized version available through Oklahoma State University Library Digital Collections.

26. Clark, Edwards, and Chouteau to the Secretary of War, St. Louis, October 1815, Potawatomi Subseries, Box 6519, GLOVE.

27. Captain Whistler, referring to the Anishinaabeg north and west of his post at Fort Wayne, shared his assessment that "all those who live in the pariarias [*sic*] and near the Lakes are" "in a starveing condition." Whistler to MacArthur, Fort Wayne, May 31, 1815. See also Talk of Nis Cau Ne Ma, a Po to Wau Ta Mie War Chief to Major John Whistler, Fort Wayne, May 22, 1815, in Potawatomi Subseries, Box 6518, GLOVE.

28. For Ninian Edwards's own account of his role in the treaty negotiations over the canal zone, see Ninian Edwards, *History of Illinois, from 1778 to 1833; and Life and Times of Ninian Edwards* (Springfield, Ill., 1870), 99; *Journal of the Senate of the General Assembly of the State of Illinois* (Springfield, Ill.: State printer, 1826), 1:77.

29. "A Treaty of Peace, Friendship, and Limits," Portage des Sioux, August 24, 1816, in *United States Statutes at Large* 7:146–148. See also Wheeler-Voegelin and Blasingham, *Indians of Western Illinois and Southern Wisconsin*, 109. For further insights into this treaty, and subsequent cessions affecting the Chicago-area Anishinaabeg, see Helen Hornbeck Tanner Papers: 1930s–2009, Box 2: Undated Folder–Treaties, BHL.

30. For an overview of the early history of the topographical engineers, later the Army Corps of Topographical Engineers, see Henry P. Beers, "A History of the U.S. Topographical Engineers, 1813–1863," *Military Engineer* 34 (June 1942): 287–291. For an older work with an important argument about how topographical engineers played a critical role in U.S. expansion, see William H. Goetzmann, *Army Exploration in the American West, 1803–1863* (New Haven, Conn.: Yale University Press, 1960). See also John W. Larson, *Those Army Engineers: A History of the Chicago District, U.S. Army Corps of Engineers* (Washington, D.C.: U.S. Army Corps of Engineers, Chicago District, 1980); Frank N. Schubert, ed., *The Nation Builders: A Sesquicentennial History of the Corps of Topographical Engineers, 1838–1863* (Washington D.C.: U.S. Government Printing Office, 1989); Theodore Catton, *Rainy Lake House: Twilight of Empire on the Northern Frontier* (Baltimore, Md.: Johns Hopkins University Press, 2017), 335–336.

31. "A Letter from L. H. Long, major of topographical engineers, and acting engineer of fortifications, addressed to George Graham, Esq. acting secretary of war," printed in the *National Register* (March 29, 1817), 3:196.

32. "Letter from L. H. Long," 3:193.

33. "Letter from L. H. Long," 3:194.

34. "Letter from L. H. Long," 3:194.

35. "Letter from L. H. Long," 3:195.

36. "Letter from L. H. Long," 3:196. See also Stephen H. Long, *A Map of the Illenois River from the Mouth to Gomo's Village 200 miles*, St. Louis, September 20, 1816, National Archives, Washington, D.C.

37. *Intelligencer*, March 11, 1818; for the legislation, see *Annals of Congress*, 15th Cong., 1st sess., 2:1677. See figure 5.1. Alès, *Illinois* (Paris, 1819), Hermon Dunlap Smith Collection. Courtesy of the Newberry Library. See also Solon Justus Buck, *Illinois in 1818* (Chicago: McClurg, 1918), 224–225; Sells, *Chicago Portage*, 103–117.

38. For this explicit logic, see Alexander Wolcott to Cass, Chicago, March 31, 1821, Potawatomi Subseries, Box 6521, GLOVE.

39. Wolcott to Cass, Chicago, January 1, 1821, Potawatomi Subseries, Box 6521, GLOVE.

40. "Proceedings of the Treaty of Chicago" (August 29, 1821), Potawatomi Subseries, Box 6521, GLOVE.

41. "Proceedings of the Treaty of Chicago" (August 29, 1821), Potawatomi Subseries, Box 6521, GLOVE.

42. Henry Schoolcraft to John C. Calhoun, Chicago, September 18, 1821, Potawatomi Subseries, Box 6521, GLOVE.

43. Pierce, *Beginning*, 26; Alfred Theodore Andreas, *History of Chicago: Ending with the Year 1857* (Chicago: A. T. Andreas, 1884), 92–93. This trading post sat near the site of the original Leigh's Place, or "Hardscrabble," strategically positioned near the portage and the early epicenter of violence leading up to the War of 1812.

44. "Journal of Jacob B. Varnum," CHM.

45. For the subsequent rivalries between the American Fur Company, independent traders, and the government factory system, see Pierce, *Beginning*, 26–29. For AFC activities, see Hubbard's reminiscences in Hurlbut, *Chicago Antiquities*, 31–32; John D. Haeger, "Business Strategy and Practice in the Early Republic: John Jacob Astor and the American Fur Trade," *Western Historical Quarterly* 19, no. 2 (1988): 183–202. For Bailly's commercial activities at this time, see "Ledger, 1825–1835," Box 4, Folder 9, Joseph Bailly Collection, Rare Books and Manuscripts, Indiana State Library, Indianapolis, Indiana.

46. If anything, the continued productivity of the fur trade around Chicago endangered traders' profits only through attracting too many would-be competitors to the trade. Running against an older trope about the fur trade's declining productivity, scholars like Susan Sleeper-Smith have demonstrated the trade's continued viability in the Lake Michigan basin into the 1840s. Sleeper-Smith, *Indian Women and French Men*. See also Hubbard's many reports from the winters throughout the 1820s where he described the fur trade: Having "secured a large number of furs . . . great success," a season being "an unusually good one," "winter's business . . . quite satisfactory," and "the hunting . . . unusually good," with "many furs collected." Hubbard, *Autobiography*, 88, 140, 149, 155. As Sleeper-Smith and, more recently, Michael Witgen have argued, the transformation of the region into settler agricultural lands threatened the fur trade more than any overhunting that took place. See Witgen, *Seeing Red*, 139–140.

47. The literature on the Erie Canal is extensive. For a foundational text, see Ronald E. Shaw, *Erie Water West: A History of the Erie Canal, 1792–1854* (Lexington: University Press of Kentucky, 1990). For more recent works, see Carol Sheriff, *The Artificial River: The Erie Canal and the Paradox of Progress, 1817–1862* (New York: Hill and Wang, 1997); Peter L. Bernstein, *Wedding of the Waters: The Erie Canal and the Making of a Great Nation* (New York: Norton, 2005). For the joint role of capitalism and state capacity in this project, see Brian Phillips Murphy, *Building the Empire State—Political Economy in the Early Republic* (Philadelphia: University of Pennsylvania Press, 2015), esp. 159–206. For an important parallel regarding canal projects and Indigenous dispossession, see Hauptman, *Conspiracy of Interests*.

48. For the social, economic, and cultural implications of this migration, see Lois (Kimball) Mathews Rosenberry, *The Expansion of New England: The Spread of New England Settlement and Institutions to the Mississippi River, 1620–1865* (Boston: Houghton Mifflin, 1909); Haeger, *Investment Frontier*. See also Ann Durkin Keating, *The World of Juliette Kinzie* (Chicago: University of Chicago Press, 2019).

49. For a parallel study of the settlement and transformation of a river valley settlement in central Illinois during this same time, see John Mack Faragher, *Sugar Creek: Life on the Illinois Prairie* (New Haven, Conn.: Yale University Press, 1986).

50. Pierce, *Beginning*, 31; John Wentworth, *Early Chicago: A Lecture, delivered before the Sunday Lecture Society, May 7, 1876* (Fergus Historical Series, No. 7: Chicago, 1876), 15–17. Tellingly, twenty-one of the thirty-five names listed remained recognizably French. For the first presidential election at Chicago, see David McCulloch, *Early Days of Peoria and Chicago, An Address Read Before the Chicago Historical Society at a Quarterly Meeting Held January 19, 1904* (Chicago: Chicago Historical Society, 1904), 112–113.

51. See, for instance, Edward Coles to D. P. Cook, State of Illinois Executive Department, December 9, 1824; Edward Coles to D. P. Cook, State of Illinois Executive Department, January 20, 1825; Edward Coles to D. P. Cook, State of Illinois Executive Department, January 30, 1826; all in *CISHL*, 4:69, 75, 95.

52. For the wider impulses of the so-called American System that outlined internal improvements and infrastructure on the regional and national level, see Maurice G Baxter, *Henry Clay and the American System* (Lexington: University Press of Kentucky, 2015); Howe, *What Hath God Wrought*.

53. In 1820, the Indian agent, Oliver Wolcott Jr., estimated the Anishinaabe population in his jurisdiction around 1,500 to 2,000 persons. See Jouett to Cass, Chicago, July 8, 1817, Potawatomi Subseries, Box 6520, GLOVE; Wolcott to Cass, Chicago, March 3, 1820, Potawatomi Subseries, Box 6521, GLOVE. For a later account that corroborates such estimates, see Schoolcraft, *Narrative*, 383.

54. Jacob Lee has effectively shown how American infiltration into Indigenous kin networks farther downstream, at St. Louis, likewise bolstered U.S. ascendancy there in similar ways. See Lee, *Middle Waters*. For studies of Chicago as a "creolized" community, see Peterson, "Prelude to Red River"; Peterson, "Goodbye, Madore Beaubien"; Keating, *Rising Up*, 203–206, 216–219; Keating, *World of Juliette Kinzie*, 64–65. Many primary sources noted the interactions and intermarriage between area Anishinaabeg, resident French, and local Anglo traders with racist disdain. For examples, see Keating, *Narrative of an Expedition*; Patrick Shirreff, *A Tour through North America* (Edinburgh, 1835), 227–228; Charles Latrobe, *The Rambler in North America*, 2 vols. (London, 1836), 2:206–210. For similar communities in Wisconsin and other parts of the Great Lakes, see Murphy, *Great Lakes Creoles*; Saler, *Settlers' Empire*; Hyde, *Empires, Nations, and Families*; Sleeper-Smith, *Indian Women and French Men*; Hubbard, *Autobiography*, 15.

55. For such complaints by white settlers, see Edmunds, *Potawatomis*, 226.

56. Edwards to Clark, Belleville, Illinois, August 20, 1827, in Edwards, *History of Illinois*, 350.

57. Hubbard, *Autobiography*, 167–168. The Chicago residents' fears were not totally unfounded, as Big Foot would conspire late that year to join the Ho-Chunk uprising in Wisconsin. See Edmunds, *Potawatomis*, 230–231; Pierce, *Beginning*, 35–36.

58. Reynolds to Dodge, Vandalia, December 13, 1832, in *CISHL*, 4:221. For further depredations, see Edmunds, *Potawatomis*, 226, 241–242.

59. Edmund Roberts to Edwards, Kaskaskia, December 4, 1829, in E. B. Washburne, ed., *The Edwards Papers: Being a Portion of the Collection of the Letters, Papers, and Manuscripts of Ninian Edwards* (Chicago: Fergus Printing Company, 1884), 464–466. See also Pierce, *Beginnings*, 46; James William Putnam, *The Illinois and Michigan Canal: A Study in Economic History* (Chicago: University of Chicago Press, 1918), 18–20.

60. For the connections between such canal projects and Indian policy in other parts of the Midwest, see Aley Ginette, "Bringing About the Dawn," in Barr, *The Boundaries Between Us* 198.

61. For Anishinaabeg engagement with Chicago's wetlands, see chapter 3, "Crossroads"; Greenberg, *Natural History*, 115–139, 216–241; Vileisis, *Discovering the Unknown Landscape*, 20–27.

62. For a fascinating work on settler migrants' cultural and bodily views on wetlands, rivers, and other geography in the Mississippi River valley, see Conevery Bolton Valencius, *The Health of the Country: How American Settlers Understood Themselves and Their Land* (New York: Basic Books, 2002).

63. Kinzie, *Wau-Bun*, 148.

64. John Tipton, quoted in Pierce, *As Others See Chicago*, 29.

65. Quoted in Vileisis, *Unknown Landscape*, 82.

66. James Herget, "Taming the Environment: The Drainage District in Illinois," *Journal of the Illinois State Historical Society* 71, no. 2 (1978): 107–118. For Indiana, see Stephen F. Strausberg, "Indiana and the Swamp Lands Act: A Study in State Administration," *Indiana Magazine of History* 73, no. 3 (1977): 191–203. For Illinois, see Margaret Beattie Bogue, "The Swamp Land Act and Wet Land Utilization in Illinois, 1850–1890," *Agricultural History* 25, no. 4 (1951): 169–180.

67. See "Statement of the 'Swamp Lands' in the State of Illinois, by Land Districts, as reported by the Surveyor General," reprinted in Herget, "Taming," 109. For a firsthand account of drainage efforts as they relate to speculation, see John Lewis Peyton, *Over the Alleghanies and across the Prairies: Personal Recollections of the Fare West One and Twenty Years Ago* (1869), 339–341. This was a protracted process, undergirded after 1850 by the federal Swamp Land Act. For more on this, see Vileisis, *Unknown Landscape*, 83–86. For the transformation of adjacent wetlands in the Calumet region, see Powell A. Moore, *The Calumet Region: Indiana's Last Frontier* (Indianapolis: Indiana Historical Bureau, 1959), 94–95, 142–143. See also Anthony E. Carlson, "The Other Kind of Reclamation: Wetlands Drainage and National Water Policy, 1902–1912," *Agricultural History* 84, no. 4 (2010): 451–478; Samuel J. Imlay and Eric D. Carter, "Drainage on the Grand Prairie: The Birth of a Hydraulic Society on the Midwestern Frontier," *Journal of Historical Geography* 38, no. 2 (2012): 109–122; Robert Michael Morrissey, "The Drains Out of Town: The Invisible Environmental History of a Flatland," *Flatland Project, New Dimensions in the Environmental Humanities*, University of Illinois (blog), https://flatlandproject.web.illinois.edu/the-drains-out-of-town-the-invisible-environmental-history-of-a-flatland/.

68. James Hall, ed., *Illinois Monthly Magazine*, vol. 1 (Vandalia, Ill.: Robert Blackwell, 1831).

69. Habitat loss as the primary driver of declining aquatic mammal populations around Chicago is further confirmed in Robert Kennicott's assessment of the local flora and fauna. See Kennicott, *Catalogue of Animals Observed in Cook County, Illinois* (Springfield: Illinois State Agricultural Society, 1855). For more on Robert Kennicott as an innovative naturalist, and his connections to Northwestern University, see Ronald S. Vasile, "The Early Career of Robert Kennicott, Illinois' Pioneering Naturalist," *Illinois Historical Journal* 87, no. 3 (1994): 150–170.

70. Vileisis, *Unknown Landscape*, 160–161. See also Greenberg, *Natural History*, 177–200, 231–232.

71. For comparable settler mindsets when it came to ecological degradation in newly developed lands, see Alan Taylor, "'Wasty Ways': Stories of American Settlement," *Environmental History* 3, no. 3 (1998): 291–310; Jon T. Coleman, *Vicious: Wolves and Men in America* (New Haven, Conn.: Yale University Press, 2004), 123–146.

72. As quoted in Ulrich Danckers and Jane Meredith, eds., *A Compendium of the Early History of Chicago: To the Year 1835 When the Indians Left* (River Forest, Ill., 1999), 163. In reality, Chicago's entire shoreline remained a space in flux, constantly shifting with water, sand, and, after incorporation, refuse. See Craig Colten, "Chicago's Waste Lands: Refuse Disposal and Urban Growth, 1840–1990," *Journal of Historical Geography* 20, no. 2 (1994): 124–142, esp. 126–129.

73. Hubbard, *Autobiography*, 39–40; Pierce, *Beginning of a City*, 90–91. The garrison may have made similar attempts earlier, with equally disappointing results. See Pierce, *As Others See Chicago*, 29.

74. See, for instance, Edmund Roberts to Edwards, Kaskaskia, December 4, 1829, in *Edwards Papers*, 464–466. For the role of the Army Engineers in Chicago's harbor development, see Forest G. Hill, *Roads, Rails & Waterways: The Army Engineers and Early Transportation* (Norman: University of Oklahoma Press, 1957), 169.

75. Cronon, *Metropolis*, 55–56; Pierce, *Beginning of a City*, 90–93. See also Robin Einhorn, "A Taxing Dilemma: Early Lake Shore Protection," *Chicago History* 18, no. 3 (Fall 1989): 34–51.

76. Charles Hoffman's observations, as quoted in Danckers and Meredith, *Compendium*, 105–106.

77. For increases in shipping, Danckers and Meredith, *Compendium*, 105–106; Pierce, *Beginning*, 92. See figure 5.2. William Howard, *Map of the Mouth Chicago River Illinois with the Plan of the Proposed Piers for Improving the Harbour* (1830), Graff Collection. Courtesy of the Newberry Library. For further optimism, see Bemsley Huntoon to Josiah Huntoon, Chicago, December 15, 1837, CHM. For the continued growth of Chicago shipping throughout the nineteenth century, see Theodore J. Karamanski, *Schooner Passage: Sailing Ships and the Lake Michigan Frontier* (Detroit, Mich.: Wayne State University Press, 2000).

78. For an account of this, see Bela Hubbard, "Field Notes, Peninsula Coast Survey, Detroit to Chicago" (July 19, 1838), Bela Hubbard Papers, Box 1, vol. 2, BHL.

79. Pierce, *Beginning*, 92. For ongoing troubles with sand, see entries in the *Chicago Democrat*, May 1, 1839, and June 5, 1844; *Chicago Daily American*, September 10, 1841; May 16, 1842; and June 25, 1842. See also Andreas, *History of Chicago*, 1:235.

80. Caroline Kirkland, "Illinois in Spring-time: With a Look at Chicago," *Atlantic Monthly* 2 (1858): 487.

81. Edwards to Thomas Neal, August 9, 1827, in Edwards, *History of Illinois*, 349.

82. As the Illinois politician Jesse Thomas noted in his assessment of Chicago's urban development in 1847, the brief conflict had "brought Chicago, and this region into notice, and laid the foundation for a permanent settlement." See Jesse B. Thomas, *Report of Jesse B. Thomas, As a Member of the Executive Committee Appointed by the Chicago Harbor and River Convention, of the Statistics Concerning the City of Chicago* (Chicago, 1847), 9.

83. The literature on Native Removal on all these fronts is extensive, but for a good overview of northern Native Removal, see Bowes, *Land Too Good for Indians*. See also Horsman, *Origins of Indian Removal*; Snyder, *Great Crossings*, 126–129; Fierst, "Rationalizing Removal"; Tiro, "View from Piqua Agency"; Carlson and Roberts, "Indian Lands"; Frymer, *Building an American Empire*; Howe, *What Hath God Wrought*, 342–357; Samantha Seeley, *Race, Removal, and the Right to Remain: Migration and the Making of the Early United States* (Chapel Hill: University of North Carolina Press, 2021). For the historical rationale of civi-

lization policies toward Native Americans during and after the American Revolution, see Griffin, *American Leviathan*; Wallace, *Jefferson and the Indians*; Reginald Horsman, *Race and Manifest Destiny: The Origins of American Racial Anglo-Saxonism* (Cambridge, Mass.: Harvard University Press, 1981).

84. Ginette, "Bringing About the Dawn," in Barr, *The Boundaries Between Us*, 207.

85. Report of the Committee on Indian Affairs, *Senate Documents*, no. 22, 21–22, serial 203, 1–4, as quoted in Ginette, "Bringing About the Dawn," in Barr, *The Boundaries Between Us*, 211.

86. Proceedings, Potawatomi and Miami Treaty Negotiations, October 5, 1826, *The John Tipton Papers*, ed. Nellie Armstrong Robertson and Dorothy Riker, 3 vols. (Indianapolis: Indiana Historical Bureau, 1942), 1:577–592; *American State Papers-Indian Affairs*, 2:679–685, as quoted in Ginette, "Bringing About the Dawn," in Barr, *The Boundaries Between Us*, 210.

87. Lewis Cass, James Ray, and John Tipton to James Barbour, October 23, 1826, *Tipton Papers*, 1:602–603.

88. Lewis Cass, *Remarks*, quoted in Witgen, *Seeing Red*, 76. For more on Cass as a leading architect behind Indian Removal as a national policy, see Tiro, "View from Piqua Agency," 41–42; Fierst, "Rationalizing Removal"; Witgen, *Seeing Red*.

89. For how the "Road Band" of Potawatomis in Indiana manipulated these road-building projects and land cession demands to their own advantage, see Ben Secunda, "The Road to Ruin? 'Civilization' and the Origins of a 'Michigan Road Band' of Potawatomi," *Michigan Historical Review* 34, no. 1 (2008): 118–149.

90. *Black Hawk's Autobiography*, 84.

91. For the Black Hawk War, and the Anishinaabeg role in aiding the United States, see John W. Hall, *Uncommon Defense: Indian Allies in the Black Hawk War* (Cambridge, Mass.: Harvard University Press, 2009). See also Patrick J. Jung, *The Black Hawk War of 1832* (Norman: University of Oklahoma Press, 2008).

92. Lewis Cass, then secretary of state, remained the chief orchestrator of the treaty behind the scenes. See Cass to Porter, Owen, and Weatherford, April 8, 1833, M21, Roll 10, 210–214, National Archives.

93. Shirreff, *Tour through North America*, 227.

94. Latrobe, *Rambler in North America*, 2:210. See also Shirreff, *Tour through North America*, 226–228.

95. James Clifton, "Chicago, September 14, 1833: The Last Great Indian Treaty in the Old Northwest," *Chicago History* (Summer 1980): 86–104, esp. 91–92.

96. For Billy Caldwell's complex background and allegiances, see James Clifton, "Personal and Ethnic Identity on the Great Lakes Frontier: The Case of Billy Caldwell, Anglo-Canadian," *Ethnohistory* 25, no. 1 (1978): 69–94. For the long-term corruption of the treaty negotiators, interpreters, and Indian traders during the treaty, see R. David Edmunds, "'Designing Men, Seeking a Fortune': Indian Traders and the Potawatomi Claims Payment of 1836," *Indiana Magazine of History* 77, no. 2 (1981): 109–122. For the overarching "political economy of plunder," prevailing at such proceedings, see Witgen, *Seeing Red*.

97. See "Journal of the Proceedings of a Treaty Between the United States and the United Tribe of Pottawottamies, Chipeways, and Ottowas," T494, Roll 3, 61–87, National Archives. For figures, see Keating, *Indian Country*, 229–230; Pierce, *Beginning*, 40–42. For a recent

overview of the treaty, see Bowes, *Land Too Good for Indians*, 149–181. For older assessments, see Cronon, *Metropolis*, 28–29; Clifton, "Last Great Indian Treaty"; Edmunds, *Potawatomis*, 250–252.

98. *Chicago Democrat*, November 26, 1833, as quoted in Pierce, *Beginning*, 40, 48.

99. For the importance of eastern investors and land speculators in Chicago's development, see Haeger, *Investment Frontier*.

100. See Pierce, *Beginning*, 44.

101. See editorials in the *Chicago Democrat*, November 26, 1933, and January 21, 1834.

102. Pierce, *Beginning*, 58.

103. For such celebrations, see *Chicago Democrat*, January 20, 1836.

104. *Chicago American*, April 23, 1836; see also Hubbard to Jackson, May 1, 1836, in Hubbard Papers, CHM; Pierce, *Beginning*, 62.

105. For good overviews of this speculative era in Chicago's early history, see Pierce, *Beginning*, 43–74; Cronon, *Metropolis*, 29–33.

106. For a description of the planned ceremonies, see *Chicago Democrat*, June 29, 1836. For "imposing ceremonies," see quote in Putnam, *Illinois and Michigan Canal*, 37. For Hubbard's role, see Hubbard, *Autobiography*, xx. The summer of 1836 was indeed a heady time across the nation for those who saw internal improvements and Indigenous dispossession as the dual cornerstones of American ascendancy. For intertwined celebrations and atrocities in other parts of the continent, see Saunt, *Unworthy Republic*, xi–xii, 231–257.

107. For the difficulties of constructing a canal through a wetland, see *Report of the Board of Canal Commissioners* (1838), 5; Putnam, *Illinois and Michigan Canal*, 37.

108. Krenkel, *Illinois Internal Improvements*, 26–46, esp. 43; Putnam, *Illinois and Michigan Canal*.

109. For the stakes of environmental exploitation for Chicago as a city, see Cronon, *Metropolis*. For Chicago as a uniquely advantageous locale for business interests and speculation, see Robin L. Einhorn, *Property Rules: Political Economy in Chicago, 1833–1872* (Chicago: University of Chicago Press, 1991).

110. Libby Hill, *The Chicago River: A Natural and Unnatural History* (Carbondale: Southern Illinois University Press, 2000). For a parallel, and wider, story of hydrological manipulation and human dominance, see John McPhee, *The Control of Nature*, reprint ed. (New York: Farrar, Straus and Giroux, 1990).

111. Robert Herrick, *The Gospel of Freedom* (New York: Macmillan, 1898), 103–104. Quoted in Cronon, *Metropolis*, 14. See also figure 5.3. George Robertson, *Chicago, Ill.* (1853). Courtesy of the Newberry Library.

Epilogue

1. John Mack Faragher, "A Nation Thrown Back Upon Itself," in *Rereading Frederick Jackson Turner*, 1–3, quote from 2. For Cody's performances at the world's fair, see Lowe, *Imprints*, 127–130. For more on Bill Cody and the performative West, see Louis S. Warren, *Buffalo Bill's America: William Cody and the Wild West Show* (New York: Vintage, 2006).

2. For the published form of Turner's talk that day, see Frederick Jackson Turner, "The Significance of the Frontier in American History," republished in Turner, *Rereading Frederick Jackson Turner*, 31–60.

3. As quoted in Faragher, "Nation," in *Rereading Frederick Jackson Turner*, 2.

4. For the fully formed historical philosophy, crafted over his career, see Frederick Jackson Turner, *The Frontier in American History* (New York: Henry Hold, 1921).

5. For more on Turner's legacies within historiography, politics, and American culture, see Weber, "Turner, the Boltonians, and the Borderlands"; Cronon, "Revisiting the Vanishing Frontier"; Martin Ridge, "The Life of an Idea: The Significance of Frederick Jackson Turner's Frontier Thesis," *Montana: Magazine of Western History* 41, no. 1 (1991): 2–13; Ridge, "Turner the Historian"; John Mack Faragher, "The Frontier Trail: Rethinking Turner and Reimagining the American West," *American Historical Review* 98, no. 1 (February 1993): 106–117; Wilbur R. Jacobs, ed., *On Turner's Trail: 100 Years of Writing Western History* (Lawrence: University Press of Kansas, 1994); Gregory H. Nobles, *American Frontiers: Cultural Encounters and Continental Conquest* (New York: Hill and Wang, 1997), esp. 3–16. For his legacy as the first "modern" American historian, beyond his thesis, see Limerick, "Turnerians All." For a contemporary attempt to square closed borders with a frontier past, see Greg Grandin, *The End of the Myth: From the Frontier to the Border Wall in the Mind of America* (New York: Metropolitan Books, 2019).

6. Quote from Frederick Jackson Turner, "Significance of the Frontier," in *Rereading Frederick Jackson Turner*, 39.

7. Turner, "Significance of the Frontier," in *Rereading Frederick Jackson Turner*, 35–37, quote on 35.

8. For the "New Indian History," see Ned Blackhawk, "Look How Far We've Come: How American Indian History Changed the Study of American History in the 1990s," *Magazine of History* 19, no. 6 (2005): 13–17; Ned Blackhawk, "Currents in North American Indian Historiography," *Western Historical Quarterly* 42, no. 3 (2011): 319–324. For continental arguments about Indigenous hegemony, see Michael Witgen, "Rethinking Colonial History as Continental History."

9. Such impulses originated from the "New Western History" of the 1970s and 1980s. See, for examples, Patricia Nelson Limerick, *The Legacy of Conquest: The Unbroken Past of the American West* (New York: Norton, 1987); Richard White, *"It's Your Misfortune and None of My Own": A History of the American West* (Norman: University of Oklahoma Press, 1991); Patricia Nelson Limerick, Clyde Milner II, and Charles E. Rankin, *Trails: Toward a New Western History* (Lawrence: University Press of Kansas, 1991). William Cronon, George A. Miles, and Jay Gitlin, *Under an Open Sky: Rethinking America's Western Past* (New York: Norton, 1992). For a more recent study historicizing the term *frontier* in early American history, see Patrick Spero, *Frontier Country: The Politics of War in Early Pennsylvania* (Philadelphia: University of Pennsylvania Press, 2016).

10. John Mack Faragher, "Nation," in *Rereading Frederick Jackson Turner*, 1.

11. See figure 6.1. For more on Turner's own engagement with portaging, and the Great Lakes ecoregion, see Terence Young, *Heading Out: A History of American Camping* (Ithaca, N.Y.: Cornell University Press, 2017), esp. 82–86. For Turner's upbringing, see Ray Allen Billington, "Young Fred Turner," *Wisconsin Magazine of History* 46, no. 1 (1962): 38–52.

12. Turner, "Problem of the West," in *Rereading Frederick Jackson Turner*, 63–65.

13. See his section on "The Middle West" in Turner, *Frontier in American History*.

14. For a contemporary of Turner's who remained much more attuned to the geographic specificity of portages like Chicago in the history of North America, see Hulbert, *Keys of the*

Continent. This lack of specificity remains an often overlooked but prescient critique of Turner and subsequent iterations of "macronarratives" that elide "local particularity." See Roney, "Containing Multitudes," 266.

15. This is especially true within the field of environmental history. See Richard White, "American Environmental History: The Development of a New Historical Field," *Pacific Historical Review* 54, no. 3 (1985): 297–335; T. R. C. Hutton, "Beating a Dead Horse? The Continuing Presence of Frederick Jackson Turner in Environmental and Western History," *International Social Science Review* 77, no. 1/2 (2002): 47–57.

16. For later scholarship that articulates the power of the state in frontier incorporation, see Griffin, *American Leviathan*; Harper, *Unsettling the West*; Abvlasky, *Federal Ground*. For the role of the state in the far west, with similar stakes, see White, *It's Your Misfortune and None of My Own*; White, *Railroaded: The Transcontinentals and the Making of Modern America* (New York: Norton, 2011); Heather Cox Richardson, *West from Appomattox: The Reconstruction of America after the Civil War* (New Haven, Conn.: Yale University Press, 2008); Elliott West, *The Last Indian War: The Nez Perce Story* (Oxford: Oxford University Press, 2011).

17. Turner, "Significance of the Frontier in American History," in *Rereading Frederick Jackson Turner*, 31.

18. For "playing Indian" as a phenomenon beyond Native peoples themselves, see Philip J. Deloria, *Playing Indian* (New Haven, Conn.: Yale University Press, 1999). For Pokagon's activities at the world's fair, see Lisa Cushing Davis, "Hegemony and Resistance at the World's Columbian Exposition: Simon Pokagon and the Red Man's Rebuke," *Journal of the Illinois State Historical Society* 108, no. 1 (2015): 32–53. For more on Indigenous involvement at the Columbian Exposition, see David Beck, *Unfair Labor? American Indians and the 1893 World's Columbian Exposition in Chicago* (Lincoln: University of Nebraska Press, 2019).

19. Copies of Pokagon's booklet can be found in the Newberry Library's Special Collections, under the title *The Red Man's Greeting*, or in the Smithsonian Cullman Library Rare Books, under the title *The Red Man's Rebuke*.

20. For the spiritual and cultural importance of birchbark scrolls in Anishinaabe society, see Frances Densmore, *Uses of Plants by the Chippewa Indians* (Washington, D.C.: U.S. Government Printing Office, 1928); Kenneth E. Kidd, "Birch-Bark Scrolls in Archaeological Contexts," *American Antiquity* 30, no. 4 (April 1, 1965): 480–483; Selwyn H. Dewdney, *The Sacred Scrolls of the Southern Ojibway* (Toronto: University of Toronto Press, 1975); Patty Loew, *Indian Nations of Wisconsin: Histories of Endurance and Renewal* (Madison: Wisconsin Historical Society Press, 2001), 84.

21. Simon Pokagon, *The Red Man's Greeting, 1492–1892* (Hartford, Mich.: C. H. Engle, 1893). Courtesy of the Newberry Library.

22. For quotes, see Pokagon, *Red Man's Greeting*.

23. Karamanski, *Schooner Passage*, 127.

24. Pokagon, *Red Man's Greeting*. For similar themes, see Pokagon, "The Chi-Kog-Ong of the Red Man," *New York Times, Sunday Magazine* (December 5, 1897), 7–10.

25. For the rhetorical and narrative problems with concluding Indigenous histories solely in the past, and the need to reemphasize persistence of Native peoples, see Jean M. O'Brien, *Firsting and Lasting: Writing Indians out of Existence in New England* (Minneapolis: University of Minnesota Press, 2010).

26. In the years following the exhibition, Pokagon would lead his Potawatomi band in claiming portions of Chicago's lakefront, which he argued had not been ceded in earlier

treaties due to changing shorelines and engineered land reclamation. See Lowe, *Imprints*. Native people exerted vital influence in Chicago throughout the twentieth century and continue to do so today. For scholarship on this, see LaGrand, *Indian Metropolis*; Straus, *Native Chicago*; Edmunds, *Enduring Nations*; Low, *Imprints*.

27. Dan Egan, "Chicago River Has Become Invasive Species Super Highway," *Milwaukee Journal Sentinel*, August 30, 2021, www.jsonline.com/in-depth/archives/2021/08/30/deep -trouble-chicago-river-has-become-invasive-species-super-highway/7881282002/; Tony Bris-coe, "Asian Carp Have Never Breached a Body of Freshwater the Size of Lake Michigan. Here's the Bizarre Way They Could Survive and Thrive in the World's Fifth Largest Lake," *Chicago Tribune*, August 13, 2019, www.chicagotribune.com/news/environment/ct-met-lake-michigan -asian-carp-study-20190812-nwanxjkymjcvteooe6ymumrroi-story.html; "Great Lakes Fishery Commission—Invasive Carps," Great Lakes Fishery Commission, 2022, www.glfc.org/asian -carp.php; Danielle Kaeding, "Efforts to Keep Asian Carp out of the Great Lakes Move Forward Under New Agreements," *Wisconsin Public Radio*, January 8, 2021, www.wpr.org /efforts-keep-asian-carp-out-great-lakes-move-forward-under-new-agreements; "If You Can't Beat 'Em, Eat 'Em: Illinois Fisheries Rebrand the Invasive Asian Carp," www.wbur.org/, www .wbur.org/hereandnow/2022/07/01/asian-carp-copi-rebrand-food; "Keeping Invasive Spe-cies Out," *Alliance for the Great Lakes* (blog), accessed October 18, 2022, https://greatlakes.org /campaigns/keeping-invasive-species-out/. For scholars' analyses of this crisis, see Justin Mando and Garrett Stack, "Convincing the Public to Kill: Asian Carp and the Proximization of Invasive Species Threat," *Environmental Communication* 13, no. 6 (August 18, 2019): 820–833; Shannon Orr, "Stakeholders and Invasive Asian Carp in the Great Lakes," *Case Studies in the Environment* 5, no. 1 (June 9, 2021): 1–9; Andrew Reeves, *Overrun: Dispatches from the Asian Carp Crisis* (Toronto: ECW Press, 2019). For the wider environmental concerns of the Great Lakes, see Dan Egan, *The Death and Life of the Great Lakes* (New York: Norton, 2017).

28. Tony Briscoe and Patrick M. O'Connell, "In 2019—the 2nd Wettest Year Ever in the U.S.—Flooding Cost Illinois and the Midwest $6.2 Billion. Scientists Predict More Water-logged Days Ahead," *Chicago Tribune*, January 16, 2020, www.chicagotribune.com/news /environment/ct-climate-disasters-cost-midwest-20200115-jubchhqe7bfdnolpw3z7cwjwvm -story.html; Tony Briscoe, "Chicago Is Sinking. Here's What That Means for Lake Michigan and the Midwest," *Chicago Tribune*, February 28, 2019, www.chicagotribune.com/news/ct-met -disappearing-glacier-chicago-sinking-20190220-story.html; Tony Briscoe, "Lake Michigan Has Swallowed up 2 Chicago Beaches This Summer. Experts Say the Worst Could Still Be on the Way," *Chicago Tribune*, August 1, 2019; Dan Egan, "A Battle Between a Great City and a Great Lake," *New York Times*, July 8, 2021, sec. Climate, www.nytimes.com/interactive/2021/07/07 /climate/chicago-river-lake-michigan.html; David Leonhardt, "Chicago's Strange Problem," *New York Times*, July 8, 2021, sec. Briefing, www.nytimes.com/2021/07/08/briefing/climate -change-chicago-lake-michigan.html; Michael Hawthorne and Morgan Greene, "Flooding in the Chicago Area Has Been So Bad in the Past Decade That Only Places Ravaged by Hurri-canes Sustain More Damage," *Chicago Tribune*, May 10, 2019, www.chicagotribune.com/news /breaking/ct-met-chicago-flooding-basement-sewage-20190506-story.html; Henry Grabar, "Tunnel Vision," *Slate*, January 2, 2019, https://slate.com/business/2019/01/chicagos-deep -tunnel-is-it-the-solution-to-urban-flooding-or-a-cautionary-tale.html; Harold L. Platt, *Sink-ing Chicago: Climate Change and the Remaking of a Flood-Prone Environment* (Philadelphia: Temple University Press, 2018).

Index

Page numbers in *italics* refer to maps and illustrations.

Printed in the USA
CPSIA information can be obtained
at www.ICGtesting.com
LVHW090953151023
759509LV00014BA/52